Fundamentals of
Smart Grid Systems

Fundamentals of
Smart Grid Systems

Abdul Gafoor Shaik
Electrical Engineering Department,
Center for Emerging Technologies for Sustainable Development,
Indian Institute of Technology, Jodhpur 342 030,
Rajasthan, India.

BSP BS Publications™
An Imprint of **BSP Books Pvt. Ltd.**
4-4-309/316, Giriraj Lane,
Sultan Bazar, Hyderabad - 500 001

Fundamentals of Smart Grid Systems
***by* Abdul Gafoor Shaik**

Published by:

BSP BS Publications™
An Imprint of **BSP Books Pvt. Ltd.**
4-4-309/316, Giriraj Lane, Sultan Bazar,
Hyderabad - 500 001
Phone : 040 - 23445688
e-mail : info@bspbooks.net
www.bspbooks.net

ISBN: 978-93-49562-05-9 (Hardback)

PREFACE

The textbook "Fundamentals of Smart Grid Systems" provides a comprehensive overview of smart grids, starting from motivation, evolution, challenges, opportunities, components, operation, management, control and protection. The book intended to serve as a textbook for undergraduate and post graduate students, faculty and also to serve as training resource by engineers from power industry. This provides essential knowledge of smart grid systems which include, basic concepts, features, the goals and the major technologies to achieve these goals in terms of renewable energy penetration, monitoring, control and protection.

This book consists of nine chapters, covering core topics such as, benefits and challenges of smart grid, introduction to renewable energy, and energy storage systems. Introduction of Plugin Electric Vehicles and their participation in grid support is covered in one of the chapters. Chapters on demand side management, monitoring, protection and power electronic converters are also included to in an effort to provide a complete operational scenario of smart grid along with the basic knowledge its components. Each chapter ends with several practical and advanced problems that instilling critical thinking and applied to industrial applications.

The first chapter is intended to provide with basic concepts of smart grid which include motivation, evolution, definitions, features and challenges of smart grids. Chapter-2 deals with the transformation of electric grid from centralized to distributed generations including renewable energy sources. The cost consideration and technical challenges are also covered in this chapter. In third chapter, history, technological developments, operation of various variable renewable energy sources are covered. The fourth chapter presents various energy storage technologies to overcome the intermittency of RE sources, their characteristics applications. Introduction of Plugin Electric Vehicles and their integration in to grid as storages on wheels, the technologies of integration and associate challenges in terms charging infrastructure are dealt in Chapter-5. Demand side management , its strategies, implementation, associated challenges and opportunities are briefed in sixth chapter. This chapter also presents mathematical formulation of demand side management optimization problem.Chapter-7 presents various monitoring technologies such as SCADA, advanced metering infrastructure and , wide area monitoring systems which deploys phasor measurement units used in smart grid. This also covers architectures of these monitoring systems and various components along with their functionalities. Various advanced schemes used to tackle challenges of smart grid protection along with their solutions are presented in Chapter-8. The ninth chapter depicts role of power electronic technology in smart grid to control the power low in the network. Some functioning of basics converters used in smart grid are also covered under this chapter.

-Author

CONTENTS

CHAPTER

1

Basics of Smart Grid

1.1 Introduction

The term *grid* is used for an electricity system that include all or some of the four operations such as generation, transmission, distribution, and control. A *smart grid (SG)*, also called smart electrical/power grid, intelligent grid, development of the 20th century power grid. In traditional grids power flows from a few central generators to a large number of users or customers. In contrast, the smart grid uses two-way power flow and information to achieve an automated and distributed advanced energy delivery network. Thus smart grid can be regarded as an electric system that uses two-way, information, cyber-secure communication technologies, and computational intelligence in an integrated way across generation, transmission, distribution and utilization to achieve a system that is clean, safe, secure, reliable, resilient, efficient, and sustainable.

1.2 Conventional Grid

A conventional grid (CG)generally consists of generation, transmission, distribution and load systems. Generation system is generally a combination of large-scale centralised generation plants such as thermal, hydro and nuclear power plants. A typical modern generating unit has rated value of over one thousand MW. Transmission system is typically designed to transfer bulk of power from generating plants to distribution network at high-voltage levels over long distances. Distribution network are designed to receive electric power from transmission system to be distributed to load centres. It is therefore important to note that the nature of distribution network in CG is passive. The main features of conventional grid can be summarized as follows:

- A conventional grid has vertical structure as shown in Figure 1.1.
- The power flow is unidirectional typically distribution networks.
- The price of electricity is dictated by the utility.

- The consumers have no choice of options in terms of electricity procurements.
- Consumer is passive stakeholder in the conventional grid.

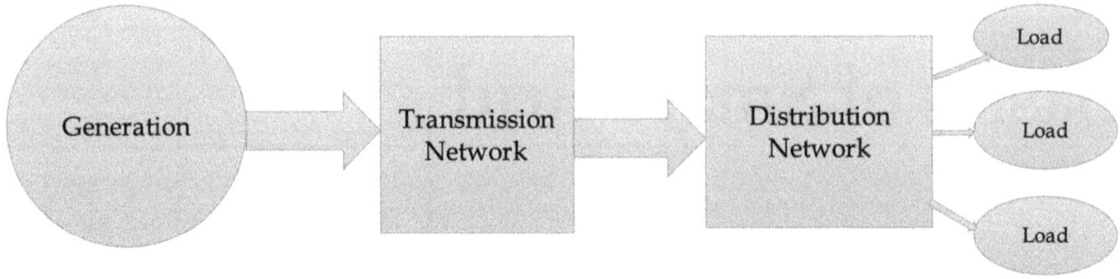

Figure 1.1 Vertical Structure of Conventional Grid.

1.3 Evolution of Smart Grid

The transformation of conventional grid in to smart grid was driven by several factors which include the technological developments in the automation industry, need to accommodate renewable energy sources due to climate change and fuel dependency, need of active customer participation in the grid operation in order to improve efficiency, reliability, stability, economy, security etc.

1.3.1 Technological Developments

Electric grids possessed some technologies since the early stage while the others have gradually been incorporated into grids over several generations as per the requirements and technological developments to improve the performance.

- Control equipment in substations driven by electro- mechanical relays replaced by digital/numerical relays.
- In 1950s the automatic telephone switchboard equipment was introduced to read and control the remote substation from the local substation which was called as supervisory control equipment(SCE)
- In the late 1960s SCE was replaced by SCADA. SCADA was extended to monitor of the majority of the transmission systems (220 kV and above) and some distribution substations in 1970s and 1980s.
- SCADA system was also used to centrally support control rooms and remote terminal units (RTUs) for data collection and control in the substations.
- Later RTUs are connected to programmable logic controllers (PLCs) originated from manufacturing industries using hardwires.
- As a course of time, hardwires are replaced by communication links.

- In the mid-1990s the RTU/PLC configuration was replaced by a different network architecture that consists of protection relays/Intelligent Electronic Devices (IEDs), PLCs and other devices communicating with each other over a network and thereby coordinating operations.
- At later stages active participation of customers was enabled by the introduction of Advanced Meter Infrastructures (AMI).

1.3.2 Changes to be Accommodated

Apart from the technological developments, CGs have found it difficult to accommodate the following changes

- Integration of distributed energy resources including renewable energy sources (RES) and Storage systems into electrical networks, specifically at distribution levels.
- Accommodation of new developments in transport industry in terms of Electric Vehicles (EV) that causes further stress on the distribution networks.
- Dynamic situation between electricity market stakeholders and electrical utilities due to liberalisation of electricity market in recent years.

These factors have adversely affected the operation, management and protection of conventional grid (CG) in number of ways. This in turn has led to initially contemplate on modernising CGs and eventually to the development of Smart Grid Concept.

1.3.3 History

The Smart Grid concept initially developed in the USA. In an article published in Wired Magazine in July 2001 a precise description of the future network, which latter on was known as a Smart Grid, states "*The best minds in electricity R&D have a plan: Every node in the power network of the future will be awake, responsive, adaptive, price-smart, eco-sensitive, real-time, flexible, humming and interconnected with everything else*".

The Smart Grid concept then promoted in Europe, North America and worldwide in countries such as India, China and South Africa. The work on Smart Grids in Europe started in 2005 by the Smart Grids European Technology Platform for Electricity Networks of the Future. The aim was to formulate and promote a vision for the development of European electricity networks at 2020 and beyond. In the USA Smart Grid concept was officially promoted by the publication of the US Energy Independence and Security Act of December 2007 whereby it is stated : "*To move the United States toward greater energy independence and security, to increase the production of clean renewable fuels, to protect consumers, to increase the efficiency of products, buildings, and vehicles, to promote research on and deploy greenhouse gas capture and storage options, and to improve the energy performance of the Federal Government, and for other purposes."* The following are some concepts of the smart grid proposed by various organization worldwide:

The IEC smart grid *as an electric power system that utilizes information exchange and control technologies, distributed computing and associated sensors and actuators, for purposes such as the integration of the behaviour and actions of the network users and other stakeholders as well as efficiently deliver sustainable, economic and secure electricity supplies.*

IEA Smart grids *are electricity network that use digital technologies, sensors and software to better match the supply and demand of electricity in real time while minimizing costs and maintaining the stability and reliability of the grid.*

IEEE Smart Grid *an electrical grid that can monitor, predict, and intelligently respond to the behaviour of all electric power suppliers and consumers connected to it in order to deliver reliable and sustainable electricity services as efficiently as possible.*

As per Swedish stand*ards it is a set of technology, regulation and market rules that are required to address, in a cost-effective way, the challenges to which the electricity network is exposed.*

1.3.4 NIST Smart Grid Conceptual Model

National Institute of Standards and Technology, Department of Commerce USA has published a Smart Grid Conceptual Model in February 2021 as shown in Figure-2. The NIST Smart Grid Conceptual Model describes the overall composition of electric grid systems and applications. It provides a high-level view of the system for better understanding of many stakeholders. Originally introduced in 2010, the Conceptual Model is updated with each Framework revision.

The visual representation of domains, interfaces, and electrical and communications flows presented in Conceptual Model provides a comprehensive understanding of various actors, roles, and responsibilities needed to ensure effective day-to-day grid operations and control. The Model allows extended concerns to be identified and understood in its context, thereby providing support for policy and planning activities from regional planning to resource adequacy assessments. Various domains and their roles/services the SG conceptual model are as follows:

Customer: The end users of electricity. May also generate, store, and manage the use of energy. Traditionally, three customer types are discussed, each with its own sub-domain: residential, commercial, and industrial.

Markets: The facilitators and participants in electricity markets and other economic mechanisms used to drive action and optimize system outcomes.

Service Providers: The organizations providing services to electrical customers and to utilities.

Generators including DERs: The producers of electricity, may also store energy for later distribution. This domain includes traditional generation sources and distributed energy resources (DER). At a logical level, "generation" includes those traditional larger scale technologies usually attached to the transmission system, such as conventional thermal generation, large-scale hydro generation, and utility-scale renewable installations usually attached to transmission. DER is associated with generation, storage, and demand response.

Transmission: The carriers of high voltage electricity over long distances. May also store and generate electricity.

Distribution: The distributors of electricity to and from customers. May also store and generate electricity.

The roles in a particular domain often interact with roles in other domains to enable smart grid functionality as shown in Figure 1.2. Moreover, as the system complexity increases, the communications and interoperability expand operational

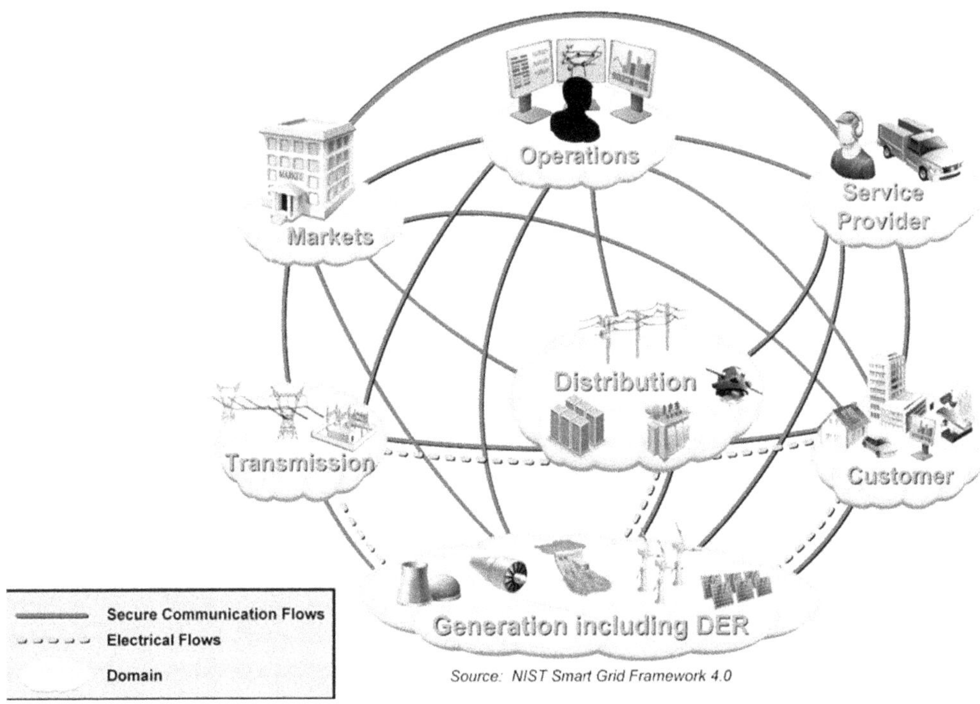

Figure 1.2 NIST Smart Grid Conceptual Model.

control beyond locational specificity of physical connections, it is likely that organizations will contain components of multiple domains. For example, the Independent System Operators (ISOs) and Regional Transmission Organizations (RTOs) in North America have roles in both the markets and operations domains. Similarly, a distribution utility is not entirely contained within the distribution domain. It is likely to contain roles in the operations domain, and perhaps also the markets domain as economic signals become more dynamic across the system.

1.4 Features and Benefit of Smart Grid

1.4.1 Features of Smart Grid

Smart Grids integrate electricity networks consisting of transmission and distribution systems and interfaces with generation, storage, and end users. The SGs employ sets of modern technologies that can be deployed globally, and guided by local commercial and legal and regulatory. The main characteristics of SGs are as follows:

- SGs have the ability to integrate different energy sources in the distributed energy system in a way that optimises power and delivered effectively to the grid
- The SG system facilitates quick electricity demand response that can ease peak load shaving, load profile shaping with high degree of effectiveness and reliability, leading to better power system stability.
- The wireless and wired communication systems and end-to-end communication management leads to integrated communication system
- Data Acquisition Systems and other technologies integrated with communication.
- New metering, system controls, and customer engagement technologies.
- The SG system uses AMI for communication between the utility company, consumers, and operators. Individual equipment and customers as well as transformers, switches, capacitor banks, voltage regulators, etc. are metered throughout the distribution system. Information from these equipment and customers is relayed to the utility in a typical two-way communication.
- The SG Applies cyber-secured communication like data supervisory control and Energy Management Systems (EMS).
- The SG enables demand profile shaping, and peak loads having to optimise power generation cost and increase consumer utility through real-time pricing technologies

- The SGs facilitate greenhouse gas emission minimisation by efficient integration of variable renewable energy sources and variable loads like electric cars unlike in conventional grid.
- The SG enables active participation of the customer who responds the requirement of the utility to reduce consumption during peak hours to relive it during times of distress.
- The SG has the infrastructure to support a two way flow of communication and of electricity from the utility to the customers which converts electricity consumers to prosumer (producer + consumer).

1.4.2 Benefits of Smart Grid

- These SG system enable seamless connection and operation of different generators in terms of size and technologies.
- Less maintenance cost due to automatic tracking and reporting compared to manual or physical tracking.
- Electricity theft is minimal due to remote monitoring of meters. Any tampering attempt raises an alert signal. Any deviations in power consumption are notified for swift action.
- They have the capability of automatic load balancing which reduces the risks of equipment failures. SGs deploy machine learning based algorithms to predict consumption patterns and manage loads dispatch which reduces stress on electrical equipment, especially during peak times.
- Electrical transmissions is regulated by intelligent technologies to mitigate electrical losses in distribution which improves the efficiency of electrical transmission.
- SGs have improved power system stability, security.
- SGs prevent transmission oscillations, overload, and resultant blackouts by facilitating automatic loading and re-routing of electricity in the event of equipment failures.
- SG consumers are timely informed about the demand and price, to plan their consumption so as to save on energy cost.
- Based on the timely data of both weather and demand SGs facilitate optimal integration of renewable energy which enables economical dispatch
- Efficient connection and operation of generators of all sizes and technologies seamlessly to a common platform.
- Consumers are allowed to participate actively in grid optimization which leads to improved overall reliability and stability of the power system

- SGs facilitate accurate and timely planning and action by availing real time data and information to create a stable, efficient, and cost-effective system.
- Reduction in the environmental impact of power systems is achieved through optimum loading of generation systems.
- Improved reliability, quality, and security of the electricity supply system.
- Improved quality and performance of energy products and services by fostering market integration and harmonious stakeholder relationship and operation.
- These grids convert electricity consumers to prosumers who do both generation and supply of grid power based on variable pricing and two-way flow of electricity and data.
- By combining or integrating weather and demand forecasts, more accurate generation planning, and dispatch can be done.
- SGs are resilient to system attacks, and disasters due to the capabilities of isolating and re-routing power flow (self-healing ability). This in turn reduces downtime and outages.

1.5 Main Features

It is very important to understand the features of Smart Grid technology so that the perfect use of Smart Grid technology can be ensured. The features of Smart Grid can be described as follows

- Smart Meters (SM)
- Distributed Generations (DG)
- Renewable Energy Integrations (REI)
- Bidirectional Communication System (BCS)
- Automatic Healing Capability (AHC)
- Data Security/Cyber Security
- Carbon Emission Reduction
- Meter Data Management (MDM)
- Field Area Networks (FAN)
- IT and Back office Computing

- Demand Response
- Electricity Storage Devices
- Distribution Automation

1.5.1 Smart Meters

Smart Meter is a unique feature in the Smart Grid which is the most dependable instrument in the field of power generations and consumption for data measurement. Smart Metering refers to usage of advanced meters in combination of communication systems to allow consumers to monitor their energy consumption and price in real time. Usually it takes the reading after a certain time and a few times a day. It also provides information of consumer loads which is utilized in Demand side Management (DMS).

Figure 1.3 Smart Meter.

1.5.2 Distributed Generation

Distributed Generation (DG) is an important part of Smart Grid Technology. Distributed Generation means generating electricity from small energy sources. Distributed Generation plays a great role in Smart Grid and is cause of some challenges in the grid operation. Some of the challenges include:

- Market oriented demand response of the Electricity. It means understanding the demand of electricity of the consumers and makes a user oriented market place for ensuring the perfect demand response.
- Ensuring the security of distributed generations and supply of electricity from generation end to the distribution end.
- Ensuring Cost Effective operations of generation and distribution so that the cost does not exceed the desired limit.
- Environmental Safety must be assured for protection of the Grid from any kind of natural disaster.

1.5.3 Renewable Energy Integration

Renewable energy integration (REI) is also an important feature of the Smart Grid. Enhancing the capability REI helps the national grid to meet the extended demand of the consumers with potential security. Environmental Impact is one of the greatest factors in case of Renewable Energy Generation and Integration. The absence of Solar Energy during the night, must be supported by storage system in the night. Voltage fluctuations due to intermittent nature of solar and wind energies are also to be countered by power electronics of the SG.

1.5.4 Bidirectional Communication Systems

Enabling the bi-directional communication system makes the Smart Grid easier to use for both consumers and suppliers. Communicating in the smart grid is like a phone conversation where users are aware of the price and usage of the electricity as well as the generation and the supplier are also aware of billing of the usage of the electricity clearly. Monitoring the whole Grid centrally is possible only through this communication system. However, the privacy must be ensured in case of communicating in the grid bidirectional or multidirectional.

1.5.5 Self-Healing Capability

Since the Smart Grid (SG) is a smart system of generating the electricity and distributing them to the consumers with a huge confirmation of data security, robustness and convenience, the smart grid technology must have one feature which is called self-healing capability (SHC). This feature includes the automatic detection of the abnormal conditions in the system like over current, surge voltage, fault current etc; sending the information to the central control room to ensure and automatic recovery/healing followed by a disturbance.

1.5.6 Data/Cyber Security

Smart Grid must be secured from considerable threats or attacks. Smart Grid can be attacked by Hackers, Cyber Terrorists, and Organized Crimes, some criminal elements, industrial competitors, careless or poorly trained employees to exploit vulnerabilities in the system. Careless operating of the system by the poorly trained employees can make the total system vulnerable to the security attacks. As SG is highly inter-connected, if one portion of the network is attacked, eventually the whole system gets exposed to serious threat and it may even led to total blackout or total system failure. Hence, cyber security must be strong enough to endure the attacks and make the system run smoothly and efficiently.

1.6 Challenges Associated with Smart Grid

SGs apply modern cutting-edge technology to address multiple challenges of the traditional grid while at the same time providing opportunities to improve their efficiency and reliability. However, several technical and socioeconomic challenges limit the development and application of the SG. SGs employ power electronics technologies like FACTS and RACDS, efficient Energy Storage Systems (ESS), etc. which have high initial cost requirements. Technical challenges in the operation of SGs include database management and cyber security. Consumers lack of understanding about the SG, hence there is a need to educate the public. For consumers to effectively participate in demand side management and other services requires training. The challenges facing the development and application of SGs are summarised Figure 1.4. The challenges also include inadequate grid infrastructure, lack of knowledge and skills by industry workforce, limited capacity of ESSs.

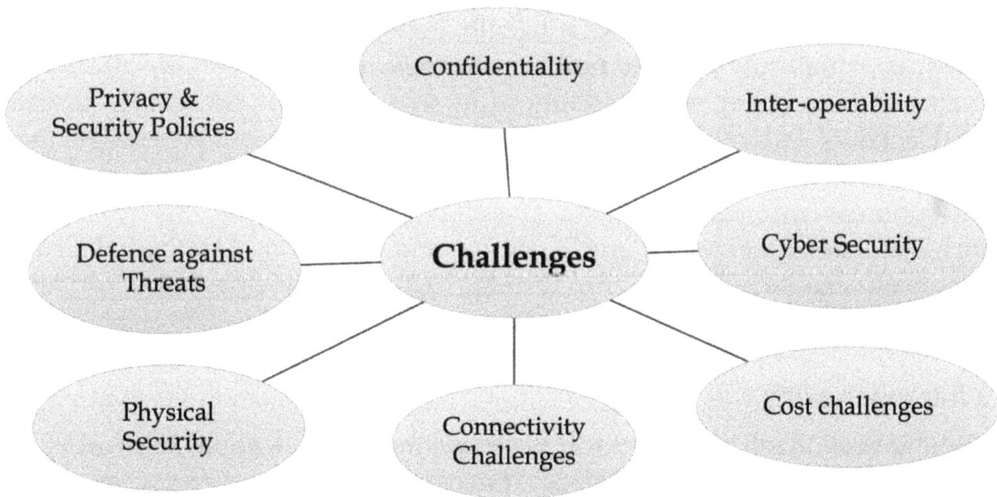

Figure 1.4 Various challenges faced by SG development.

1.6.1 Cost Challenges

The biggest challenge to the development and implementation of SGs is cost. The main cost elements in SG development are associated with the transmission and distribution system, metering, and critical technologies. As per the economic analysis and financial feasibility, studies the cost of the development for SG infrastructure is paramount which is ultimately disadvantageous to the developing countries. Smartening of the traditional grid requires the integration of additional sensors to the infrastructure which is an expensive investment to be undertaken by distribution grid operators.

1.6.2 Cyber Security

There are increased possibilities of remote operation of smart grids since the grids are large-scale system with participants widely spread between power plants and electricity consumers and devices geographically. This requires protection of resources against theft, abuse, and malicious activities. The Electric Power Research Institute has identified power system cyber security as one of the greatest issues of smart grids. It is necessary to identify loopholes in smart grids, exploited by attackers and carryout Threats Analysis by integrating Systems Security Threat Models. There are various approaches used to address cyber-physical security issue for Wide-Area Monitoring and protection and control which are vulnerable to a coordinated cyber-attack. Hence, it is necessary to assess the SG security by review of its methodology to identify weaknesses. Since SGs are deployed across the value chain, there is an increase in points of interconnection with other information networks which exposes the system to the potential threat of cyber-attack on important power management systems and infrastructure. The SG is heavily dependent on ICT which ultimately increases the complexity and the number channels for possible cyber-attacks. The cyber-attacks may target the distribution infrastructure not necessarily to cause power blackouts but other destructive disruptive activities like energy meter manipulation, blackmail, power supply disruptions, damage of equipment, damage of distribution or customer equipment. Therefore, it is necessary to ensure that cyber security is an essential aspect of any SG design and operations. Interoperability and cyber security can be achieved only through rigorous implementation of various standards.

1.6.3 Interoperability

The communication infrastructure for SG should meet specifications in terms of time synchronisation, data delivery efficiency, latency reliability, and multicast support. Additionally, the main challenge in networking communications in SG is interoperability. Rigorous implementation of applicable standards is an effective measure to achieve interoperability. The NIST has put in place interoperability standards in areas like advanced metering infrastructure (AMI), end-to-end security, metering, communication between control centres, substation automation and protection, EMS interfaces, power system control operations information security, PMU communications, electrical and physical interconnections between utility and DG, security for intelligent electronic devices (IEDs), cyber security standards and guidelines for federal information systems and bulk power systems, HAN device communication, measurement, and information model and SGIP catalogue of standards.

Each country should develop their standards for smooth implementation of SG considering their existing standards and for easy implementation. However, internationally accepted standards and security should also be put in by all countries involved in the SG implementation to make the power grid a homogeneous global grid.

1.6.4 Confidentiality Challenges

Access Control and Identity Management is needed for the confidentiality of data transmitted via SGs. There is a need for authentication to verify the identity of the receivers and avoid any disruption or exploitation. Only authenticated users should be allowed access to the control centre for transmission, and distribution grids.

1.6.5 Privacy and Security Policies

With wide scale interconnectivity and system access including the growth of IoT associated with SG development, privacy, and development of far policies for all stakeholders is a real challenge as objectives and needs may differ and even conflict among stakeholder. Therefore, there is a huge necessity for suitable security policies to establish relationships between the consumers, utilities, generating companies, transmission firms, and third parties, although applying security and privacy policies in a manner that does not lead to dissatisfaction which is a complex affair.

1.6.6 Defence against Threat

The SGs have inherent vulnerabilities against which it must be protected. This can be achieved by building an effective, layered defence system for the protection of the SG infrastructure. This defence or protection against threats or vulnerabilities gives the network segmentation and access control to defend against denial-of-service (DoS). Technologies used include firewall, virtual private network (VPN), intrusion prevention system (IPS). This leads to extra investment in technology and resources adding to system costs.

1.6.7 Physical Security

SG systems have thousands or millions of widely spread remote points and networks. This the challenge of policing or physical protection from theft, sabotage and illegal or unauthorised access. Therefore, physical protection is a real challenge for SGs. Additionally, the wider geographical dispersion of the many assets and systems makes it difficult to access all of the terminals for both maintenance and protection

1.6.8 Connectivity Challenges

This refers to the communications connectivity and interchange within the SG infrastructure and between users or stakeholders. The current trend in communication technology is undergoing a transition from analog to digital and to an Internet-like distributed environment where large number of devices are interconnected. Not all regions, devices or users have access to modern and effective internet and communication infrastructure especially where the services a public accesses and not limited to the grid system.

References

1. X. Fang, S. Misra, G. Xue and D. Yang, "Smart Grid — The New and Improved Power Grid: A Survey," in *IEEE Communications Surveys & Tutorials*, vol. 14, no. 4, pp. 944-980, Fourth Quarter 2012, doi: 10.1109/SURV.2011.101911.00087. keywords: {Energy management; Smart grids; Security ;NIST; Power grids; Privacy; Smart grid; power
2. F. R. Yu, P. Zhang, W. Xiao and P. Choudhury, "Communication systems for grid integration of renewable energy resources," in IEEE Network, vol. 25, no. 5, pp. 22-29, September-October 2011, doi: 10.1109/MNET.2011.6033032.
3. S. K. Salman, "Evolution of Conventional Power Systems to Smart Grids," 2019 54th International Universities Power Engineering Conference (UPEC), Bucharest, Romania, 2019, pp. 1-6, doi: 10.1109/UPEC.2019.8893444.
4. Moses Jeremiah Barasa Kabeyi & Oludolapo Akanni Olanrewaju (2023) Smart grid technologies and application in the sustainable energy transition: a review, International Journal of Sustainable Energy, 42:1, 685-758, DOI: 10.1080/14786451.2023.2222298
5. U.S Department of Energy. 2022. "The Smart Grid." US Department of Energy. Accessed September 22, 2022.https:// www.smartgrid.gov/ the_smart_grid/smart_grid.html.
6. Panda, D. K., and S. Das. 2021. "Smart Grid Architecture Model for Control, Optimization and Data Analytics of Future Power Networks with More Renewable Energy." Journal of Cleaner Production 301: 126877, 2021/06/10. doi:10.1016 /j.jclepro.2021.126877.
7. Kiran, P. A. S., and B. L. Rao. 2018. "Smart Grid Functionalities at Distribution Level." International Journal of Pure and Applied Mathematics (Special Issue): 17, April 17, 2018. [Online]. https://acadpubl.eu/hub/2018-118-24/1/ 177.pdf.
8. S. Paul, M. S. Rabbani, R. K. Kundu and S. M. R. Zaman, "A review of smart technology (Smart Grid) and its features," 2014 1st International Conference on Non Conventional Energy (ICONCE 2014), Kalyani, India, 2014, pp. 200-203, doi: 10.1109/ICONCE.2014.6808719.
9. https://www.smartgrid.gov/the_smart_grid/smart_grid.html
10. https://nvlpubs.nist.gov/nistpubs/SpecialPublications/NIST.SP.1108r4.pdf
11. https://www.iea.org/energy-system/electricity/smart-grids

CHAPTER

2

Electric Grid and Transformations

2.1 Introduction

In the late nineteenth century, electricity was first made available through central stations serving a group of nearby customers. Generation and distribution were localized, and long-distance transmission was not yet necessary. As demand for electricity service increased, distribution networks expanded in capacity and geography. Systems once isolated from one another were becoming interconnected. Out of this emerged the basic operating structure of the grid still in place today:

- Large geographically dispersed power plants generate electricity at low voltage levels.
- The voltage is then stepped up via transformers in transmission substations and the same is transmitted over long distances on interconnected transmission lines to distribution substations, where transformers step down the voltage to low levels.
- The electricity at low voltage levels is then distributed over relatively short distances via a network of lines using smaller transformers, which step the voltage down to further safe and appropriate levels to suit the consumer needs.

These elements are illustrated in Figure 2.1, including (1) a power plant, (2) a transmission substation, (3) a transmission line, (4) a distribution substation, (5) a distribution line/transformer, and (6) an end user. The basic operating structure of the grid has largely remained the same over the decades, the practices used to plan and operate the grid and the regulatory structures that govern the industry have evolved substantially since that time.

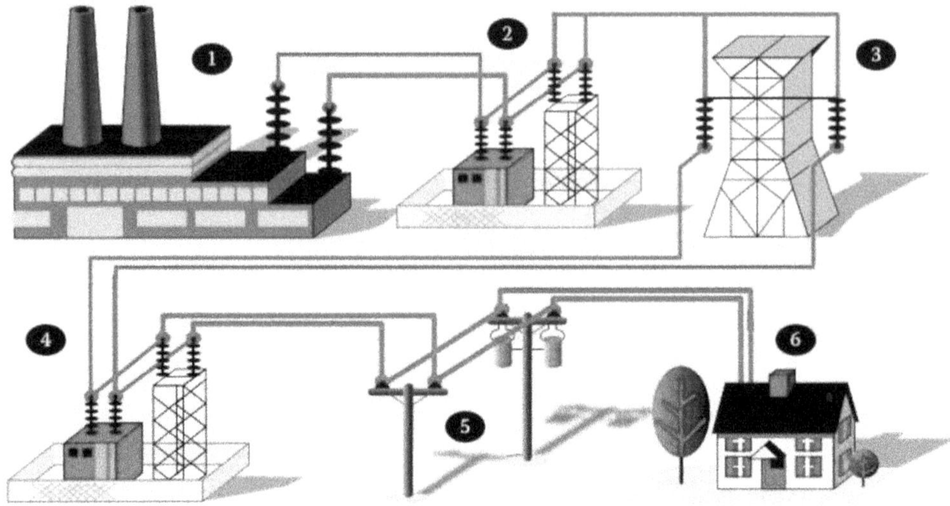

Figure 2.1 Electric utility interconnection overview. (Courtesy of the Advisory Board of the Utility Executive Course, University of Idaho, Moscow, ID).

Electricity is now available to approximately 7.2 billion people around the world, more than 0.7 billion people are left in the dark with no, or very limited, access to electricity. Developing nations continue to lag in the provision of electricity to their citizenry. A full understanding of the new challenges and opportunities, the industry will face over the coming decades requires an understanding of the factors that have shaped the utility industry's history. The regulatory structures that exist today will also fundamentally shape the development of the smart grid. The complexity of the energy infrastructure, coupled with its social, economic, regulatory, and political operating environments, directly impacts the understanding, acceptance, and ultimate promotion of innovative technologies, solutions, and new service models described through smart grid and many other definitions.

The demand for electrical energy is increasing drastically. It was estimated that electricity demand will double between 2000 and 2030, with an annual growth rate of 2.4 faster than the increase of any non-renewable energy source. Hence more renewable energy sources are needed in future energy systems. In Europe, the target is to raise the renewable energy penetration percentage from the current level of 20% to 50% by 2050. The renewable energy sources are usually located in a distributed manner, e.g., photovoltaic (PV) and wind, in contrast to the conventional large centralized power plants. This leads to a larger and more complex networked energy system. It is a large nonlinear highly structured system consisting of a number of interconnected distributed generators (DGs) or subsystems. In view of these changes, the configuration of power system is moving towards more distributed generation. Several articles and reports have discussed the technical advantages and economic benefits of this new distributed paradigm. Some of them are:

- DG enhances system resilience and supply reliability by reducing the dependence on the distribution network availability. They also reduces transmission and distribution network investments needs.
- DG avoids the large amounts of transmission losses.
- DG aggregation could be used to provide local ancillary services such as voltage support or even to avoid overloads in networks.
- DG aggregation facilitates fulfilling market commitments since their global operation allows reallocating these commits in case of individual failure.
- The economic investment is diluted among several small agents making an active participation of end-users possible. Each agent can take its own decision and choose an appropriate technology
- Deployment of DG and renewable energy sources has significant environmental impacts such as reduction of greenhouse gases, higher energy efficiencies, reduced damages to human health.

2.2 Centralized and Distributed Generations

2.2.1 Centralized Generation

Centralized Generation is the bulk electric power generating system with central power plants. It includes large fossil-fired (gas or coal) boilers or nuclear boilers to produce steam that drives turbine generators. In some cases, large hydro is also used. The widespread concentration of CG for power generation was driven by several factors which include:

Economies of scale: With the advent of steam turbines, it was possible to increase the size of the turbines which lead to decrease in cost of electricity production.

The search for high energy efficiency: Efficiency gains can be achieved through handling higher pressures and temperatures of steam used in electricity generation.

Innovation in electricity transmission: Usage AC in place of DC was made possible to transmit electricity over long distances with a significant loss reduction.

Environmental constraints: It was possible to relocate the generation facilities outside the city centres thus replacing the exhaust from plants.

The search for reliability: Reliability of supply was increased as the failure of one plant was compensated by the other plants in the interconnected system.

2.2.2 Drive for Distributed Generators

The following drawbacks of centralized generation brought the concept of mixed generation wherein the distributed generation compensate for the same:

- **Transmission and distribution costs**: In CG, transmission and distribution costs amount for up to 30% of the cost of delivered electricity on average. These costs are mainly made up of line losses, unaccounted for electricity losses and conversion losses. In addition to the cash cost, these billions of kWh losses have an implicit cost in terms of greenhouse gas emissions.
- **Rural electrification**: In CG, rural electrification over long distance to serve small consumptions become uneconomical.
- **Investment in transmission and distribution networks**: Significant investment is required to upgrade the transmission and distribution networks to meet the ever-growing energy demand by CG.
- **Thermal efficiency:** Continuous efforts to increase the thermal efficiency in steam turbines, causes super-heated pressurized steam pressed against the turbine blades and boiler tubes, metallurgical fatigue increased substantially, decreasing the reliability of huge power plants and increasing maintenance costs.
- If the efficiencies are not pushed utilities could no longer achieve significant cost declines from this aspect of their industry's technological progress.

2.2.3 Simple Benefits of Distributed Generators

- Shorter construction times
- Reduced financial risk of over- or under-building
- Reduced project cost-of-capital over time due to better alignment of incremental demand and supply
- Lower local impacts of smaller units may qualify for streamlined permitting or exempted permitting processes, reducing fixed costs per kW
- Significantly reduced exposure to technology obsolescence
- Local job creation for manufacturing, technician installers/operators
- Higher local, small-business development and taxes vs. overseas manufacturing
- Lower unit-cost, automated manufacturing processes shared with other mass-production enterprises
- Shorter lead times reduce risk of exposure to changes in regulatory climate

- Significant reduction in fuel disruption risk (portfolio of locally produced fuels and "fuel-less" technologies— solar, wind)
- Reduced fuel-forward price risk
- Reduced trapped equity
- Reduced exposure to interest-rate fluctuations
- Potential for more modular, routine analysis for capital expansions
- Multiple off ramps for discontinued projects, without same level of risk
- Ability to redeploy portable resources as demand profiles change
- Portability = Higher capacity utilization
- Reduced site remediation costs after decommissioning
- Higher system efficiency reduces ratio of fixed-to-variable costs (fuel)
- Potential for lower unit costs for replacement parts when mass produced
- Displaces that portion of customer load with highest line losses
- Displaces that portion of customer load with greatest reactive power requirements
- Displaces that portion of customer load with highest marginal energy costs
- Weather-related (solar, wind) interruptions more easily predicted and of shorter duration than equipment failures at central plants
- "Hot swap" capability - when one DG module (panel, tracker, inverter, turbine) is unavailable, all other modules continue operating
- Load siting reduces or eliminates line losses on electric transmission and distribution lines
- Inherently improved system stability due to multiplicity of inputs
- Reduced regional consequences of system failure
- Improved transmission and distribution reliability due to reduced peak loading, conductor and transformer cooling
- Fast ramping within the distribution system, ability to reduce harmonic distortions at customer's site.

2.3 Benefits and Costs of Renewable Energy Sources

The International Renewable Energy Agency (IRENA) has analysed the socio-economic benefits of renewable energy since 2011. An initial focus on employment creation and skills was subsequently extended to cover aspects such as gross domestic product (GDP), broader measures of welfare, local economic value creation, improved livelihoods, gender and other benefits. The assessments include present-day global, regionall and selected national impacts, as well as projections to 2030 and 2050. The benefits associated with the Renewable energy can be broadly categorized in to the following areas which are detailed in succeeding subsections.

2.3.1 Benefits of Renewable Energy Sources

Employment: China, Brazil and the US are leading employers in the renewable energy sector (see Figure 2.2). Recent years have seen a considerable shift towards Asian countries, whose share of global renewable energy employment rose from 50% in 2013 to 62% in 2016. This shift is the result of two factors. Strong deployment policies have led to the emergence of dynamic domestic markets, and industrial policies have supported the growth of globally competitive manufacturing facilities, especially in the solar photovoltaics (PV) industry.

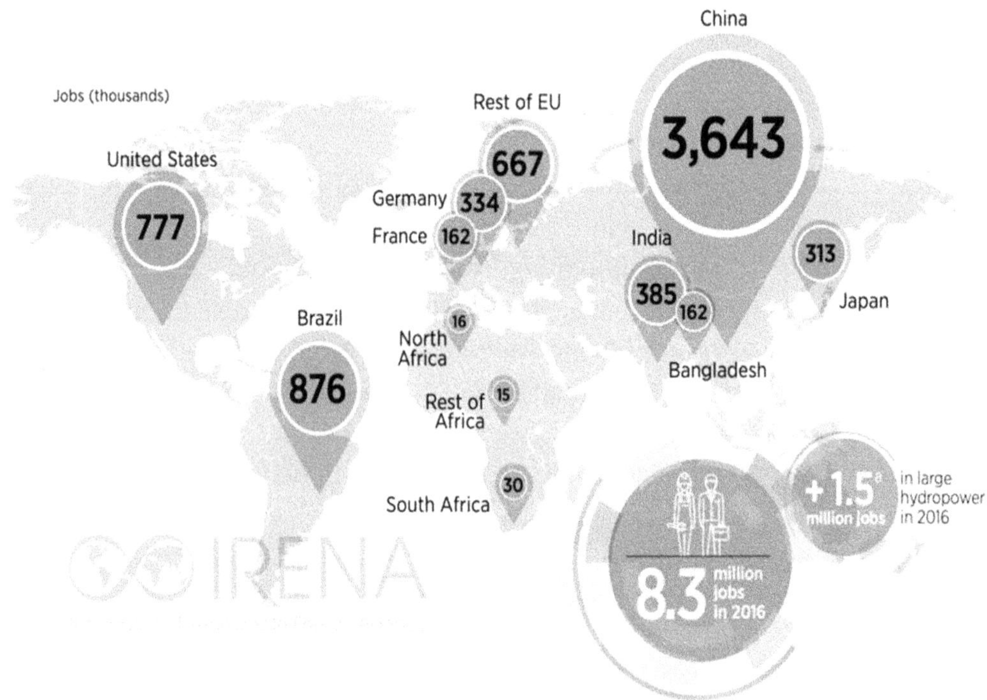

Source: IRENA, 2017c.

Figure 2.2 Renewable energy employment in selected countries, 2016.

IRENA's Leveraging Local Capacity for Solar PV (IRENA, 2017d) analysed the occupational patterns and skills needs of a typical 50 megawatt (MW) solar PV project. In total, some 230 000 person days are needed along the value chain as shown in Figure 2.3. Operations and maintenance account for 56%, manufacturing for 22%, and construction and installation for 17%. Construction workers (35 500 person days) and factory workers and technicians (32 000 person days) are among the most prominent occupations. Many of the occupations that can be filled locally - especially in construction - do not require highly renewables-specific skills.

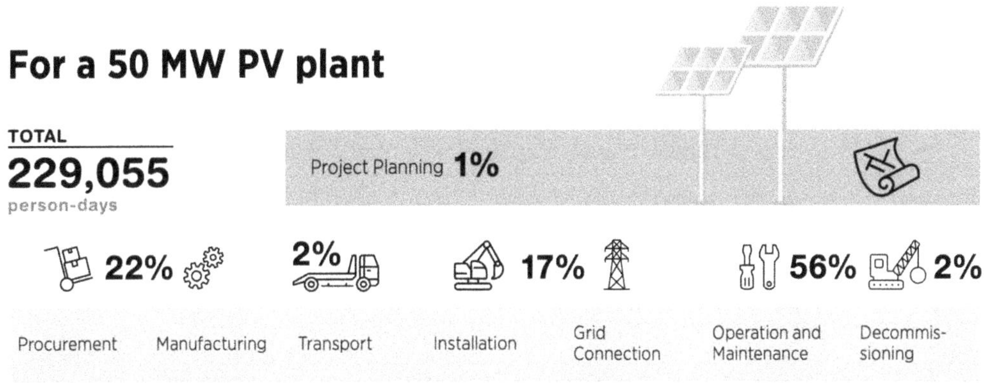

Source: IRENA, 2017c.

Figure 2.3 Employment impacts in the solar PV value chain.

In case of Wind energy, a strong pace of new installations (particularly in China, the US and Germany) resulted in a 7% increase in global employment to reach 1.2 million jobs. IRENA's Leveraging Local Capacity for Onshore Wind (IRENA, 2017e) found that a typical 50 MW project requires a total of 144 000 person days, with operations and maintenance representing 43%, construction and installation 30%, and manufacturing 17% as depicted in Figure 2.4. Construction workers (26 600 person days) are the single largest occupational contingent, followed by factory workers (close to 12 500 person days).

Health: Wind, solar and hydropower produce little or no air pollution. Technologies, such as biomass and geothermal, do emit air pollutants, but at much lower rates than most conventional fuels. According to the World Health Organization (WHO), about 99% of the people in the world breathe air that exceeds air quality limits and threatens their health, and more than 13 million deaths around the world each year are due to avoidable environmental causes, including air pollution. The unhealthy levels of fine particulate matter and nitrogen dioxide originate mainly from the burning of fossil fuels. In 2018, air pollution from fossil fuels caused $2.9 trillion in health and economic cost, about $8 billion a day. Switching to clean sources of energy, such as wind and solar, thus helps address not only climate change but also air pollution and health.

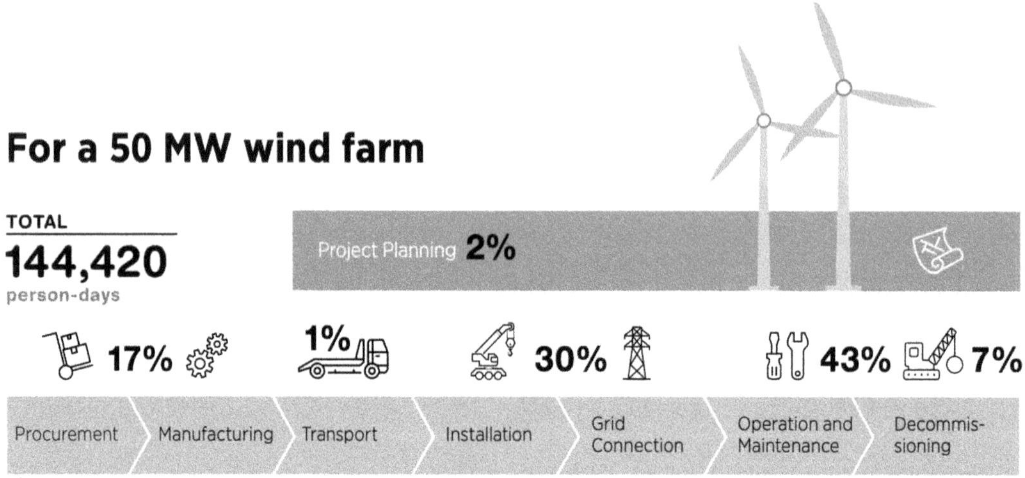

Source: IRENA, 2017d.

Figure 2.4 Employment impacts in the onshore wind value chain.

Resilience: Renewable energy systems are more resistant to global shocks than the systems based on fossil fuels. This is because renewable energy is locally produced, managed, and maintained, so it doesn't rely on imports. Renewable energy projects are less likely to be damaged or disrupted by extreme weather events. For example, during Hurricane Sandy in 2012, renewable energy projects continued to operate with minimal disruption, while fossil fuel-based electric generation was damaged. Solar energy can be distributed to individual buildings, allowing them to generate their own power. This can help create microgrids that can provide power to critical infrastructure during outages. Solar energy can be stored and discharged without the need for fuel deliveries. This means that solar energy can continue to deliver power even during long outages.

Access to Energy: Having an electricity source that can provide very basic lighting, and charge a phone or power a radio for 4 hours per day. Over one billion people lack access to electricity, while a further one billion have an unreliable supply. Stand-alone and mini-grid renewable electricity solutions viable for the 80% of those without access in rural areas or small developing island states. Off-grid solutions are decentralised and local projects and hence, local jobs are created.

2.3.2 Cost of Renewable Energy

The cost advantage of fossil fuels over renewable energy sources has been decreasing in recent years, and certain renewables can already compete with fossil fuels solely on financial terms. Renewable energy costs are expected to decline further in the future, while fossil fuel prices will likely to rise. Thus even without policies to promote a transition toward renewables, economic factors are currently moving us in that direction.

Cost comparisons between different energy sources are made by calculating the levelized cost of energy (LCOE). Levelized costs represent the present value of building and operating a plant over an assumed lifetime, expressed in real terms to remove the effect of inflation. For energy sources that require fuel, assumptions are made about future fuel costs. The levelized construction and operations costs are then divided by the total energy obtained to allow direct comparisons across different energy sources.

Levelized Cost Of Energy (LCOE): The LCOE is also an important calculation to allow financial experts to compare different energy-producing technologies, such as wind, solar, and nuclear power sources. It allows for these comparisons regardless of unequal life spans, differing capital costs, size of the projects, and the differing risk associated with each project. This is because the LCOE reflects a per-unit cost of electricity generated, and the risk of each project is an implication of the specific discount rate used for each power-generating asset.

The LCOE can be calculated by first taking the net present value of the total cost of building and operating the power generating asset. This number is then divided by the total electricity generation over its lifetime. The total costs associated with the project generally will include:

- The initial cost of investment expenditure (I)
- Maintenance and operations expenditures (M)
- Fuel expenditures (if applicable) (F)

The total output of the power-generating asset will include:

- The sum of all electricity generated (E)

The last two important factors to be considered in the equation are:

- The discount rate of the project (r)
- The life of the system (n)

$$LCOE = \Sigma[(I_t + M_t + F_t) / (1 + r)^t] / \Sigma[(E_t/(1 + r)^t]$$

Trend in LCOE of Renewable Energy Sources: The cost of electricity from renewable energy technologies has fallen steadily, and even dramatically, in recent years. This is especially the case since 2000, with the rise of solar and wind power generation as viable commercial options. Today, power generation from renewable sources and technologies has become increasingly competitive with, or indeed, lest costly than, fossil-based or nuclear power. Where untapped and economical resources exist, bioenergy, hydropower and onshore wind all offer new, low-cost power generation. Recent and often rapid cost declines for electricity from solar photovoltaics (PV), offshore wind and concentrating solar power (CSP) mean that these technologies, too, can offer competitive electricity, either now or in the next few years when contracted plants are commissioned.

For plants commissioned in 2016, the global weighted average cost of electricity from bioenergy was USD 0.066 per kilowatt-hour (kWh), from hydropower USD 0.048/kWh, from onshore wind USD 0.07/kWh, from geothermal USD 0.064/kWh, from solar PV USD 0.11/kWh, from offshore wind USD 0.152/kWh and from CSP USD 0.27/kWh. The global weighted average costs of electricity from all renewable technologies except CSP and offshore wind now fall within the range of fossil fuels (USD 0.45/kWh to USD 0.14/kWh).

Solar PV and onshore wind have seen very rapid cost reductions in recent decades. PV modules have experienced learning rates of 18% to 22%, and module prices have fallen by around 80% since 2010. Onshore wind has experienced a learning rate of 15% for the cost of electricity delivered, as installed cost reductions (wind turbine prices have fallen 38% on average since 2009) and performance improvements have raised yields. Between 2010 and 2016, the global weighted average cost of electricity from utility- scale solar PV plants commissioned in those years fell 69%, from USD 0.36 to USD 0.11/kWh. At the same time, the 1st and 99th percentile values fell from a range of USD 0.13 to USD 0.49/kWh to a range of USD 0.07 to USD 0.26/kWh. Recent auction and tender results suggest that the lower range could fall to just USD 0.04/kWh by 2019. Where excellent solar resources exist in the sunbelt, low-cost financing is available, and competitive installed costs and operation and maintenance (O&M) costs can be achieved. Auction results in Abu Dhabi, Chile, Dubai, Mexico and Saudi Arabia all suggest that USD 0.03/kWh will be achievable from 2019 and beyond. Where good resources exist, onshore wind now represents one of the least-cost sources of new electricity generation capacity. The global weighted average cost of electricity from onshore wind fell 18% between 2010 and 2016, from USD 0.085 to USD 0.07/kWh, with projects regularly delivering electricity for just USD 0.04/kWh. Recent auction results suggest costs as low as USD 0.03/kWh within the next two to three years in areas with excellent wind sites. This decline has been driven by falling wind turbine costs since 2009, as well as by increasing hub heights and larger swept areas that allow today's turbines to harvest much more electricity from the same site than turbines from five to ten years ago. Over the space of 17 years, the average capacity factor of new wind farms commissioned in Denmark doubled, while those in Brazil increased 83%, in the United States 46% and in Germany 41%.

Although recently commissioned offshore wind and CSP projects have been relatively expensive, 2016 and 2017 represented watershed years for these technologies. Both are still in the infancy of their deployment, at 13 gigawatts (GW) and 5 GW, respectively. In 2016 and 2017 the results of auctions for offshore wind delivery up until around 2022 in Denmark, Germany, the Netherlands and the United Kingdom saw the cost of electricity fall to the range of USD 0.06 to USD 0.10/kWh, which is highly competitive for new generation capacity in Europe. Similarly, CSP tenders in Australia and Dubai yielded highly competitive prices of around USD 0.07/kWh for a dispatchable, renewable generation technology that has less than 2% of the cumulative installed capacity of solar PV.

While solar and wind power technologies are commercially mature, they still have significant potential for cost reduction. By 2025 the global weighted average cost of electricity from solar PV could fall by as much as 59%, and from CSP by up to 43%. Onshore and offshore wind could see cost declines of 26% and 35%, respectively. Figure 2.5 illustrates the trends in Levelized cost of electricity by project and renewable generating technology in 2010 and 2016.

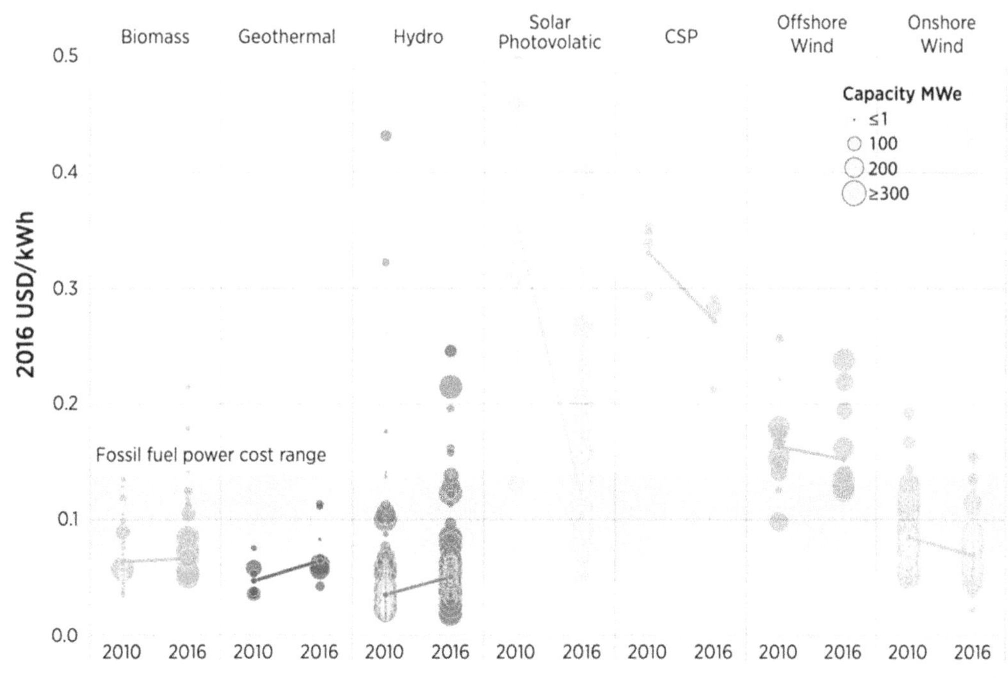

Figure 2.5 Levelized cost of electricity by project and renewable generating technology in 2010-2016.

Facilitating Further Penetration Renewable Energy Sources: Significant progress is being made in decarbonising the electricity sector, notably with the growth of the variable renewable electricity technologies, meaning solar and wind power. Experience in leading markets has demonstrated the viability of integrating high shares of these sources smoothly into existing power grids, with smart technologies helping to manage daily and seasonal variations in power supply. The focus must now increasingly shift to how to integrate renewables in the end-use sectors (heating, cooling, transport, etc.).

This need to accommodate variable energy supply while providing undisrupted output in the electricity sector, as well as efforts to integrate renewables into the end-use sectors has brought into sharp relief the significant potential, as well as crucial importance, of electricity storage to facilitate deep decarbonisation.

Electricity storage that is based on rapidly improving batteries and other technologies will permit greater system flexibility, a key asset as the share of variable renewables increases. More directly, electricity storage makes possible a transport sector dominated by electric vehicles; enables effective, 24-hour off-grid solar home systems; and supports 100% renewable mini-grids.

Cost reduction potential until 2030: Total electricity storage capacity could triple in energy terms by 2030, in tandem with rapid uptake of renewable energy. This assumes sufficient uptake to double the share of renewables in the global energy mix in less than a decade and a half. With growing demand for electricity storage from stationary and mobile applications, the total stock of electricity storage capacity will need to grow more than 150%, from an estimated 4.67 terawatt-hours (TWh) in 2017 to 12 TWh or more by 2030.

The cost reduction potential for new and emerging electricity storage technologies is significant. The total installed cost of a lithium-ion battery could fall by at least an additional 54% by 2030 in stationary applications. This will open up new economic markets for electricity storage, as well as increase their role in providing flexibility services to the grid.

Other battery storage technologies also offer large cost reduction potential. The total installed cost of "flow batteries" could drop two-thirds by 2030. High-temperature sodium sulphur (NaS) and sodium nickel chloride batteries also will become much more affordable. Their installed cost could fall 56% to 60% by 2030, at the same time that their performance improves. The installed cost of flywheels could fall 35% by 2030. Compressed air energy storage (CAES), although based on a combination of mature technologies, could see a 17% cost decline by 2030.

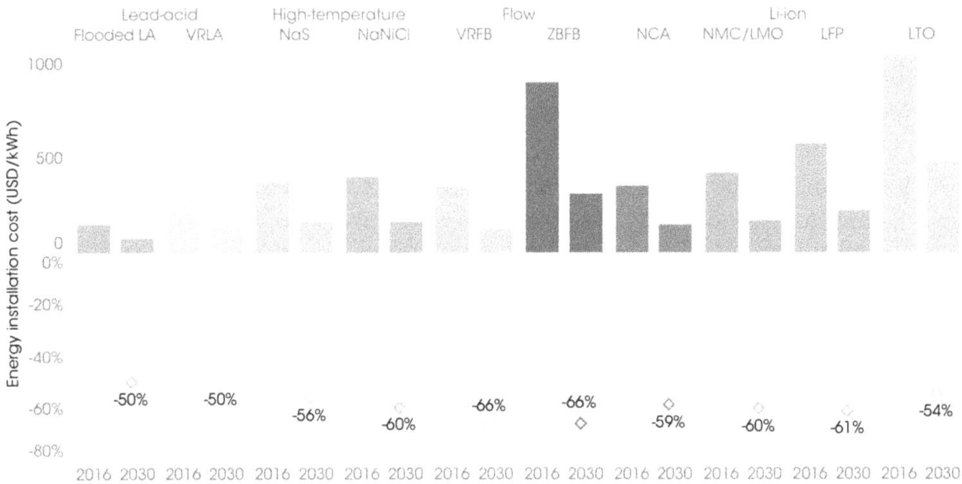

Figure 2.6 Battery electricity storage systems: Installed energy cost reduction potential, 2016-2030.

2.4 Renewable Energy Integration

The share of renewables in overall power generation is rapidly increasing, both in developed and developing countries. Furthermore, many countries have ambitious targets to transform their power sector towards renewables. To achieve these objectives, the structure and operation of existing power grid infrastructures will need to be revisited as the share of renewable power generation increases. Renewable energy technologies can be divided into two categories:

- Dispatchable (i.e. biomass, concentrated solar power with storage, geothermal power and hydro).
- Non-dispatchable, also known as Variable Renewable Energy or VRE (i.e. ocean power, solar photovoltaics and wind).

VRE has four characteristics that require specific measures to integrate these technologies into current power systems:

- Variability due to the temporal availability of resources.
- Uncertainty due to unexpected changes in resource availability.
- Location-specific properties due to the geographical availability of resources.
- Low marginal costs since the sources are freely available.

A transition towards high shares of VRE requires a re-thinking of the design, operation and planning of future power systems from a technical and economic point of view. In such a system, supply and demand will be matched in a much more concerted and flexible way. From a technical perspective, VRE generation can be ideally combined with smart grid technologies, energy storage and more flexible generation technologies. From an economic perspective, the regulatory framework will need to be adjusted to account for the cost structure of VRE integration, to allow for new services and revenue channels, and to support new business models. There are several technological options that can help to integrate VRE into the power system grid: system-friendly VREs, flexible generation, grid extension, smart grid technologies, and storage technologies. New advances in wind and solar PV technologies allow them to be used over a wider range of conditions and provide ancillary services like frequency and voltage control. Flexible generation requires changes in the energy mix to optimise production from both dispatchable and non-dispatchable resources.

The integration of a significant share of variable renewables into power grids requires a substantial transformation of the existing networks to:

- Allow for a bi-directional flow of energy; that is top-down (from generators to users) and bottom-up (with end-users contributing the electricity supply) aimed at ensuring grid stability while installing distributed generation.
- Establish an efficient electricity-demand and grid management mechanisms aimed at reducing peak loads, improving grid flexibility, responsiveness and security of supply to deal with increased systemic variability.
- Improve the interconnection of grids at the regional, national and international level, aimed at increasing grid balancing capabilities, reliability and stability.
- Introduce technologies and procedures to ensure proper grid operation stability and control (e.g. frequency, voltage, power balance) in the presence of a significant share of variable renewables
- Introduce energy storage capacity to store electricity from variable renewable sources when power supply exceeds.

2.5 Smart Grid Technologies

Smart grid technologies can act as an enabler for VRE integration, given their ability to reduce the variability in the system by allowing the integration of renewables into diverse electricity resources, such as

- Load control (e.g. Demand Side Management (DSM),
- Advanced Metering Infrastructure (AMI) by implementing advanced technologies (e.g. smart inverters, Phasor Measurement Unit (PMU) and Fault Ride Through (FRT) capabilities).

- Energy storage technologies can alleviate short-term variability (up to several hours), or longer-term variability through
  - Pumped-storage hydroelectricity
  - Thermal energy storage
  - Conversion of electricity into hydrogen or gas.

Smart grids are meant to improve reliability and service, enhance markets, reduce costs and improve the efficiency of power systems. Consequently, smart grids can play a crucial role in facilitating the smooth integration of high shares of variable renewables and supporting distributed generation [SG&RE]. They involve a new integrated architecture for transmission and distribution with smart metering of bi-directional flows of energy and communication technologies that provide central operators with the information needed for the efficient operation and monitoring of the grid (e.g. fault detection, isolation and system restoration), efficient control of electricity supply (from both centralised and distributed power plants) and demand (i.e. smoothing demand peaks and/or meeting peak demand with variable renewables as well as possible).

This new architecture may eventually include automated "intelligent" management of end-use appliances to shift electricity demand towards off-peak periods (e.g. overnight) or make use of energy as and when available from variable renewables. Of key importance is the involvement of variable renewables and distributed generation plants in several tasks aimed at ensuring efficient and reliable operations (e.g. voltage and frequency regulation, reactive power regulation, active power reservation, congestion management, optimisation of grid losses, network restoration). By reducing peak demand and enabling more efficient energy management, smart grids can also reduce the need for new and back-up capacity. Smart grids require the use of power electronics components, such as advanced inverters, which are crucial for the integration of variable renewables, especially when higher penetrations (around 15%) are reached. Besides the main task of feeding renewable-based power into the grid, inverters provide additional functions, such as the balance of active and reactive powers. They can participate in frequency control through active power regulation (with proper control algorithms) and can provide voltage control, fault ride-through capability, and a reduction of grid losses that depend mainly on reactive power control.

2.5.1 Cost Considerations

The high costs for power generation in these markets make VREs and grid integration technologies economically attractive since they can simultaneously improve the reliability, efficiency and performance of these power systems. For example, the case of the Smart Grid demonstration project in Jeju Island, South

Korea. Furthermore, the right assessment and understanding of VRE integration costs are relevant for policy making and system planning.

Any economic analysis of the transition towards renewables-based power systems should, therefore, consider all different cost components for VRE grid integration, such as

- Grid costs (e.g. expansion and upgrading),
- Capacity costs
- Balancing cost.

Integration costs are due not only to the specific characteristics of VRE technologies but also to the power system and its adaptability to greater variability. Therefore, these costs should be carefully interpreted and not entirely attributed to VRE, especially when the system is not flexible enough to deal with variability (i.e. in the short-term). Moreover, RE integration delivers broader benefits beyond purely economic ones, such as social and environmental benefits. Even though not straightforward, these externalities should be considered and quantified to integrate them into the decision-making process and maximise socio-economic benefits.

Due to the rapid technological progress and multiple grid integration options available, policy makers should build a framework for RE grid integration based on the current characteristic of the system, developing technological opportunities and long-term impacts and targets. Policy makers should adopt a long-term vision for their transition towards renewables and set regulatory frame-works and market designs to foster both RE development and management of greater system variability. Such regulatory frameworks could include new markets for ancillary services and price signals for RE power generators that incentivise the reduction of integration costs.

2.5.2 Technical Challenges

Technical issues to be addressed for renewables integration into power grids also include the connection of renewable power plants to distribution grids, Fault Ride-Through (FRT) operations, their impact on protection systems and bi-directional communication.

Capacity of VRE Source: The connection of renewable electricity generation plants to distribution grids requires the analysis of several factors which may impact the grid's operation. A major criterion for plant connection is the impact on the grid voltage during normal operations. In conventional distribution networks with radial configuration (i.e. centralised power generation), a node voltage decreases, because of a voltage drop in network elements. In networks with distributed generation, either negative or positive deviations of node voltage may occur. The RE to be connected, is required to keep the voltage increase in an acceptable range of typically 2%–3%. Voltage changes can be determined by complex load-flow calculations. The voltage increase depends on the network short-circuit capacity and the impedance angle at the connection point. It also depends on equipment characteristics (i.e. transformers, distribution lines) and is also affected by the phase angle (ϕ) of the plant, which can be adjusted by reactive power control.

An approximate calculation of the expected relative voltage increase is given by the formula:

$$\Delta U_{aV} = [S_{Amax} \times \cos(\psi_{kV} + \phi)] / S_{kV}$$

qhere:

S_{Amax}: Maximum apparent power of the generating plant

S_{kV}: Short-circuit power at the connection point

ψ_{kV}: Network impedance angle at the connection point

Φ: Angle between generating plant current & voltage

A study in 2011 calculated for various cases that the operation of PV plants at a power factor of 0.9 could increase the maximum allowable PV power that could be connected to low-voltage radial distribution networks by a factor of 1.5 to more than 2. However, the thermal limits of transformers and cables must be considered. In general, a high withdrawal of inductive reactive power can reduce the voltage rise. However, a trade-off must be accepted as this also leads to increased losses and reduced transmission capacity. A further criterion for connecting a renewable electricity generation plant to the distribution grid is the thermal limit of the grid components (mainly electric lines). At the Medium Voltage (MV) level, the loading limits of the lines are determined by their short-circuit capacity. The network short-circuit current is increased by the power plant short-circuit current capacity, particularly in the vicinity of the connection point.

The following rules of thumb can be used to estimate the contribution of the power plant to the short-circuit current:

- Eight times the rated current for synchronous generators.
- Six times the rated current for asynchronous generators and double-fed asynchronous generators.
- One times the rated current for generators with inverters; (this depends on the timescale and could be 3–5 times for events under one second).

A major requirement is that the total fault level, which is determined by the combined short-circuit contribution of the upstream grid and the distributed electricity source, remain below the network rated value. This constraint is often the main inhibiting factor for connecting new distributed electricity sources to existing grids. Distributed electricity sources coupled with inverters can relieve this constraint.

Fault Ride-Through Operation (FRT): In conventional power grids, inverter-based distributed generation plants must be quickly disconnected when the grid voltage or frequency exceeds the allowable operating range. In grids with a high share of distributed renewable units, the simultaneous loss of a large number of generation plants and capacity due to short-term voltage or frequency fluctuations (which may, for example, result from a fault in the transmission network) can threaten the grid's overall stability. Under such circumstances, appropriate management of the generation plants can help react to the voltage and frequency deviations and avoid grid instability or even collapse. The ability of generation plants to remain connected to the network during such short-term fluctuations referred to as Fault Ride-Through (FRT) capability. This is crucial for large-scale renewables integration into the power grids. FRT requirements for the connection of generation plants to transmission and distribution networks have already been included in several national grid regulations to solve the problems associated with the connection of wind parks and a large number of PV systems. This means that generation plants must remain connected for a certain length of time in the case of network faults and support the network voltage by injections of reactive power.

Effect on Protection System: The operation of a power grid requires protection systems to detect abnormal conditions and restore normal operations through corrective action. In general, distribution networks are overcurrent protected. Protection systems are usually designed assuming a mono-directional power flow and proper coordination of overcurrent devices based on the available fault current. The introduction of distributed electricity sources into the distribution network may cause unwanted impacts on the protection systems, mainly due to load flow changes and increased fault current contribution. For example, the introduction of a distributed source and its associated current may cause an erroneous operation of distance relays. Furthermore, in contrast to the basic rule of the protection systems that disconnect only the faulted parts, the overcurrent protection device of distributed sources may also be activated by faults in adjacent sections (sympathetic tripping). Therefore, to ensure safe and selective protection, the impact of distributed electricity sources on protection systems should be considered in planning grid operations by considering new relays, as well as the development of new directional protection algorithms.

Communication Technology: The introduction of variable renewable power generation and distributed electricity generation in power grids requires significant advances in monitoring and control systems to obtain optimal performance. As a consequence, information and communication technologies (ICT) are key elements of smart grids. To allow local renewable and distributed generation units to interact with each other and with the grid management systems, a harmonisation of communication methods and/or physical media employed by different vendors and users is needed. In this context, developers and manufacturers all over the world propose harmonisation through specific standards, with the IEC 61850 as the most representative. This standard permits inter-operation between different systems from different vendors, thus increasing benefits to all owners, operators and users of distributed generation systems. The IEC 61850 models cover all operational aspects of distributed generation systems. However, they do not address market operations.

References

1. F. Blaabjerg and J. M. Guerrero, "Smart grid and renewable energy systems," 2011 International Conference on Electrical Machines and Systems, Beijing, China, 2011, pp. 1-10, doi: 10.1109/ICEMS.2011.6073290.
2. Y. Sabri, N. El Kamoun and F. Lakrami, "A Survey: Centralized, Decentralized, and Distributed Control Scheme in Smart Grid Systems," 2019 7th Mediterranean Congress of Telecommunications (CMT), Fez, Morocco, 2019, pp. 1-11, doi: 10.1109/CMT.2019.8931370.
3. https://www.statista.com/statistics/856703/global-population-with-access-and-without-to-electricity/#:~:text=In%202022%2C7.2%20billion%20people,over%2090%20percent%20in%202022.
4. F. Martín-Martínez, A. Sánchez-Miralles, M. Rivier, C.F. Calvillo, “Centralized vs distributed generation. A model to assess the relevance of some thermal and electric factors. Application to the Spanish case study”, Energy, Volume 134, 2017, Pages 850-863, ISSN 0360-5442, https://doi.org/10.1016/j.energy.2017.06.055
5. Md. Rejwanur Rashid Mojumdar1 , Mohammad Sakhawat Hossain Himel , Golam Kayes, “A Distinctive Analysis between Distributed and Centralized Power Generation” International Journal of Recent Research in Electrical and Electronics Engineering (IJRREEE) Vol. 2, Issue 4, pp: (1-6), Month: October - December 2015
6. A report on “The Potential Benefits of Distributed Generation and Rate-Related Issues That May Impede Its Expansion” by department of Energy June 2007.
7. Lovins, A., Datta, K. and T. Feiler, A. Lehmann, K. Rabago, J. Swisher, K. Wicker, 2002. Small is Profitable: The Hidden Economic Benefits of Making Electrical Resources the Right Size. Rocky Mountain Institute, Snowmass, Colorado.
8. https://www.irena.org//media/Files/IRENA/Agency/Publication/2017/Nov/IRENA_Understanding_Socio_Economics_2017.pdf
9. https://www.ise.fraunhofer.de/content/dam/ise/en/documents/publications/studies/EN2024_ISE_Study_Levelized_Cost_of_Electricity_Renewable_Energy_Technologies.p

df&tbm=ilp&ctx=atr&sa=X&ved=2ahUKEwjZjMv054aLAxWZ4TgGHe6MOLYQv5AHegQIABAD

10. https://www.bu.edu/eci/files/2019/06/RenewableEnergyEcon.pdf
11. https://corporatefinanceinstitute.com/resources/valuation/levelized-cost-of-energy-lcoe/
12. https/www.irena.org//media/Files/IRENA/Agency/Publication/2017/Nov/%20IRENA_Sharply_falling_costs_2017
13. https://www.stannescet.ac.in/cms/staff/qbank/EEE/Notes/EE8703-RENEWABLE%20ENERGY%20SYSTEMS-644988498-RES%20UNIT%203.pdf
14. https://www.irena.org/-/media/Files/IRENA/Agency/Publication/2015/IRENA-ETSAP_Tech_Brief_Power_Grid_Integration_2015.pdf

CHAPTER

3

Renewable Energy Sources and Integration

3.1 Introduction

The basic principle behind both solar thermal and solar photovoltaic (PV) is the same. They absorb raw energy from the sun and use it to create usable energy. In solar PV systems this is through the creation of electricity, whereas thermal systems are used directly for heating water or air. The amount of solar radiation on the earth surface can be instrumentally measured using Pyrheliometer, Pyranometer, Photoelectric sunshine recorder and many instruments. Solar thermal power plants collect and concentrate sunlight to produce the high temperature heat needed to generate electricity. Thermal energy storage (TES) is a technology that stocks thermal energy by heating or cooling a storage medium so that the stored energy can be used at a later time for heating and cooling applications and power generation. A photovoltaic module consists of multiple PV cells connected in series to provide a higher voltage output. A photovoltaic array is a system composed of multiple PV modules. They can be connected in one or more series circuits, which are connected to a combiner box to provide a single direct-current output.

3.2 Solar Energy

The basic principle behind both solar panels of solar photovoltaic (PV) and solar thermal is the same. They absorb raw energy from the sun and use it to create usable energy. In solar PV systems the solar energy is converted to electricity, where in thermal systems the solar energy is used directly for heating water or air. Solar thermal power plants collect and concentrate sunlight to produce the heat needed to generate electricity. Thermal energy storage (TES) is used to store the heat produced by heating or cooling a storage medium so that the stored energy can be used later for heating and cooling applications and power series/parellal generation. A photovoltaic module consists of multiple PV cells connected in series to provide a higher voltage output. A photovoltaic array is a system composed of multiple PV modules. The amount of so radiation on the earth surface can be instrumentally measured using Pyrheliometer, Pyranometer, Photoelectric sunshine recorder and many instruments.

Solar radiation, often called the solar resource, is a general term for the electromagnetic radiation emitted by the sun. The sun emits electromagnetic radiations as a black body having a surface temperature of about 6000 K. This is because of the internal nuclear reaction, where hydrogen is converted into helium.

The solar power intensity without the effect of the atmosphere or ground surface is called the Air Mass Zero where one AM0 = 1 SC = 135.3 mW/cm^2. The solar power intensity after crossing one air mass perpendicular to the earth is called the air mass 1, AM1. It represents the area under the spectral irradiance curve and amounts to 92.5 mW/cm^2. If θ is the angle of incidence with normal to the earth surface, then the optical path in units of the air mass will be larger. The air mass AM = 1/(cos θ).

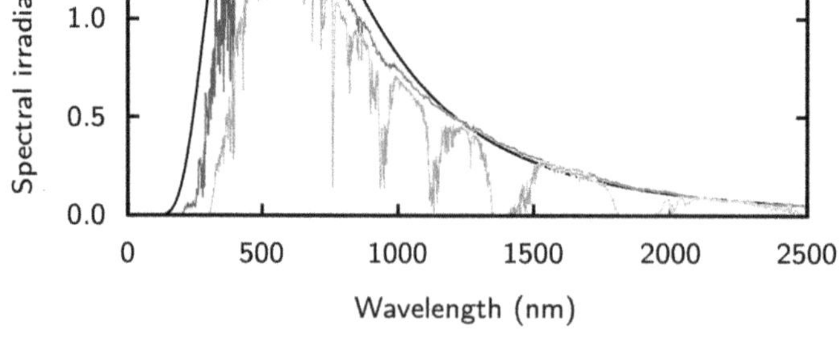

Figure 3.1 The power spectral distribution of the sunlight.

As the solar radiation passes through the atmosphere, it gets absorbed, scattered, reflected, or transmitted. All these processes result in reduction of the energy flux density. Actually, the solar flux density is reduced by about 30% compared to extraterrestrial radiation flux on a sunny day and is reduced by as much as 90% on a cloudy day. The major losses include:

- Absorbed by particles and molecules in the atmosphere: 10-30%
- Reflected and scattered back to space: 2-11%
- Scattered to earth (direct radiation becomes diffuse): 5-26%

As a result, the direct radiation reaching the earth surface (or a device installed on the earth surface) never exceeds 83% of the original extraterrestrial energy flux. This radiation that comes directly from the solar disk is defined as beam radiation. The scattered and reflected radiation that is sent to the earth surface from all directions (reflected from other bodies, molecules, particles, droplets, etc.) is defined as diffuse radiation. The sum of the beam and diffuse components is defined as total (or global) radiation. The beam radiation can be concentrated, while the diffuse radiation, in many cases, cannot be concentrated. Short-wave radiation, in the wavelength range from 0.3 to 3 µm, comes directly from the sun. It includes both beam and diffuse components. Long-wave radiation, with wavelength 3 µm or longer, originates from the sources at near-ambient temperatures - atmosphere, earth surface, light collectors, other bodies. The solar radiation reaching the earth is highly variable and depends on the state of the atmosphere at a specific locality. Two atmospheric processes can significantly affect the incident irradiation: scattering and absorption.

Scattering is caused by interaction of the radiation with molecules, water and dust particles in the air. The amount of light scattered depends on the number of particles in the atmosphere, particle size and the total air mass the radiation comes through.

Absorption occurs upon interaction of the radiation with certain molecules, such as ozone (absorption of short-wave radiation - ultraviolet), water vapour, and carbon dioxide (absorption of long-wave radiation - infrared). Due to these processes, out of the whole spectrum of solar radiation, only a small portion reaches the earth surface. Thus, most of x-rays and other short-wave radiation is absorbed by atmospheric components in the ionosphere, ultraviolet is absorbed by ozone and not-so abundant long-wave radiation is absorbed by Carbonioxide. As a result, the main wavelength range to be considered for solar applications is from 0.29 to 2.5 µm.

Insolation is the incident solar radiation onto some object. Specifically, it is a measure of the solar energy that is incident on a specified area over a set period of time. Generally, insolation is expressed in two ways. One unit is kilowatt-hours per square meter (kWh/sq.m) per day which represents the average amount of energy hitting an area each day. Another form is watts per square meter (W/sq.m) which represents the average amount of power hitting an area over an entire year. It is important to have values for insolation at certain positions on the Earth as these figures are used to help determine the size and output of solar power systems. Values for insolation can help to determine the expected output for solar panels and determine where on Earth solar panels would be most effective.

3.2.1 Solar Photo Voltaic Cells – Working and Construction

A photovoltaic cell is comprised of many layers of materials, each with a specific purpose. The most important layer of a photovoltaic cell is the specially treated semiconductor layer. It is comprised of two distinct layers of p-type and n-type which converts the Sun's energy into useful electricity through a process called the photovoltaic effect. On either side of the semiconductor there exists a conducting layer which "collects" the electricity produced. The backside or shaded side of the cell can be completely covered in the conductor, whereas the front or illuminated side uses the conductors sparingly to avoid blocking radiation from reaching the semiconductor. The final layer of the illuminated side of the cell has anti-reflection coating. Since all semiconductors are naturally reflective, reflection loss can be significant.

Light is composed of photons, which are simply small bundles of electromagnetic radiation or energy. When light of a suitable wavelength is incident on these cells, energy from the photon is transferred to an electron of the semiconducting material, causing it to jump to a higher energy state known as the conduction band. In their excited state in the conduction band, these electrons are free to move through the material, and it is this motion of the electron that creates an electric current in the cell. Figure 3.2 illustrates Working Principle of Solar PV Cell.

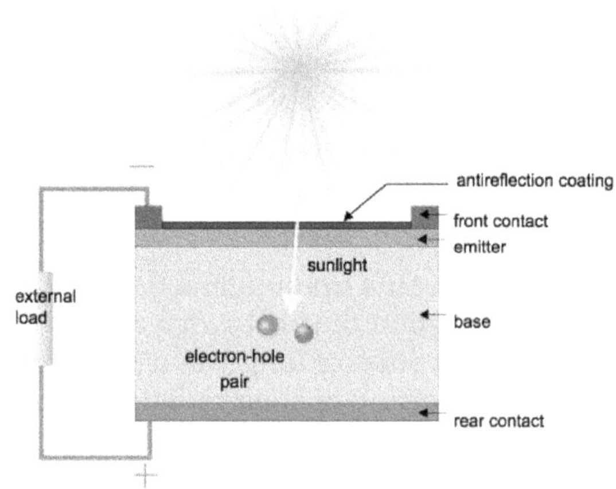

Figure 3.2 Working Principle of Solar PV Cell.

Cell: A photovoltaic cell is the most basic unit of a solar PV system - solar cells can be either monocrystalline or polycrystalline, and their key characteristic is that they produce a voltage output when exposed to light. It is important to note that although they are normally called "solar cells", they can respond to any type of light. Each cell produces approximately 0.5 volt, and a solar module can have any number of solar cells.

Module: A photovoltaic module consists of multiple PV cells connected in series to provide a higher voltage output. PV modules are manufactured in standard sizes. A solar module designed for charging a 12-volt battery will typically have 36 solar cells while the typical residential grid connected system uses solar modules with 60 solar cells. For large commercial and utility scale solar systems, solar modules will have typically 72 solar cells. By increasing the number of solar cells, the module voltage and wattage increases. The term solar panel is sometimes used interchangeably with solar module. The main difference is that some solar panels models are composed of multiple modules mounted together.

Array: A photovoltaic array is a system composed of multiple PV modules. They can be connected in one or more series circuits, which are connected to a combiner box to provide a single direct-current output. This output can be used to charge batteries, power DC loads, or fed to an inverter to provide an AC voltage for home appliances or exporting to the electric grid. Figure 3.3 demonstrates various levels of Solar PV

Types of Solar Cells: There are different types of photovoltaic cells available to buy, but mainly they are manufactured from silicon (Si). The use of silicon in the manufacture of photovoltaic cells produces the stereo typical uniform blue coloured PV cell which we see on roof tops and the sides of buildings. The two major types of photovoltaic cell materials used are crystalline silicon and thin film deposits, which vary from each other in terms of light absorption efficiency, energy conversion efficiency, manufacturing technology and cost of production. Crystalline silicon PV cells are the most common type of photovoltaic cell in use today and are also one of the earliest successful PV devices. The three general types of photovoltaic cells made from silicon are:

- Mono-crystalline Silicon – also known as single-crystal silicon
- Poly-crystalline Silicon – also known as multi-crystal silicon
- Thin Film Silicon

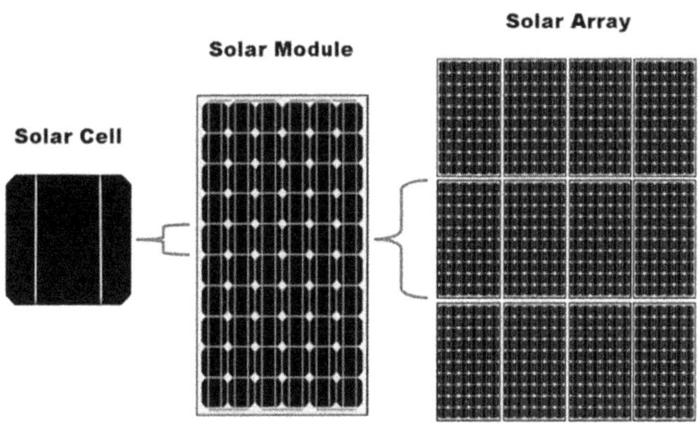

Figure 3.3 Cell, Module and Array.

Crystalline Silicon (c-Si) is the most common technology used to produce photovoltaic cells representing about 90% of the market today. Crystalline photovoltaic cells are made from silicon which is first melted and then crystallised into ingots or castings of pure silicon. Thin slices of silicon called wafers, are cut from a single crystal of silicon or from a block of silicon crystals to make individual cells. The conversion efficiency for these types of photovoltaic cell ranges between 10% and 20%.

Mono-crystalline Silicon is a type of photovoltaic cell material manufactured from a single-crystal silicon structure which is uniform in shape because the entire structure is grown from the same crystal. High purity silicon is melted in a crucible. A single-crystal silicon seed is dipped into this molten silicon and is slowly pulled out from the liquid producing a single-crystal ingot. The ingot is then cut into very thin wafers or slices which are then polished, doped, coated, interconnected and assembled into modules and arrays. These types of photovoltaic cells are also widely used in photovoltaic panel construction. Compared to non-crystalline cells, the uniform molecular structure of the silicon wafer makes it ideal for transferring loose electrons through the material resulting in a high energy conversion efficiency. The conversion efficiency for a mono-crystalline cell lies between 15 to 20%. Not only are they energy efficient, mono-crystalline photovoltaic cells are highly reliable for outdoor power applications due to their wafer thickness. However, to make an effective PV cell, the silicon has to be "doped" with other elements to make the required N-type and P-type conductive layers. Figure 3.4. shows Monocrystalline solar cell and its panel.

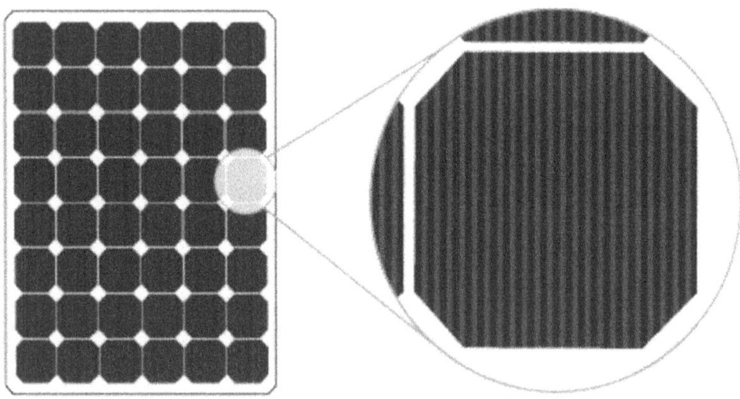

Figure 3.4 Monocrystalline solar panels.

Poly-crystalline Silicon also known as multi-crystalline silicon, is cast to produce a silicon ingot. The silicon molecular structure consists of several smaller groups or grains of crystals, which introduce boundaries between them. Poly-crystalline PV cells are less energy efficient than the mono- crystalline silicon PV cells because these boundaries restrict the flow of electrons through it by encouraging the negative electrons to recombine with the positive holes reducing the power output of the cell. The result of this means that a poly-crystalline PV cell only has an energy conversion efficiency of between 10 to 14%. However, these types of photovoltaic cell are much less expensive to produce than the equivalent single mono-crystalline silicon due to their lower manufacturing costs. A Polycrystalline solar cell and its panel is shown Figure 3.5.

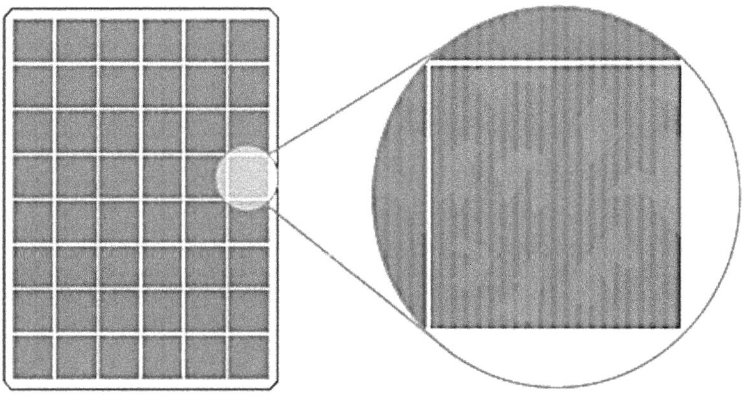

Figure 3.5 Polycrystalline solar panels.

Thin Film Solar Cells are another type of photovoltaic cell which were originally developed for space applications with a better power-to-size and weight ratio compared to the other crystalline silicon devices. Thin film photovoltaics are produced by printing or spraying a thin semiconductor layer of PV material onto a glass, metal or plastic foil substrate. By applying these materials in thin layers, the overall thickness of each photovoltaic cell is substantially smaller than an equivalent cut crystalline cell, hence the name "thin film". As the PV materials used in these types of photovoltaic cells are sprayed directly onto a glass or metal substrate, the manufacturing process is therefore faster and cheaper making thin film PV technology more viable for use in a home solar system as their payback time is shorter. However, although thin film materials have higher light absorption than equivalent crystalline materials, thin film PV cells suffer from poor cell conversion efficiency due to their non-single crystal structure, requiring larger sized cells. Semiconductor materials used for the thin film types of photovoltaic cell include Cadmium Telluride, Amorphous Silicon and Copper Indium Di Selenide

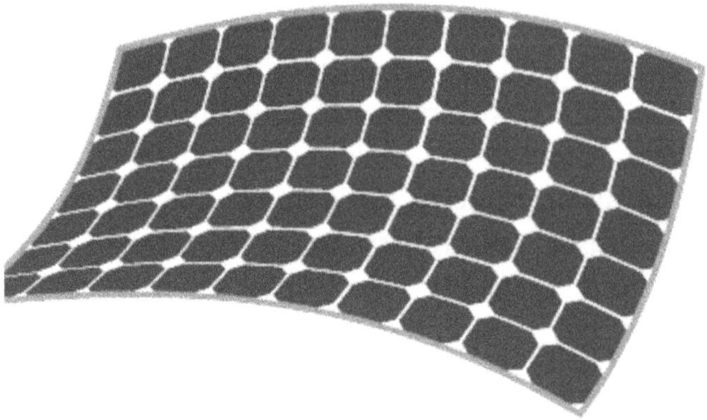

Figure 3.6 Thin film solar panels.

Cadmium Telluride (CdTe) is a poly-crystalline semiconductor material made from cadmium and tellurium. Thin film cadmium telluride has a high light absorption level so the amount of CdTe required can be quite minimal with less than 1.0 microns of semiconductor material is needed to effectively absorb sunlight for the solar device to perform. Although the process of spraying or printing the thin film is relatively easy making it cheap to manufacture these types of photovoltaic cell, the main material, cadmium is a toxic heavy metal can pollute the environment if the cell is damaged or broken. Another disadvantage of these types of photovoltaic cells is that the conversion efficiency for a cadmium telluride PV cell can be low at less than 10%.

Amorphous Silicon, (a-Si) is a non-crystalline form of silicon that is widely used in calculators, consumer electronics and solar garden products that require a small current at a low voltage. Of the different types of photovoltaic cell available, amorphous silicon has the highest light absorption of over 40 times higher than crystalline silicon. The advantage of this is that a much thinner layer of amorphous silicon material is required to make a thin film PV cell reducing manufacturing costs and price. Amorphous silicon cells have various advantages and disadvantages. On the plus side, amorphous silicon can be deposited on a variety of low-cost rigid and flexible substrates such as polymers, thin metals and plastics as well as tinted glass for building integration. However, on the minus side, two of the main disadvantages of amorphous silicon (a-Si) is its very low conversion efficiency ranging from between 7 to 9% when new, degrading down within a few months of exposure to sunlight to less than 5%.

Copper Indium diSelenide, (CIS) is another type of poly-crystalline semiconductor material composed of Copper, Indium and Selenium, ($CuInSe_2$). Thin film CIS types of photovoltaic cell can produce conversion efficiencies of nearly 10%, almost double that of amorphous silicon without suffering from the same outdoor degradation problems due to their thicker film. Also, CIS cells are one of the most light-absorbent semiconductor compounds absorbing up to 90% of the solar spectrum. Although Copper Indium diSelenide, CIS cells are efficient, the complexity of the formulation of the semiconductor compound makes them difficult to manufacture and expensive. Also, Indium is a relatively expensive material due to its limited availability with manufacturing safety issues a concern as hydrogen selenide is an extremely toxic gas. Table-3.1 presents the comparison of various technologies in different aspects.

Table 3.1 Comparison various technologies.

Solar panel type	Advantages	Disadvantages
Monocrystalline	• High efficiency/performance • Aesthetics	• Higher costs
Polycrystalline	• Low cost	• Lower efficiency/performance
Thin-film	• Portable and flexible • Lightweight • Aesthetics	• Lowest efficiency/performance

3.2.2 V-I Characteristics

Solar Cell I-V Characteristic Curves show the current and voltage (I-V) characteristics of a particular photovoltaic (PV) cell, module or array giving a detailed description of its solar energy conversion ability and efficiency. Knowing the electrical I-V characteristics (more importantly Pmax) of a solar cell, or panel is critical in determining the device's output performance and solar efficiency. The main electrical characteristics of a PV cell or module are summarized in the relationship between the current and voltage produced on a typical solar cell I-V characteristics curve. The intensity of the solar radiation (insolation) that hits the cell controls the current (I), while the increases in the temperature of the solar cell reduces its voltage (V). Solar cells produce direct current (DC) electricity and current times voltage equals power, so we can create solar cell I-V curves representing the current versus the voltage for a photovoltaic device. Solar Cell I-V Characteristics Curves are basically a graphical representation of the operation of a solar cell or module summarising the relationship between the current and voltage at the existing conditions of irradiance and temperature. I-V curves provide the information required to configure a solar system so that it can operate as close to its optimal peak power point (MPP) as possible.

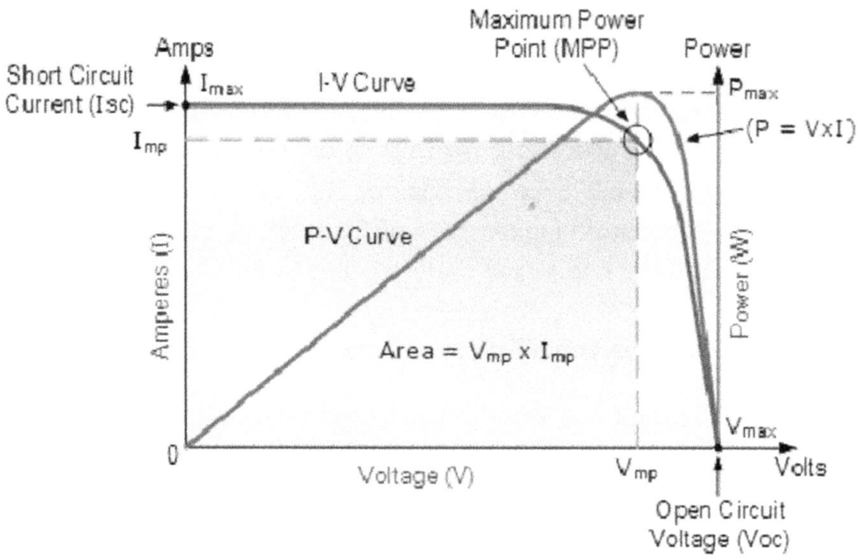

Figure 3.7 Solar Cell I-V Characteristic Curve.

Figure 3.7 shows the current-voltage (I-V) characteristics of a typical silicon PV cell operating under normal conditions. The power delivered by a solar cell is the product of current and voltage. If the multiplication is done, point for point, for all voltages from short-circuit to open-circuit conditions, the power curve above is obtained for a given radiation level. With the solar cell open-circuited, that is not connected to any load, the current will be at its minimum (zero) and the voltage across the cell is at its maximum, known as the solar cells, open circuit voltage, or V_{oc}. At the other extreme, when the solar cell is short circuited, the voltage across the cell is at its minimum (zero) but the current flowing out of the cell reaches its maximum, known as the solar cells short circuit current, or I_{sc}.

Then the span of the solar cell I-V characteristics curve ranges from the short circuit current (I_{sc}) at zero output volts, to zero current at the full open circuit voltage (V_{oc}). In other words, the maximum voltage available from a cell is at open circuit, and the maximum current at closed circuit. Of course, neither of these two conditions generates any electrical power, but there must be a point somewhere in between where the solar cell generates maximum power. However, there is one particular combination of current and voltage for which the power reaches its maximum value, at I_{mp} and V_{mp}. In other words, the point at which the cell generates maximum electrical power, and this is shown at the top right area of the rectangle. This is the "maximum power point" or MPP.

Hence, the ideal operation of a photovoltaic cell (or panel) is defined to be at the maximum power point. The maximum power point (MPP) of a solar cell is positioned near the bend in the I-V characteristics curve. The corresponding values of V_{mp} and I_{mp} can be estimated from the open circuit voltage and the short circuit current: $V_{mp} \cong (0.8–0.90)V_{oc}$ and $I_{mp} \cong (0.85–0.95)\ I_{SC}$. Since solar cell output voltage and current both depend on temperature, the actual output power will vary with changes in ambient temperature.

Photovoltaic panels can be wired or connected together in either series or parallel combinations, or both to increase the voltage or current capacity of the solar array. If the array panels are connected together in a series combination, then the voltage increases and if connected together in parallel then the current increases. The electrical power in Watts, generated by these different photovoltaic combinations will still be the product of the voltage times the current, ($P = V \times I$). However, the solar panels are connected together, the upper right-hand corner will always be the maximum power point (MPP) of the array. The electrical characteristics of a photovoltaic array are summarised in the relationship between the output current and voltage. The amount and intensity of solar insolation (solar irradiance) controls the amount of output current (I), and the operating temperature of the solar cells affects the output voltage (V) of the PV array. Solar cell I-V characteristic curves that summarise the relationship between the current and voltage are generally provided by the panels manufacturer as follows:

Open-circuit voltage (V_{OC}): This is the maximum voltage that the array provides when the terminals are not connected to any load (an open circuit condition). This value depends upon the number of PV panels connected in series.

Short-circuit current (I_{SC}): The maximum current provided by the PV array when the output connectors are shorted together

Maximum power point (MPP): This relates to the point where the power supplied by the array that is connected to the load. The maximum power point of a photovoltaic array is measured in Watts (W) or peak Watts (Wp).

Fill factor(FF): The fill factor is the relationship between the maximum power that the array can actually, provide under normal operating conditions and the product of the open-circuit voltage times the short-circuit current, ($V_{oc} \times I_{sc}$) This fill factor value gives an idea of the quality of the array and the closer the fill factor is to 1 (unity), the more power the array can provide. Typical values are between 0.7 and 0.8.

Percent efficiency: The efficiency of a photovoltaic array is the ratio between the maximum electrical power that the array can produce compared to the amount of solar irradiance hitting the array. The efficiency of a typical solar array is normally low at around 10-12%, depending on the type of cells used. Photovoltaic I-V characteristics curves provide the information needed to configure a solar power array so that it can operate as close as possible to its maximum peak power point. The peak power point is measured as the PV module produces its maximum amount of power when exposed to solar radiation equivalent to 1000 watts per square metre, 1000 W/m2 or 1kW/m2

Conversion Efficiency of the Cell: The efficiency is the most used parameter to compare the performance of one solar cell to another. Efficiency is defined as the ratio of energy output from the solar cell to input energy from the sun. In addition to reflecting the performance of the solar cell itself, the efficiency depends on the spectrum and intensity of the incident sunlight and the temperature of the solar cell. Therefore, conditions under which efficiency is measured must be carefully controlled in order to compare the performance of one device to another. Terrestrial solar cells are measured under AM1.5 conditions and at a temperature of 25°C. Solar cells intended for space use are measured under AM0 conditions. The efficiency of a solar cell is determined as the fraction of incident power which is converted to electricity and is defined as:

$$P_{max} = V_{OC} I_{SC} FF$$

$$\eta = \frac{V_{OC} I_{SC} FF}{P_{in}}$$

where,

V_{oc} is the open circuit voltage.

I_{sc} is the short-circuit current.

FF is the fill factor and

η is the efficiency.

The conversion efficiency is affected by the following factors

Wavelength: The sunlight that reaches the earth's surface has wavelengths from ultraviolet, through the visible range, to infrared. When light strikes the surface of a solar cell, some photons are reflected, while others pass right through. Some of the absorbed photons have their energy turned into heat. The remainder have the right amount of energy to separate electrons from their atomic bonds to produce charge carriers and electric current.

Re-combination: It is one of the fundamental factors that limits efficiency. Indirect recombination is a process in which the electrons or holes encounter an impurity, a defect in the crystal structure, or interface that makes it easier for them to recombine and release their energy as heat.

Temperature: Solar cells generally work best at low temperatures. Higher temperatures cause the semiconductor properties to shift, resulting in a slight increase in current, but a much larger decrease in voltage. Extreme increases in temperature can also damage the cell and other module materials, leading to shorter lifetimes.

Reflection: A cell's efficiency can be increased by minimizing the amount of light reflected away from the cell's surface. For example, untreated silicon reflects more than 30% of incident light. Anti- reflection coatings and textured surfaces help decreased reflection. A high-efficiency cell will appear dark blue or black.

The efficiency of PV cells can be improved by:

- Decreasing the number of semiconductor impurities and crystal structure deformations. This can be achieved through the production of monocrystalline, or "single crystal" cells. A purer and more uniform cell has a higher chance of interacting with incoming photons.
- Another method is to use a more efficient semiconducting material such as Gallium Arsenide. Although it's much rarer and more expensive than silicon, gallium arsenide has an optimal bandgap of 1.4 electron volts, which allows higher percentage of the Sun's energy to be harnessed.
- Multiple layers of semiconductor material called p-n junctions can also be used to increase cell efficiency. These multi-junction cells harness energy from multiple sections of the solar spectrum as each junction has a different band gap energy.
- Efficiency can also be increased through concentrated photovoltaics. This method involves concentrating the Sun's energy through various methods to increase the intensity of energy hitting the solar cell.

Quality of the Cell: The Fill Factor (FF) is essentially a measure of quality of the solar cell. The "fill factor", more commonly known by its abbreviation "FF". The FF is defined as the ratio of the maximum power from the solar cell to the product of V_{oc} and I_{sc}. A larger fill factor is desirable and corresponds to an I-V sweep that is more square-like. Typical fill factors range from 0.5 to 0.82. The manufacture of photovoltaic modules is governed by several standards required by the IEC (International Electrotechnical Commission) in order to be marketed on the international market. These standards are as follows:

- The standard IEC 61215 (crystalline) or IEC 61646 (amorph) certifies a guarantee of quality in terms of electrical parameters and mechanical stability. The requirements of this standard refer to the qualification of the design and approval of photovoltaic modules for land application and for long-term use.
- The standard IEC 61730 specifically addresses the topics of prevention against electric shock, fire hazard and bodily injury due to mechanical and environmental constraints.

- This standard, whose specificities concern the safety aspects of the modules, complements the IEC 61215 standard, which fixes the electrical performance.

3.2.3 Maximum Power Point Tracking

MPPT or Maximum Power Point Tracking is an algorithm that is included in charge controllers used for extracting maximum available power from PV module under certain conditions. The voltage at which PV module can produce maximum power is called maximum power point (or peak power voltage). Maximum power varies with solar radiation, ambient temperature and solar cell temperature. Typical PV module produces power with maximum power voltage of around 17 V when measured at a cell temperature of 25°C, it can drop to around 15 V on a very hot day and it can also rise to 18 V on a very cold day.

The major principle of MPPT is to extract the maximum available power from PV module by making them operate at the most efficient voltage (maximum power point). MPPT checks output of Module, compares it to battery voltage then fixes what is the best power that PV module can produce to charge the battery and converts it to the best voltage to get maximum current into battery. The commonly used methods are classified in to

(a) Indirect methods or open loop methods and

(b) Direct methods or closed loop methods.

Indirect methods include

(a) Fixed operating point method,

(b) Fractional Open Circuit Voltage method, and

(c) Fractional Short Circuit method.

Some of the popular direct methods include

(a) Perturb and Observe method,

(b) Incremental Conductance method and

(c) Hill Climbing method

Indirect Methods

Fixed Operating Point method: This method finds the voltage V_{MPP} or the current I_{MPP} at which the PV array delivers the maximum power under a given temperature T and irradiance G. Then the method automatically puts the PV in this condition. The reason is that MPP varies in a small region and that on the left part of the P-V characteristic, the slope P/V is roughly constant

Fractional Open-circuit Voltage method: The method uses the ratio of the voltage at maximum power V_{MPP} to its open circuit voltage V_{OC} which is nearly a constant. $\frac{V_{MPP}}{V_{OC}} \approx$ k. The factor k_1 is not constant and ranges between 0.71 and 0.78. Once the constant k_1 is known, V_{MPP} can be computed. This method consists in measuring V_{OC} periodically and then fixing V_{MPP}. The implementation is simple and cheap, but the tracking efficiency is relatively low due to inaccurate values of the constant $k_1$in the computation of V_{MPP}.

Fractional Short-circuit Current method: This method is based on the fact that the current at maximum power point I_{MPP} is approximately proportional to the short circuit current I_{SC} of the PV array. $I_{MPP} \approx k_2 . I_{SC}$. The factor k_2 is not constant but, varies between 0.78 and 0.92. The accuracy of this method and its tracking efficiency depend on the accuracy of knowledge of k_2 and the periodic measurement of short circuit current I_{SC}.

Direct Methods

Perturb and Observe Method: The most commonly used MPPT algorithm is the Perturbation and Observation (P&O) due to its ease of implementation. It uses the P-V characteristics P_{PV} = f(V) of the PV module. At the point of maximum power given by P(n)=V(n) × I(n) the gradient $\frac{dP}{dV}$ = 0 is accomplished, regardless of the sun irradiance magnitude. In actual experiments, the system oscillates around the MPP. To minimize the oscillations amplitude, the perturbation step size is reduced. However, smaller step size slows down the convergence of the MPPT. The Modified Enhanced Perturb and Observe (MEPO) algorithm uses and adaptive step adjustment gain and simplify the implementation using on commutation functions. This algorithm have been revisited, its rationale behind have been clarified and then implementation obviated using commutation functions.

Incremental Conductance Method: The incremental conductance method is based on the fact that the slope (or the PV conductance G = dI/dV) of the PV array, is zero at the MPP and it is positive (constant) on the left of the MPP. The slope becomes negative on the right of the MPP.

Hill Climbing Method: The basic idea of the Hill Climbing method is the same as P&O method. It tests if P(n) is greater than P(n-1) or not, to reach MPP. The P&O method uses instead a test on dP/dV to determine whether the maximum power point has been found or not. However, the Hill Climbing method uses a test condition on P(n)-P(n-1).

3.2.4 Solar PV Systems

Solar cells produce direct current therefore they are only used for DC equipment. If alternating current is needed for AC equipment or backup energy is needed, solar photovoltaic systems require other components in addition to solar modules. These components are specially designed to integrate into solar PV system. The components of solar photovoltaic system are:

- **Solar or Photovoltaic Module** is the essential component of any solar PV system that converts sunlight directly into DC power.
- **Solar Charge Controller** regulates voltage and current from solar arrays, charges the battery, prevents battery from overcharging and performs controls discharges.
- **Battery** stores the energy produced from solar arrays for using when sunlight is not visible, nighttime or other purposes.
- **Inverter** is a critical component of any solar PV system that converts DC power output of solar arrays into AC for AC appliances.
- **Lightning protection** prevents electrical equipment from damages caused by lightning or induction of high voltage surge. It is required for the large size and critical solar PV systems.

Photovoltaic power systems are generally classified according to their functional and operational requirements, their component configurations, and how the equipment is connected to other power sources and electrical loads. The two principal classifications are grid-connected or utility interactive systems and stand-alone systems. Photovoltaic systems can be designed to provide DC and/or AC power service, can operate interconnected with or independent of the utility grid, and can be connected with other energy sources and energy storage systems.

PV Direct System: These are the simple most type of solar PV systems, with the fewest components; the Solar Panels and the load. Because they don't have batteries and are not hooked up to the grid, they only power the loads when the sun is shining as shown in Figure 3.8. They are appropriate for a few applications e.g. water pumping or attic ventilation fan.

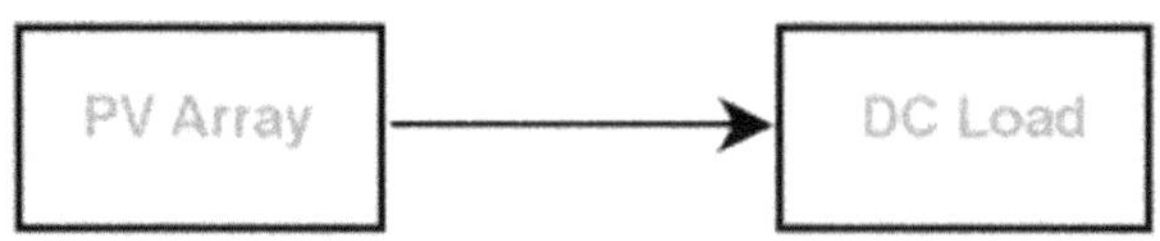

Figure 3.8 Direct PV System.

Solar Off Grid System: Also referred to as stand-alone systems, it is designed to be independent of the power grid. Batteries are used to store energy when the sun is not available during cloudy days or at night. This type of system will require regular attention to battery electrolyte levels and terminal corrosion. The features are:

- Independence from the utility grid
- Not subjected to the terms/policies of the utility company
- Rate increases, blackouts, or brownouts do not apply
- In remote areas, it is cost effective than extending a grid

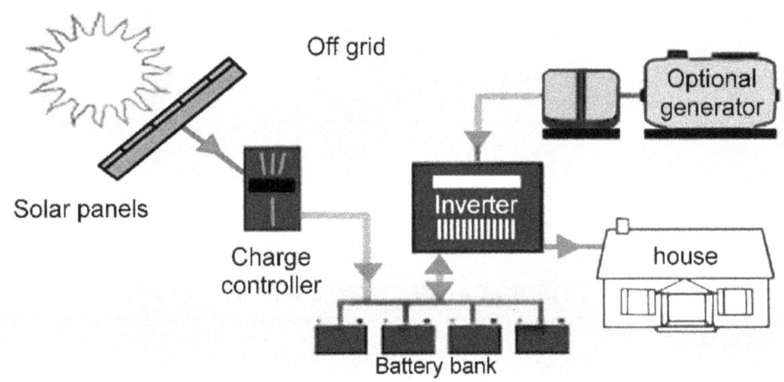

Figure 3.9 Solar Off Grid System.

- Batteries require maintenance and has limited life
- Batteries decrease system efficiency
- It is more expensive than a grid-direct system
- When the batteries are fully charged, l power from the PV array is not utilized
- They are expensive, noisy, dirty, and require fuel and regular maintenance

Grid Tied System: These are most common type of PV systems. They are also known as on-grid, grid-tied, grid-intertied, or grid-direct systems. They generate solar electricity and route it to the loads and to the grid, offsetting some of electricity usage. System components comprise of the PV array and inverter. Grid-connected system is similar to regular electric powered system except that some or all of the electricity comes from the sun. The drawback of these battery less systems is that they provide no outage protection when the utility grid fails, these systems cannot operate. Grid-tied-system has increased design flexibility since the system does not have to power all of the home's loads. It is less expensive compared to stand-alone or grid-tied with battery backup systems. It requires the least amount of maintenance. If the system produces more than the loads need, then the extra energy is exchanged with the grid. Grid-direct systems have a higher efficiency because batteries are not part of the system. Electricity costs are fixed for the life of your system. There is no power to the home when the grid goes down. Paperwork requirements for interconnection, incentives, and rebates. Figure 3.10 depicts the inter-connectivity of various components and the power flow in Grid-Tied Solar PV system.

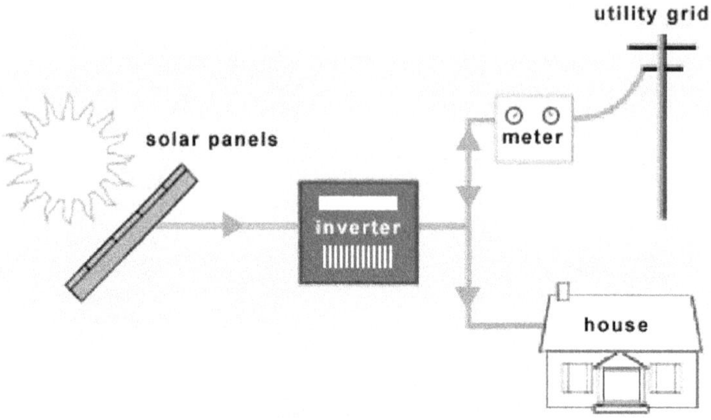

Figure 3.10 Power flow in Grid-Tied Solar PV system.

Hybrid Solar PV System: This type is very similar to an off-grid system in design and components, but adds the utility grid, which reduces the need for the system to provide all the energy all the time. This grid-tied system has a battery backup to energize designated loads to have power when the grid goes down. If the system produces more than the home needs, then the extra energy is sold back to the utility after the batteries get fully charged. This system requires rewiring circuits from main service panel to a separate subpanel. Batteries decrease system performance because of their efficiency losses. This system is more expensive than a grid-direct system. It requires paperwork for interconnection, incentives, and rebates. Figure 3.11 presents a typical hybrid solar system.

3.2.5 Solar Thermal Power Plant

Solar thermal power plant is a combination of solar energy and thermal energy. The sun's radiation is used as fuel in the power plant. Solar energy is converted into heat or thermal energy using ponds or collectors which is further converted to mechanical energy using turbine and electrical energy using generators. Further categories are based upon the power cycles i.e. low, medium and high temperature cycles. These cycles are based upon the solar radiations and type of collectors used for collecting the solar radiations. In low temperature cycles the temperature is limited to about 100^0 C, in medium temperature, range varies from 150^0 C to 300^0 C whereas in high temperature cycles temperature may go above 300^0 C. For different temperature cycle different thermodynamic cycles are used in the power plant. Generally, for low and medium temperature ranges, Rankine cycle is preferred whereas for high temperature range Brayton cycle is used.

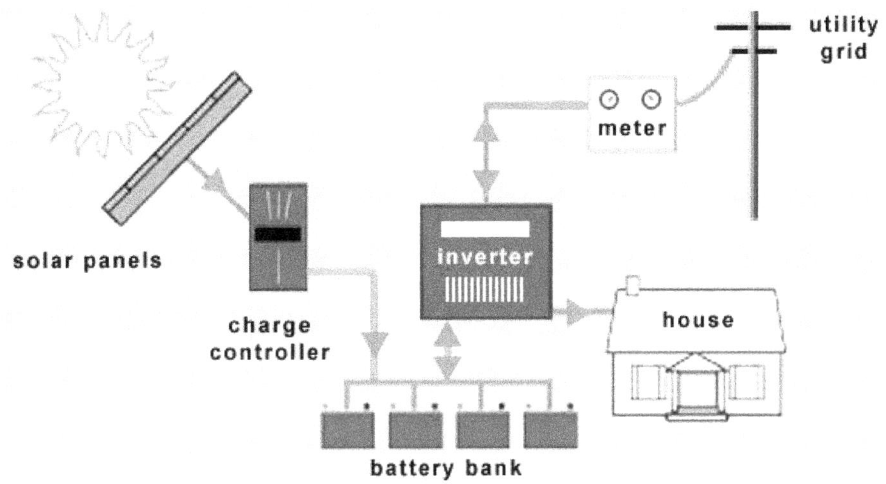

Figure 3.11 Typical hybrid solar system.

3.3 Wind Energy

Wind is used to produce electricity by converting the kinetic energy of air in motion into electricity. In modern wind turbines, wind rotates the rotor blades, which convert kinetic energy into rotational energy. This rotational energy is transferred by a shaft which to the generator, thereby producing electrical energy.

3.3.1 History

Wind is a motion of a mass of air, blowing across the Earth from high- pressure areas to low-pressure areas, and bringing with it inexhaustible natural motive power. Mankind has used wind energy for thousands of years, with early applications including pumping water and grinding grains. The construction techniques of the time discovered in the Middle East were mainly based on using vertical axes to extract drag force from the wind. Early use of wind energy in Europe dates back to 1191 AD, when countries with rich wind resources, such as Holland and England, built horizontal axis windmills with tilted wings or sails. By the 18th century AD, multi-sail Dutch windmills were extensively used in Europe, with numbers estimated at about 8000 windmills in operation in Holland and 10,000 in England and Germany around 1750 AD. European settlers also built windmills in North America, mainly along the east coast area. The use of wind power for sailing ships and the mastering of sea navigation brought enormous fortunes to the European colonial powers until the age of steam. Then with the intermittent nature and uncertain availability of wind combined with the slowness of wind powered vessels, these gradually gave way to fossil-powered commercial shipping. Nevertheless, yachting and small boat sailing remain important recreational sports throughout the world.

The development of wind turbines with aerodynamically formed rotor blades for electricity generation only started around 80 years ago, including projects by German engineers (Kleinhenz and Honnef), and by the American Palmer Putnam (for the Smith company) in the 1940s which produced a turbine of 1250 kW. Wind powered machines were mass produced in the early 1950s, mainly by the German construction company Allgaier, to supply electricity to farms lying far from the public grid. However, these were small-scale generators having power ratings of 6 kW to 10 kW. Large megawatt class wind turbines re-appeared only in the 1970s due to rising fuel prices. Particularly in the USA, Sweden and Germany, megawatt machines were produced such as the American MOD-2 with 2.5 MW nominal power and 91 m rotor diameter, the 61m Swedish- American 4 MW WTS-4, Swedish 3 MW WTS-75 AEOLUS model, and the German 3 MW GROWIAN having a 100 m rotor diameter and 100 m hub height.

Independently of large turbine development, the use of wind-generated power to supply energy to the grid on a large scale started in the 1980s in the USA state of California. Other countries including Denmark, Holland and Germany were also developing grid connected wind converters. Turbines of initially 50 kW, steadily scaled up to megawatt ranges, were used. This development has led to mass production of wind turbines and has resulted in a considerable improvement in performance. Figure 3.12 shows the trend in installed capacity of wind generation across the world installed capacity along vertical axis and the year of installation along horizontal axis.

3.3.2 Construction of Wind Turbines

The main element is a wind turbine or wind rotor which converts the kinetic energy of the wind into mechanical energy by rotating the rotor. The main elements of a typical wind generator include a rotor with blades, main shaft, gearbox, brake, coupling, generator, wind vane/anemometer, housing, tower, and yaw motor/gear as shown in Figure 3.13.

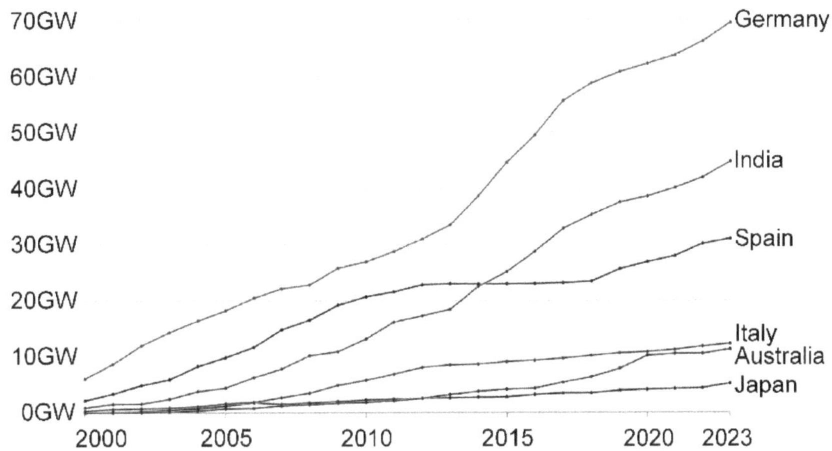

Figure 3.12 Installed capacity of wind generation across the world.

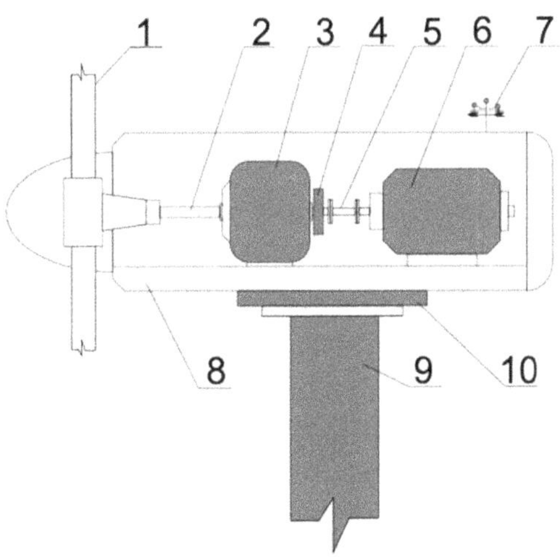

Figure 3.13 Components of a wind turbine: 1 – rotor with blades, 2 – main shaft, 3 – gearbox, 4 – brake, 5 – coupling, 6 – generator, 7 – wind vane/anemometer, 8 – housing, 9 – tower, 10 – yaw motor/gear.

The tower supports the rotor of the wind turbine at the desired height. The main types of towers used in wind turbines are the tubular steel tower, lattice tower, and the guyed tower.

Rotor: The portion of the wind turbine that collects energy from the wind is called the rotor. The rotor usually consists of two blades which rotate about an axis (horizontal or vertical) at a rate determined by the wind speed and the shape of the blades. The blades are attached to the hub, which in turn is attached to the main shaft.

Transmission: The number of revolutions per minute (rpm) of a wind turbine rotor can range between 40 rpm and 400 rpm, depending on the model and the wind speed. Generators typically required rpms of 1,200 to 1,800. As a result, most wind turbines require a gear-box transmission to increase the rotation of the generator to the speeds necessary for efficient electricity production. Some DC-type wind turbines do not use transmissions. Instead, they have a direct link between the rotor and generator. These are known as direct drive systems.

Generator: The generator converts the turning motion of a wind turbine's blades into electricity. Inside this component, coils of wire are rotated in a magnetic field to produce electricity. Different generator designs produce either alternating current (AC) or direct current (DC), and they are available in a large range of output power ratings. The generator rating, or size, is dependent on the length of the wind turbine's blades because more energy is captured by longer blades. It is important to select the right type of generator to match intended use. Generators that produce AC are generally equipped with features to produce the rated voltage and frequency, even when the windspeed is fluctuating. DC generators can be used to produce AC electricity with the use of an inverter, which converts DC to AC.

Tower: The tower on which a wind turbine is mounted is not just a support structure. It also raises the wind turbine so that its blades safely clear the ground and so it can reach the stronger winds at higher elevations. Maximum tower height is optional in most cases, except where zoning restrictions apply. The decision of what height tower to use will be based on the cost of taller towers versus the value of the increase in energy production resulting from their use. Studies have shown that the added cost of increasing tower height is often justified by the added power generated from the stronger winds. Large wind turbines are usually mounted on towers ranging from 40 to 70 meters tall. Towers for small wind systems are generally "guyed" designs. This means that there are guy wires anchored to the ground on three or four sides of the tower to hold it erect. These towers cost less than freestanding towers, but require more land area to anchor the guy wires. Some of these guyed towers are erected by tilting them up. This operation can be quickly accomplished using only a winch, with the turbine already mounted to the tower top. This simplifies not only installation, but maintenance as well. Towers can be constructed of a simple tube, a wooden pole or a lattice of tubes, rods, and angle iron. Large wind turbines may be mounted on lattice towers, tube towers or guyed tilt-up towers. Towers must be strong enough to support the wind turbine and to sustain vibration, wind loading and the overall weather elements for the lifetime of the wind turbine. Their costs will vary widely as a function of design and height.

Classification: Although there are several ways to categorize wind turbines, they are broadly classified into horizontal axis machines and vertical axis machines, based on their axis of rotation.

Horizontal Axis Wind Turbine: Horizontal axis wind turbines (HAWT) have their axis of rotation horizontal to the ground and almost parallel to the wind stream (Figure 3.14). Most of the commercial wind turbines fall under this category. Horizontal axis machines have some distinct advantages such as low cut-in wind speed and easy furling. In general, they show relatively high power coefficient. However, the generator and gearbox of these turbines are to be placed over the tower which makes its design more complex and expensive. Another disadvantage is the need for the tail or yaw drive to orient the turbine towards wind.

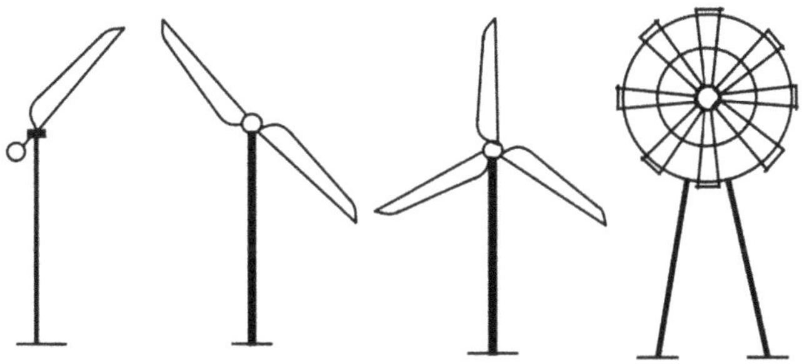

Figure 3.14 Horizontal Axis Wind Turbines wit different number of blades.

Vertical Axis Wind Turbine: The axis of rotation of vertical axis wind turbine (VAWT) is vertical to the ground and almost perpendicular to the wind direction as seen from Figure 3.15. The VAWT can receive wind from any direction. Hence complicated yaw devices can be eliminated. The generator and the gearbox of such systems can be housed at the ground level, which makes the tower design simple and more economical. Moreover the maintenance of these turbines can be done at the ground level. For these systems, pitch control is not required synchronous applications. The major disadvantage of some VAWT is that they are usually not self-starting. Additional mechanisms may be required to 'push' and start the turbine, once it is stopped. As the rotor completes its rotation, the blades have to pass through aerodynamically dead zones which will result in lowering the system efficiency. There are chances that the blades may run at dangerously high speeds causing the system to fail, if not controlled properly. Further, guy wires are required to support the tower structure which may pose practical difficulties.

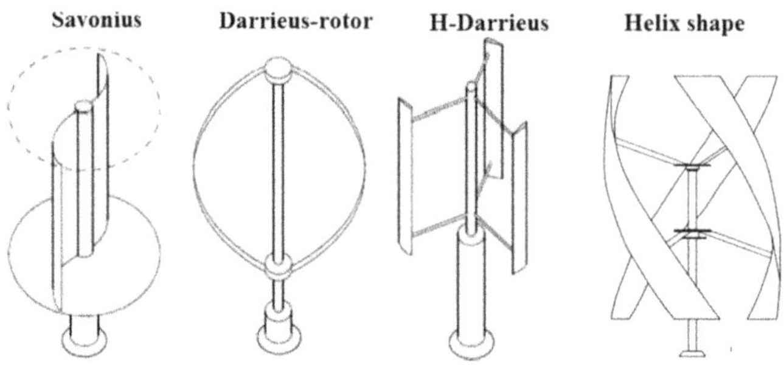

Figure 3.15 Various types of Vertical Axis Wind Turbines.

3.3.3 Energy Calculation in Wind Stream

Let us consider a wind flows perpendicular to a vertical disc of radius "R" and surface area "A". let us visualize a cylindrical mass of air of exactly the right size so that it can pass snugly through the disc. This cylinder of air has length of "l" and hence a volume "V"= A x l ". The kinetic energy"E" of the wind can be expressed as

$$E = \frac{1}{2}\rho\, Alv^2 \qquad .....(3.1)$$

where

A: Cross-sectional area of the disc ($\Pi\, R^2$)

v: Velocity of wind

If this kinetic energy is extracted in an amount of time "t", then the average power that can be extracted is given by equation

$$P = E/t \qquad .....(3.2)$$

However $\quad l = v \times t \qquad .....(3.3)$

The power extracted "P" is given by

$$P = \frac{1}{2}\rho\, Av^3 .... \qquad .....(3.4)$$

This is really convenient, because this expression works for any arbitrary time length "t", so we can let "t" become infinitesimally small (t → 0) and we can thus interpret this expression as a continuous rate of energy extraction from the wind, expressed in Joules per second, or Watts.

BETZ Limit: When the kinetic energy is extracted from wind by wind turbine, the wind speed eventually decreases. If v_1 and v_2 are maximum and minimum wind speeds on upstream and downstream respectively as shown in Figure 3.16. The average speed of the stream "v_d" is the average of v_1 and v_2 i.e., $\frac{v_1+v_2}{2}$.

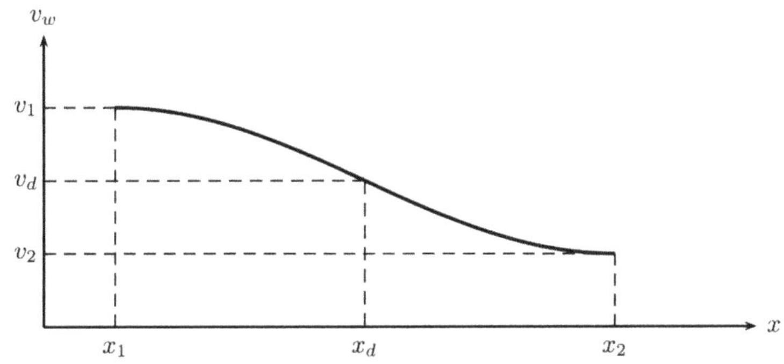

Figure 3.16 Variation of wind speed from far upstream to far downstream.

As per the law of conservation of momentum (when applied to the flow), the difference between the input and output momentum rates is nothing but the drag force due to the disc actuator, which can be expressed as

$$F=\rho A v_d (v_1 - v_2) \quad .....(3.5)$$

Hence, the power extracted from the wind by the disc actuator can be given by

$$P=Fv_d = P = \frac{1}{4}\rho A (v_1 - v_2)(v_1 + v_2)^2 \quad .....(3.6)$$

For given upstream wind velocity the power extracted by the turbine is function of downstream speed v_2. For the maximum output extraction the condition to be satisfied is $(dP/dv_2)=0$; which yields to a relation $v_1 = \frac{1}{3}v_1$. This leads to $v_d = \frac{2}{3}v_1$.

Thus the maximum possible power that can be extracted can be obtained as

$$P_{max} = \left(\frac{16}{27}\right) * \frac{1}{2}\rho A v_1^3 \quad .....(3.7)$$

The factor$\left(\frac{16}{27}\right)\approx$ 0.59% is called the maximum power coefficient, and P_{max} is commonly called the Betz limit, named after the German physicist Albert Betz.

Power Characteristics: In practice, it can be shown using more advanced principles of aerodynamics that the power extracted by a wind turbine can be expressed as

$$P = \frac{1}{2} C_p(\lambda, \theta)\rho A v_1^3 \quad .....(3.8)$$

The function C_p is called the performance coefficient, and it depends on the tip-speed ratio λ and the blades' pitch angle θ. The tip-speed ratio is defined as

$$\lambda = \frac{\omega R}{v_1} \quad .....(3.9)$$

where "ω" is angular velocity of the blades, in rad/s. The explicit dependence on θ is due to the fact that modern wind turbines have the capability to alter the pitch of the blades by rotating them. The pitching the blades away from the optimal point where ∂Cp/∂θ = 0 in effect reduces the aerodynamic torque on the blades and the power extracted from the wind. This functionality is useful when a turbine needs to limit its power output, or when it needs to slow down. A typical variation of Cp with different pitch angles and tip speed ratio is shown in Figure 3.17. These curves help to harvest maximum power from the wind; the turbine depending upon the wind speeds.

Another interesting plot is the one shown in Figure 3.18, which depicts the power extracted from the wind as a function of the blades' rotational speed, for different wind speeds (and assuming that the blades are not pitched). The red dashed line connects the peak points of all the power curves, and represents the maximum power extraction curve. This curve is cubic with respect to v and $\boldsymbol{\omega}$ (because these two are proportional at the peaks), and so it rises very fast as the wind speed increases.

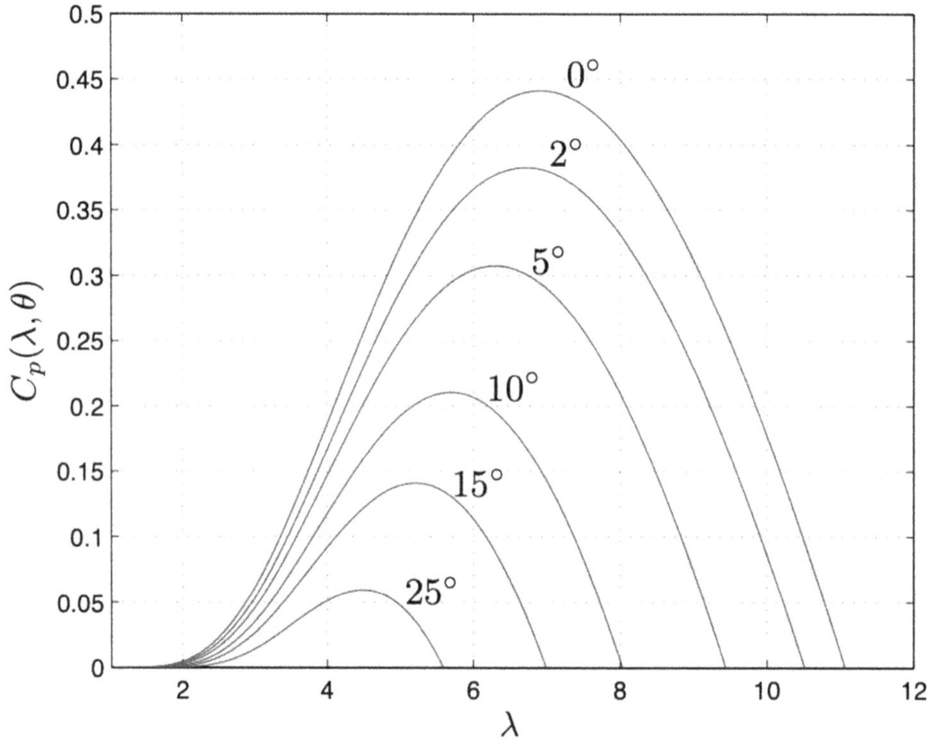

Figure 3.17 A typical performance coefficient variation for arrange of blade pitch angles.

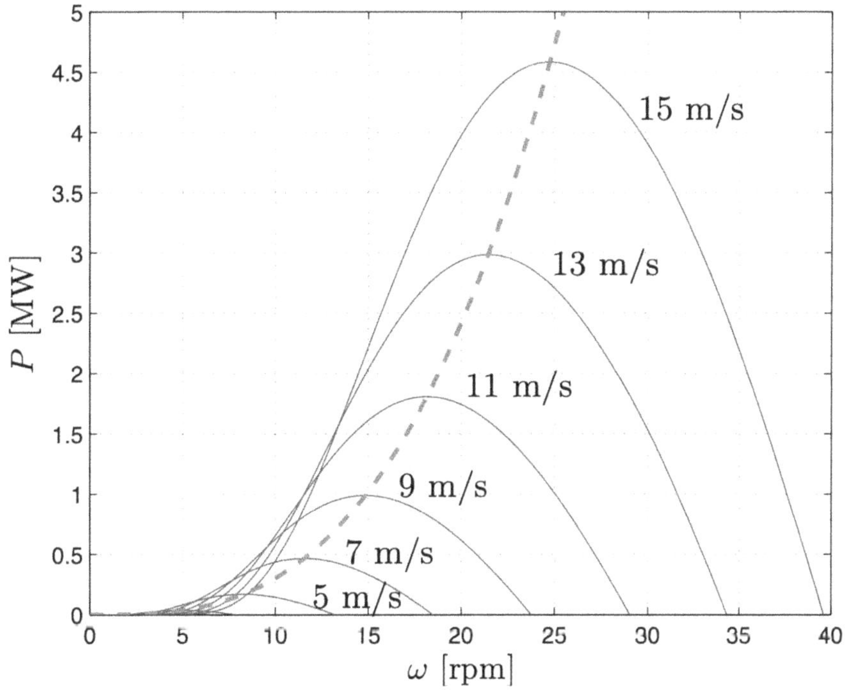

Figure 3.18 Turbine power vs. blade speed for wind speeds for a turbine with an 80-m wide swept area.

The wind speed at which a turbine reaches its nominal power output is called the **rated wind speed**. Above this value, pitching of the blades occurs, and the power output is maintained constant at rated value. Some wind turbines do not have active pitch control systems. These turbines are stall-regulated, meaning that their speed is restrained from reaching dangerously high values because their blades are designed to stall above a certain speed threshold. The wind turbine stops producing power at a maximum wind speed called the **cut-out speed**. For very low speeds for instance less than 3 m/sec the power production is insignificant and is almost indistinguishable in the plot, having a peak power around 35 kW. So there is a minimum wind speed, called the **cut-in speed**, below which a turbine just shuts itself down because there is not enough energy in the wind to supply its own mechanical losses and other auxiliary systems. A typical power production curve vs. wind speed of a variable-speed wind turbine is sketched in Figure 3.19. It should be noted that the output power is equal to the aerodynamic power that we have calculated minus all mechanical and electrical losses incurred from the blades to the electrical terminals.

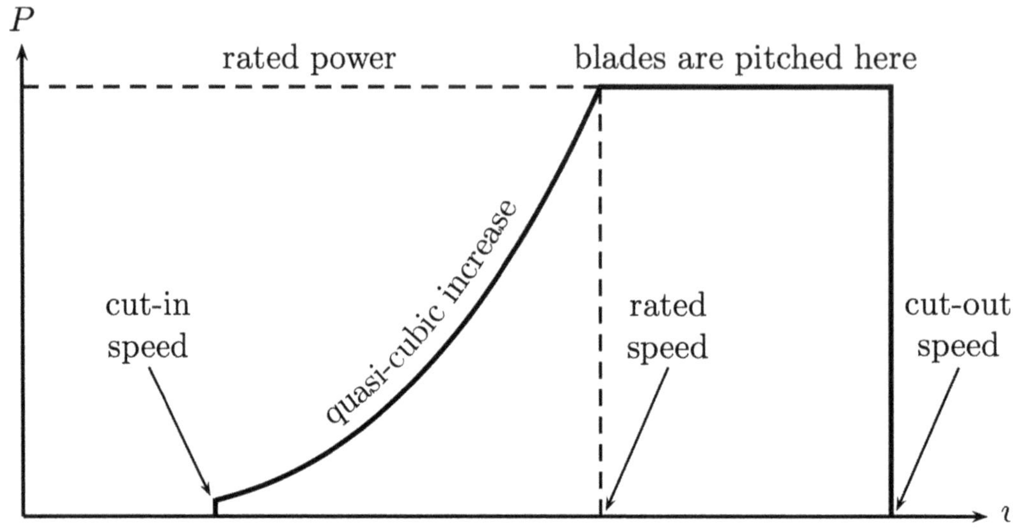

Figure 3.19 Wind turbine power output as a function of wind speed.

3.3.4 Types of Wind Turbine Generators

Beyond mechanical power regulation, turbines are further divided into fixed speed (Type-1), limited variable speed (Type-2), or variable speed with either partial (Type-3) or full (Type-4) power electronic conversion. The different speed control types are implemented via different rotating AC machines and the use of power electronics. There is one other machine type that will be referred to as Type-5 in which a mechanical torque converter between the rotor's low-speed shaft and the generator's high-speed shaft controls the generator speed to the electrical synchronous speed. This type of machine then uses a synchronous machine directly connected to the medium voltage grid.

Type-1: The Type-1 WTG is implemented with a squirrel-cage induction generator (SCIG) and is connected to the step-up transformer directly as shown in Figure 3.20. The turbine speed is fixed (or nearly fixed) to the electrical grid's frequency, and generates real power (P) when the turbine shaft rotates faster than the electrical grid frequency creating a negative slip. Type-1 turbines typically operate at or very close to a rated speed. A major drawback of the induction machine is the reactive power that it consumes for its excitation field and the large currents the machine can draw when started "across-the-line." To ameliorate these effects the turbine typically employs a soft starter and discrete steps of capacitor banks within the turbine.

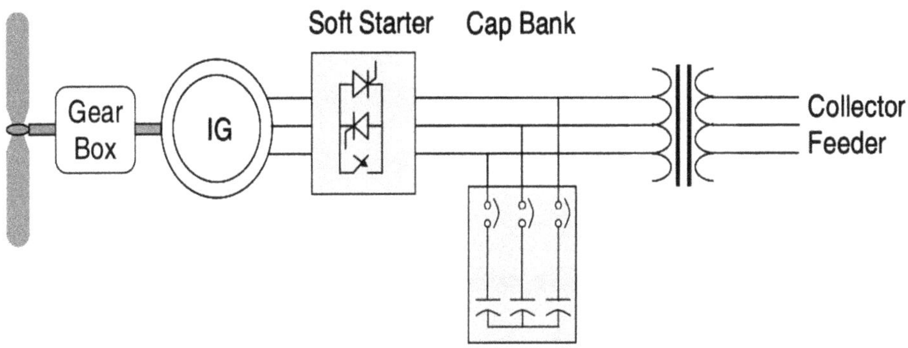

Figure 3.20 Typical Configuration of a Type-1 WTG.

Type-2: In Type 2 turbines, wound rotor induction generators are connected directly to the WTG step-up transformer in a fashion similar to Type 1 with regards to the machines stator circuit, but also include a variable resistor in the rotor circuit as shown in Figure 3.21. This can be accomplished with a set of resistors and power electronics external to the rotor with currents flowing between the resistors and rotor via slip rings. Alternately, the resistors and electronics can be mounted on the rotor, eliminating the slip ring. The variable resistors are connected into the rotor circuit softly and can control the rotor currents quite rapidly so as to keep constant power even during gusting conditions, and can influence the machine's dynamic response during grid disturbances.

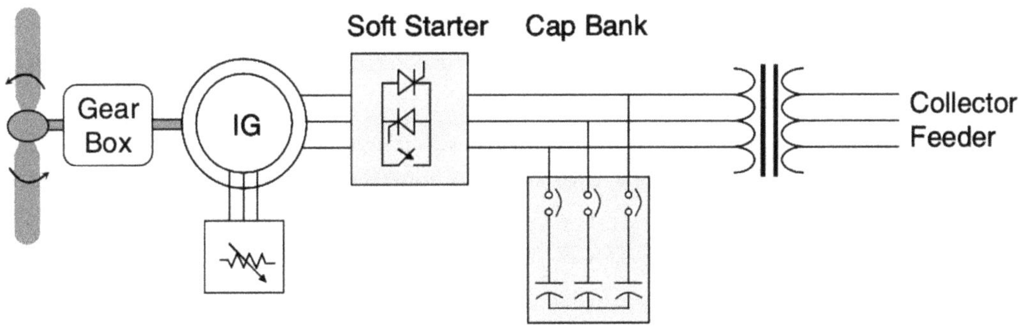

Figure 3.21 Typical Configuration of a Type 2 WTG.

By adding resistance to the rotor circuit, higher slip and higher speed ranges and hence higher power range are made feasible. That is to say that the turbine would have to spin faster to create the same output power, for an added rotor resistance.

This allows some ability to control the speed, with the blades' pitching mechanisms and move the turbines operation to a tip speed ratio (ratio of tip speed to the ambient wind speed) to achieve the best energy capture. It is typical that speed variations of up to 10% are possible, allowing for some degree of freedom in energy capture and self-protective torque control.

Type-3: The Type 3 turbine, known commonly as the Doubly Fed Induction Generator (DFIG) or Doubly Fed Asynchronous Generator (DFAG), takes the Type 2 design to the next level, by adding variable frequency ac excitation to the rotor circuit. The additional rotor excitation is supplied via slip rings by a current regulated, voltage-source converter, which can adjust the rotor currents' magnitude and phase nearly instantaneously. This rotor-side converter is connected back-to-back with a grid side converter, which exchanges power directly with the grid as shown in Figure 3.22.

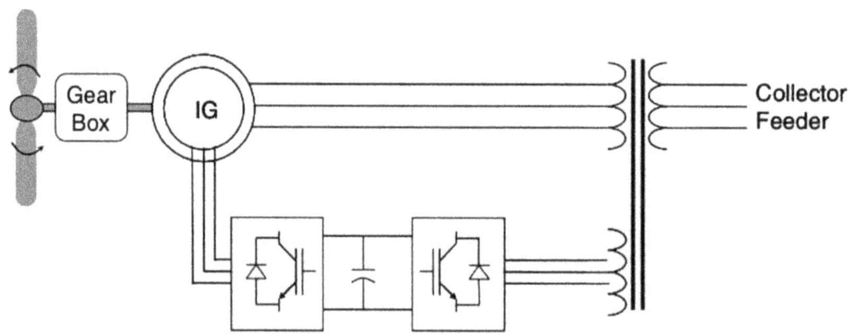

Figure 3.22 Typical Configuration of a Type 3 WTG.

A small amount power injected into the rotor circuit can effect a large control of power in the stator circuit. This is a major advantage of the DFIG – a great deal of control of the output is available with the presence of a set of converters that typically are only 30% of the rating of the machine. In addition to the real power that is delivered to the grid from the generator's stator circuit, power is delivered to the grid through the grid-connected inverter when the generator is moving faster than synchronous speed. When the generator is moving slower than synchronous speed, real power flows from the grid, through both converters, and from rotor to stator. These two modes, made possible by the four-quadrant nature of the two converters, allows a much wider speed range, both above and *below* synchronous speed by up to 50%, although narrower ranges are more common.

The greatest advantage of the DFIG, is that it offers the benefits of separate real and reactive power control, much like a traditional synchronous generator, while being able to run asynchronously. The field of industrial drives has produced and matured the concepts of vector or field oriented control of induction machines. Using these control schemes, the torque producing components of the rotor flux can be made to respond fast enough that the machine remains under relative control, even during significant grid disturbances. Indeed, while more expensive than the Type 1 or 2 machines, the Type 3 is becoming popular due to its advantages.

Type-4: The Type 4 WTG shown in Figure 3.23 offers a great deal of flexibility in design and operation as the output of the rotating machine is sent to the grid through a full-scale back- to-back frequency converter. The turbine is allowed to rotate at its optimal aerodynamic speed, resulting in a "wild" AC output from the machine. In addition, the gear box may be eliminated, such that the machine spins at the slow turbine speed and generates an electrical frequency well below that of the grid. The rotating machines of this type have been constructed as wound rotor synchronous machines, similar to conventional generators found in hydroelectric plants with control of the field current and high pole numbers, as permanent magnet synchronous machines, or as squirrel cage induction machines. However, based upon the ability of the machine side inverter to control real and reactive power flow, any type of machine could be used. Advances in power electronic devices and controls in the last decade have made the converters both responsive and efficient. It does bear mentioning, however, that the power electronic converters have to be sized to pass the full rating of the rotating machine, plus any capacity to be used for reactive compensation.

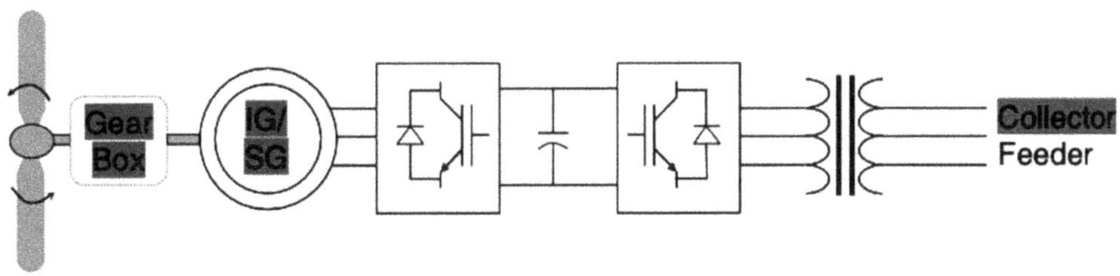

Figure 3.23 Typical Configuration of a Type 4 WTG.

Type-5: Type 5 turbines (Figure 3.24) consist of a typical WTG variable-speed drive train connected to a torque/speed converter coupled with a synchronous generator. The torque/speed converter changes the variable speed of the rotor shaft to a constant output shaft speed. The closely coupled synchronous generator, operating at a fixed speed (corresponding to grid frequency), can then be directly connected to the grid through a synchronizing circuit breaker. The synchronous generator can be designed appropriately for any desired speed (typically 6 pole or 4 pole) and voltage (typically medium voltage for higher capacities). This approach requires speed and torque control of the torque/speed converter along with the typical voltage regulator (AVR), synchronizing system, and generator protection system inherent with a grid-connected synchronous generator.

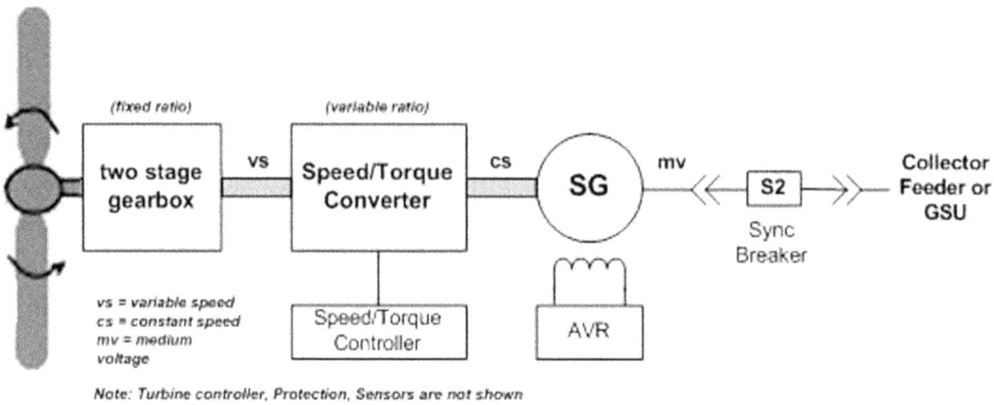

Figure 3.24 Typical Configuration of a Type 5 WTG.

3.4 Hydro Power

Hydropower is generated from water moving in the hydrological cycle, which is driven by solar radiation. Incoming solar radiation is absorbed at the land or sea surface, heating the surface and creating evaporation where water is available. A large percentage close to 50% of all the solar radiation reaching the Earth's surface is used to evaporate water and drive the hydrological cycle. The potential energy embedded in this cycle is therefore huge, but only a very limited amount may be technically developed. Evaporated water moves into the atmosphere and increases the water vapour content in the air. Global, regional and local wind systems, generated and maintained by spatial and temporal variations in the solar energy input, move the air and its vapour content over the surface of the Earth, up to thousands of kilometres from the origin of evaporation. Finally, the vapour condenses and falls as precipitation, about 78% on oceans and 22% on land. This creates a net transport of water from the oceans to the land surface of the Earth, and an equally large flow of water back to the oceans as river and groundwater runoff. It is the flow of water in rivers that can be used to generate hydropower, or more precisely, the energy of water moving from higher to lower elevations on its way back to the ocean, driven by the force of gravity

3.4.1 Types of Hydro Power Plants

Hydropower plants come in a wide range of sizes. The largest have generating capacities of thousands of megawatts, the smallest a few kilowatts. The techniques used to build these plants vary too. Large plants will always be built to suit a particular site and all the components are likely to be tailored for this specific project. Smaller plants, in contrast, will often use off-the-shelf components to keep costs down. Given these variations, hydropower plants are traditionally broken down into categories depending on their size. Categories can vary, depending on the defining agency but the most common categorisation is shown in Table 3.1. The smallest plants, with capacities of between 1 kW and 100 kW are called micro hydropower plants. Between100 kW and 1 MW a plant is described as a mini hydropower plant. Small hydropower plants are generally those with capacities of between 1 MW and 10 MW but this upper limit can vary from country to country and in some cases may be as high as 30 MW. Plants with capacities larger than 10 MW (or up to 30 MW depending on jurisdiction) are classified as large hydropower plants.

Sometimes an intermediate category for medium hydropower plants is also introduced between small and large hydropower. If used, this is typically for plants between 5 and 50 MW; those above are large and those below are small. From a global perspective, large hydro- power is the most important category and this accounts for most of the hydropower capacity in operation today. However the smaller categories are important locally.

Large hydropower plants are the most technically sophisticated and are generally individually designed for each site using turbines that have also been made specifically for the power plant. Small hydro- power plants are similar to large plants but some use off-the-shelf turbines and other components rather that bespoke components. Mini and micro hydropower installations usually employ standard turbines and many involve novel, often cost-effective, designs not used in larger plants

Table 3.2 Categories of Hydro Power Plants.

Micro	1 kW–100 kW
Mini	100 kW–1 MW
Small	1 MW to 10–30 MW
Large	Above 10–30 MW
Source: *UNDP/World Bank*[9]	

Table 3.2 provides an indication of the potential for these different categories of hydropower plant across the globe. In this table large hydropower is defined as being greater than 10 MW, small hydropower covers 100 kW to 10 MW and micro hydropower is below 100 kW. As would be expected, large hydropower potential dominates in most regions. Globally it accounts for 16,644 GW out of a total of 19,343 GW or 83% of the total. Small hydropower is estimated to potentially offer a further 2949 GW and micro hydropower a further 396 GW. It is worth noting, however, that small hydropower offers the largest potential in Oceania, 30 GW out of a total of 52 GW. In Europe too, the small hydropower contribution is large, with 217 GW (27%) out of a total of 792 GW.

Table 3.3 Global Potential of Hydro Power.

Region	Total Potential (GW)	Large Hydropower Potential (GW)	Small Hydropower Potential (GW)	Micro Hydropower Potential (GW)
Africa	2111	1729	355	27
Asia	13701	11,714	1775	212
Oceania	52	7	30	15
Europe	792	540	217	35
North America	2038	1653	327	58
South America	1296	1001	245	50
World	19,990	16,644	2949	397

[a] *Totals have been corrected where there were inconsistencies.*
Source: *Delft University of Technology*[7].

3.4.2 Dams and Barriers

Modern hydropower plants designed to generate electrical power use similar strategies and the water mill provides an example of two of the most important ways of harnessing the water in a river or stream. The first and perhaps the most important is the dam and reservoir power plant. In this type of hydropower plant a large physical structure, a dam, is built across the river. Once in place, this dam will cause a lake the reservoir of water to build up behind it. As its name implies, the reservoir provides a reserve of water that is stored and ready to be used to generate power as necessary while also damping the peaks and troughs of the flow. Channels within the dam allow water to be taken from the reservoir and delivered to one of more hydro-turbines which each provide the mechanical power to drive a generator and provides electrical energy. Once through the turbine(s) the water is released back into the original waterway. A run-of-river hydropower scheme is a type of power plant that takes the water flowing down a waterway, diverts it so that it passes through a hydro-turbine generator to generate electric power and then returns the water to the waterway at a point lower down the watercourse.

Dam Types: There are several different types of dam, each suited to a different situation. Four principle types are used for large and small hydropower plants: embankment dams, gravity dams, buttress dams and arch dams.

Embankment dams: An embankment dam is made from natural materials and they are usually constructed using material sourced locally. Ideally the material will be taken from land that will be submerged by the reservoir once it has formed. The dams come in two principle varieties, earth fill embankment dams and rockfill embankment dams. An earth fill embankment dam is built around a core of an impermeable material such as clay or concrete which is keyed into the rock of subsurface to prevent water flowing beneath it. This core is then buttressed by banks of earth, on both sides, to create a gradually sloping wall on both the upstream and the downstream face of the dam as shown in Figure 3.25. The earth provides the mass that holds the dam in place and resists the pressure from the water in the reservoir. A rockfill dam is similar to an earth fill dam but the body of the dam is constructed from rock. This is also permeable to water and a rockfill dam may have an impermeable core, like an earth fill dam. Alternatively it may have its upstream face sealed with concrete to render it impermeable. One of the main dangers with an embankment dam is that over-topping of the dam can lead to erosion of the material from which it is constructed and to eventual failure. To prevent this, the structure will be fitted with a concrete spillway to control the water level.

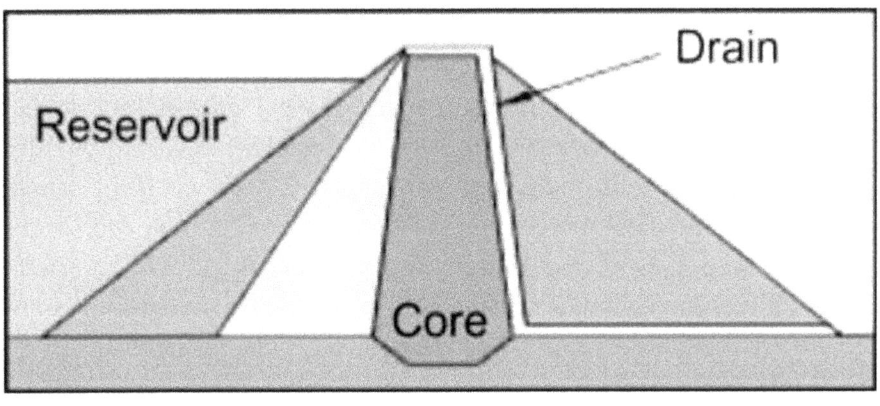

Figure 3.25 Cross section of an embankment dam. Source: The British Dam Society.

Gravity dams: A gravity dam is a massive dam constructed from concrete, masonry or both. As its name implies, it holds in position as a result of gravity acting on its mass; i.e it is held in position by its weight. Gravity dams normally have a vertical or near-vertical upstream face, while the downstream side is sloping to create a triangular structure, as shown in Figure 3.26. This type of dam must be built on secure rock foundations because of the pressure it exerts on those foundations. It will usually be keyed into the rock both to increase its resistance to the pressure from the water and to prevent any water seeping underneath the dam. The dams can be built in both wide and narrow valleys and they can be. much taller than embankment dams. The tallest concrete gravity dam, the Grande Dixence dam in Switzerland, is 285 m high.

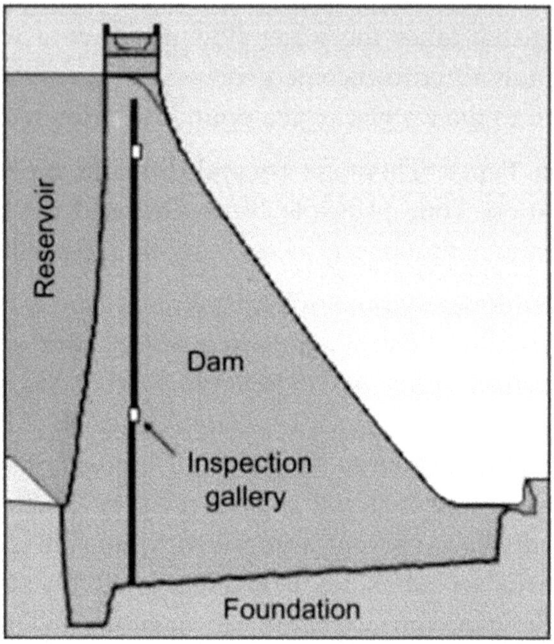

Figure 3.26 Cross section of a gravity dam. Source: British Dam Society.

Buttress dams: The buttress dam, sometimes called a hollow dam, is a variant of the concrete gravity dam. It too is built from concrete and has a vertical upstream face but downstream it is supported by massive triangular buttresses walls built into the rock below. The buttresses help the dam resist the pressure of the water on the structure. There are spaces between the buttresses, spaces that would be filled in the case of a conventional gravity dam hence the name hollow dam. In consequence, these dams require much less material than a conventional gravity dam and this has made them attractive. Like a gravity dam, a buttress dam can be built across both narrow and wide river valleys and like the gravity dam it requires sound rock footings to ensure it is stable. Figure 3.27 presents a typical buttress dam

Figure 3.27 Typical buttress dam.

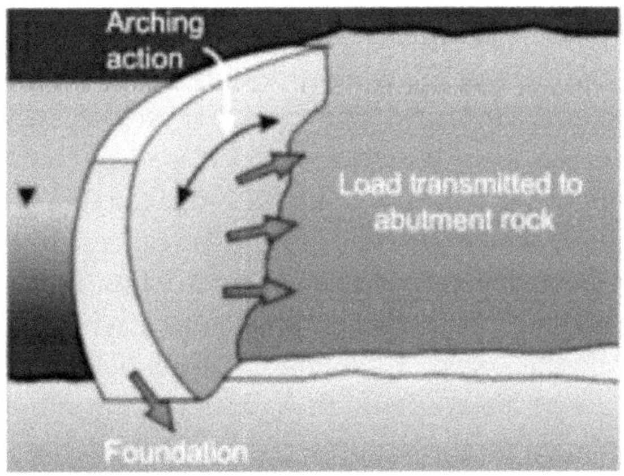

Figure 3.28 Schematic of an arch dam. Source: US Department of Homeland Security.

Concrete arch dam: The arch dam is the most elegant of all the dam types and relies on the principle of the arch, a parabolic structure, which redirects all forces applied to it along the line of the parabola. In consequence all these forces are resolved into compressive stress and there are no tensile stresses. The forces it supports are, by this means, transmitted to the base of the arch (see Fig. 3.28). In the case of an arch dam this will be the rock faces of a valley wall. The earliest known arch dams were built by the Romans. An arch dam can only be built in a steeply sided rock ravine where both the sides of the ravine and the underlying substrata are sound rock formations. The arch is built on its back, bowing upstream, with the sides bedded into the rock faces of the side of the ravine.

The arch then relies on the rock of the valley walls to support the weight of water pressing down on it. The concrete arch dam is the most precisely engineered of all dam types and requires less material than other types and where it can be built it is the strongest of dams. They are also among the tallest of dams. Most of the world's dams over 250 m high are concrete arch dams. In addition to the single arch dam, there are also multiple arch dams that rely on but-tresses to support the forces on the arches in intermediate positions. In the United States, arch dams are classified according to their height with low dams less than 30 m tall, medium dams 30-91 m tall and high dams more than 90 m tall.

3.4.3 Hydro Power Turbines

Development of the water wheel during the 18th century allowed energy efficiency as high as 60% to be achieved. However the water wheel had limitations and during the 19th century even higher efficiency was achieved with the development of the turbine. The name turbine is derived from the Latin word turbo meaning 'spinning top' or 'that which swirls' and the name was introduced by the French engineer Charles Burdin. The key to these early turbines was that the device was enclosed and water was introduced tangentially, creating a swirling motion that improved energy efficiency. The most important of the early turbines was that designed by Benoît Fourneyron, a student of Burdin. His prototype was built in 1827 and he perfected it over the next decade. The design is still in use.

Impulse Turbine: A tall column of water creates a high pressure at its base by virtue of the mass of the water pressing down on it and if this water under high pressure is released through a fine nozzle it will create a high- speed jet of water. If this jet is directed onto a bucket-shaped paddle on the circumference of a wheel, it will generate an impulse that will cause the wheel to turn. This is the basis for the impulse turbine. The main type of impulse turbine in use today is the Pelton turbine, patented by the American engineer Lester Allen Pelton in 1889, which operate by firing a jet of water at buckets mounted onto a wheel. It operate at its highest efficiency, 95%, when the speed of movement of the bucket on the wheel rim is half that of the water in the jet directed into it. A modern Pelton turbine has a set of spoon-shaped buckets mounted on the circumference of a wheel as shown in Figure 3.29.

The largest Pelton turbines can provide a generating capacity of up to 400 MW while the smallest are only a few centimetres in diameter and generate 5 kW or less.

Reaction Turbine: Impulse turbine exploits the kinetic energy and momentum of flowing water, where as a reaction turbine relies on a combination of the kinetic energy and the pressure of the water generated from a head to drive it.. A typical modern reaction turbine will be enclosed within a casing and the water is introduced around the sides of the casing, entering tangentially from all sides. The casing feeding the water into the turbine forms a spiral to create a swirl that helps animate the turbine. As the water enters the turbine, the pressure will act on the blades, forcing them to turn and the shape of the blade then forces the water to exit axially. The water is carried away through a channel called the draft tube which is designed to add suction to the forces acting on the turbine rotor, helping to transfer more energy to the machine. There are several different types of reaction turbine. The most popular, is the Francis turbine. For all heads up to 450 m a reaction turbine is often the preferred design although there is some overlap of reaction and impulse turbines between 200 and 600 m. Figure 3.30 represents the construction of a typical Francis turbine.

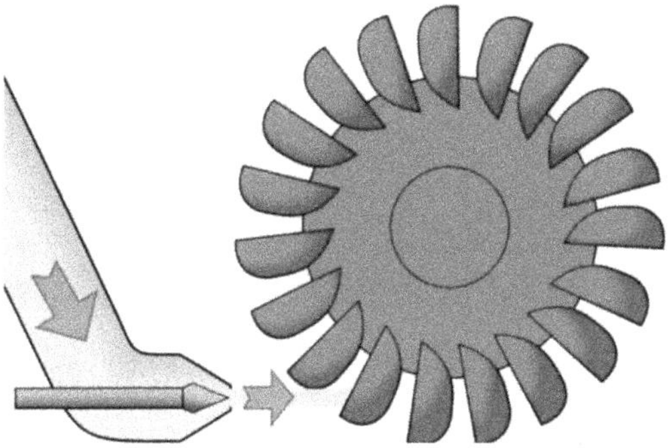

Figure 3.29 A typical Pelton turbine.

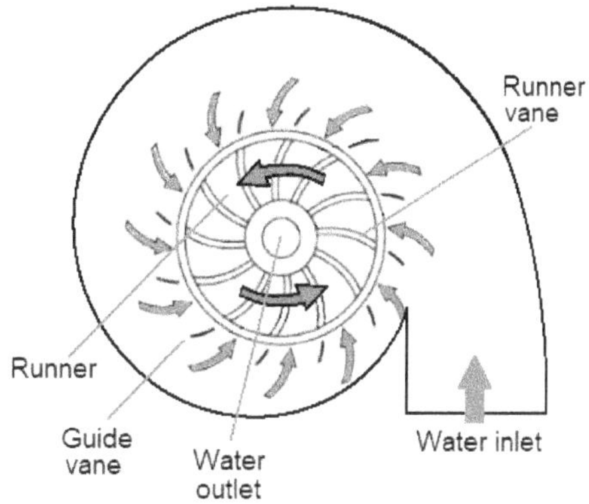

Figure 3.30 A typical Francis turbine.

Propeller and Kaplan turbines: If the head of water becomes too low, the rotational speed of a Francis turbine falls and hence its efficiency. For low-head applications an alternative turbine, called as propeller turbine is employed. A propeller turbine looks much like the screw of a ship but its mode of operation is the reverse of the ship's propulsion unit. In a ship a motor turns the propeller which pushes against the water, forcing the ship to move. Whereas in the hydropower plant, moving water drives the propeller turbine to generate power. A cutaway of a typical propeller turbine generator is shown in Figure 3.31(a)

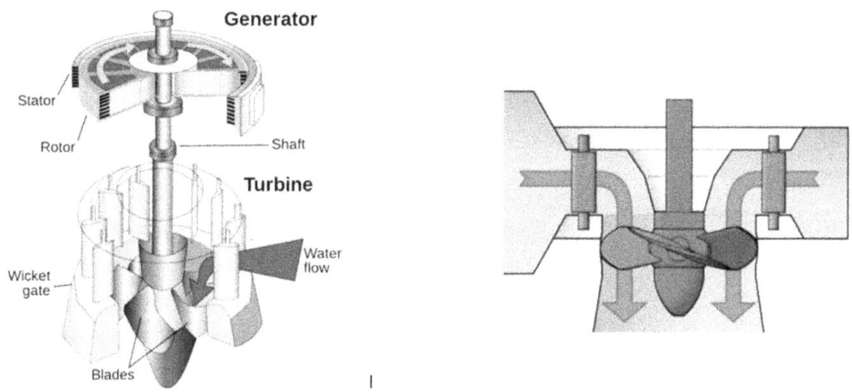

(a) (b)

Figure 3.31(a) Cutaway of a typical propeller turbine generator
(b) Cross-sectional view of a typical Kaplan turbine.

In 1915, Viktor Kaplan has developed a turbine, which is called as Kaplan turbine used for very low heads. A schematic of a Kaplan turbine is shown in Figure 3.31(b). Its primary feature is a set of blades that can be adjusted to maximise efficiency under different flow conditions. These turbines can operate at higher flow levels than conventional propeller turbines and are suited to heads of between 10 and 50 meters.

Deriaz Turbine: The Deriaz turbine invented by Paul Deriaz, has blade shapes similar to a Francis turbine but these blades are adjustable and hence different flow rates can be adapted. The Deriaz turbine is another mixed flow turbine in which water enters from the side but exits along the turbine axis. It is best suited for head heights between 20 and 100 meters.

3.4.4 Hydro Power Generators

The generator is the part of a power plant that converts rotary motion provided by a device such as a hydro-turbine into electrical energy. The machine exploits a phenomenon discovered during the 19th century by the English scientist Michael Faraday; when a conductor moves through a magnetic field a current is generated in the conductor. The design of generators based on this principle has been refined over a period of more than 140 years and modern generators are extremely sophisticated and efficient electromechanical devices. The earliest devices of this type were called dynamo-electric machines, named by Werner Siemens, the inventor of the first generator, in 1867. This name was soon shortened to dynamo. A dynamo is a machine that generates a direct current (DC) from mechanical rotation. Later, machines that could provide an alternating current (AC) were designed because it proved easier to distribute over large distances. These AC devices were originally called alternators but today they are usually called generators.

In order to reduce the speed of rotation of a directly coupled generator the number of rotor poles increased. If the number of poles is doubled to four then there will be two complete AC cycles induced in the stator of the generator for each rotation of the rotor and therefore the rotor need only rotate at half the speed, 1500 rpm or 1800 rpm, to synchronise with the grid. The number of poles can be increased further in order to match slower rotational speeds. For example a 50-Hz generator coupled to a turbine rotating at 166.4 rpm will have 36 poles.

A feature of these low-speed hydropower generators is salient poles as shown in Figure-3.27. This arrangement is more convenient for generators that have multiple poles. Most hydropower plants are connected as synchronous generators which means that they supply AC power at exactly (or very close to) the grid frequency. In order to maintain the speed accurately, the rotation of the turbine is controlled by a device called a governor. The governor, in turn, controls the opening and closing of gates through which water enters the turbine. Opening the gates further allows more water through and the turbine will speed up. Increasing the closure of the gates will restrict the water flow further and the turbine will slow down. The governor is essentially a feedback system that monitors the rotational speed of the shaft and then applies a correction if the speed is too high or too low.

3.4.5 Components of Hydropower Plant

The main components of a Hydroelectric power plant include, Power House, Penstock, Water Reservoir, Turbine, Spillway, Dam, Surge Tank, Draft Tube, Tail Race Level, Gate, and Pressure Tunnel as shown in Figure 3.32.

Power House: Power house contains generator, water turbine, with transformer and control room. When the water rushes through the turbine, it turns the turbine shaft, which is coupled to electric generator. Generator has a rotary electromagnet called as rotor with a stationary element called as stator. Rotors generates magnetic fields that create an electric charge in stator. Charge is transmitted as electricity. Step up transformer increase the voltage coming from the stator. Electricity is than dispersed through transmission lines.

Penstock: Penstock pipe is used to convey water from the dam to hydraulic turbine. Penstock pipes are made of steel or reinforced material. Turbine is installed at a lower level from the dam. Penstock is connected by a gate valve at inlet to totally close the water supply. It has a control valve to control water flow rate into turbine.

Water Reservoir: In reservoir, water is collected at the catchment area during raining period and is stored at the dam. Catchment area obtains its water from rains and streams. Permanent accessibility of water is an essential necessity for hydroelectric power plant. The stage of water surface in reservoir is call head.

Turbine: Hydraulic turbines change potential energy of water into kinetic energy. Mechanical energy (revolution) accessible on turbine shaft is attached to shaft of an electric generator where electricity is created. Water after performing work on turbine blade is discharge through draft tube.

Spillway: Overload addition of water endanger the strength of dam construction. Also in order to avoid the overflow of water out of dam mainly during raining seasons spillways are provided. This prevents the increase of water level in dam. Spillways are passage which allows excess water to flow to a dissimilar storage area away from the dam.

Dam: The function of dam is to store water and control the outgoing flow of water. Dam helps to store all incoming water. It also helps to raise the head of water. In order to make a necessary quantity of power, it is needed that an enough head is available.

Surge Tank: Surge tank is a little tank or reservoir installed on penstock in which water level rise or fall due to unexpected changes in pressure. This rapid rise of pressure in penstock pipe is called as water hammer. Surge tank serves the follow reason:

- To decrease the distance among the free water surface in dam and turbine, thus dropping the water hammer cause.
- To provide as a supply tank to turbine while the water in pipe accelerates during amplified load situation and as a storage tank while the water is decelerating during reduced load situation.

Draft Tube: Draft tube is joined to outlet of turbine. It changes the kinetic energy available in water in pressure energy in diverge section. Therefore, it retains a pressure of just above the atmospheric level at the end of draft tube to travel the water into a tail race. Water from the tail race is free for irrigation.

Tail Race Level: Tail race is a water path to guide the water discharged from the turbine to river or canal.

Gate: Gate is use to adjust or control the flow of water from the dam.

Pressure Tunnel: It carries the water from the reservoir to surge tank.

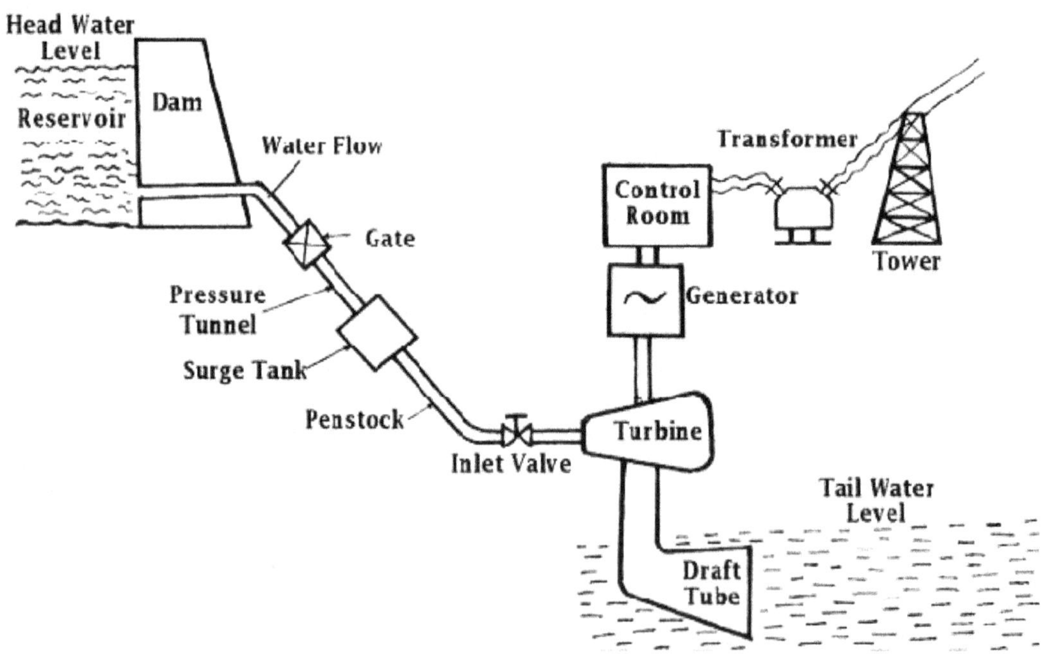

Figure 3.32 A typical Hydropower Plant layout.

The global gross theoretical resource is estimated at about 3.7 TW. 3.5 TW is the resource computed excluding areas with a benign wave climate (areas with less than 5 kW/m) and the net resource (where also areas with potential ice cover is excluded) is about 3 TW; the total reduction from gross to net resource is then about 20 %. In Europe there is a decrease of 25 % from gross to net resource, mostly a result of ice coverage, the gross and net values being 381 and 286 GW, respectively.

3.5 Wave Energy

Wave energy technologies consist of a number of components: 1) the structure and prime mover that captures the energy of the wave, 2) foundation or mooring keeping the structure and prime mover in place, 3) the power take-off (PTO) system by which mechanical energy is converted into electrical energy, and 4) the control systems to safeguard and optimise performance in operating conditions.

There are two ways in which wave energy technologies can be categorised, one is by the way the wave energy is converted into mechanical energy or by the technology used. A broad classification could be **1**. Oscillating Water Column, 2. Wave Activated Bodies and 3. Overtopping Devices, as shown in Figure 3.33.

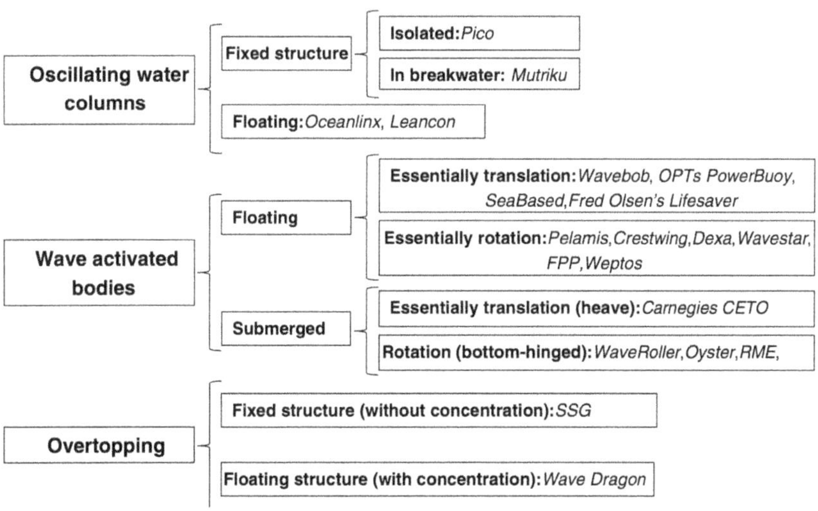

Figure 3.33 Categorization of wave energy technologies.

3.5.1 Oscillating Water Column

Oscillating Water Columns are conversion devices with a semi-submerged chamber, keeping a trapped air pocket above a column of water. Waves cause the column to act like a piston, moving up and down and thereby forcing the air out of the chamber and back into it. This continuous movement generates a reversing stream of high-velocity air, which is channelled through rotor blades driving an air turbine-generator group to produce electricity. The main advantages of these systems are their simplicity (essentially there are no moving parts other than the air turbine) and the fact that they are usually reliable. Conversely, the performance level is not high, although there are new control strategies and turbine concepts under development, which are notably increasing the power performance. Figure 3.34 shows a power plant as an example of an OWC technology.

3.5.2 Wave Activated Bodies

These are either floating (usually) or submerged (sometimes fixed to the bottom). They exploit the more powerful wave regimes that normally occur in deep waters where the depth is greater than 40 metres. In general, they are more complex than OWCs, particularly with regards to their Power Take-off (PTO) systems. In fact, the many different concepts and ways to transform the oscillating movement into electricity has given rise to various PTO systems, e.g., hydraulic generators with linear hydraulic actuators, linear electric generators, piston pumps, etc. As waves pass down the length of the machine and the sections bend in the water, the movement is converted into electricity via hydraulic power take-off systems housed inside each joint of the machine tubes, and power is transmitted to shore using standard subsea cables and equipment. The advantages of oscillating body converters include their size and versatility since most of them are floating devices. A distinct technology has yet to emerge and more research, to increase the PTO performance and avoid certain issues with the mooring systems, needs to be undertaken. Figure 3.35 shows some representative devices of oscillating bodies.

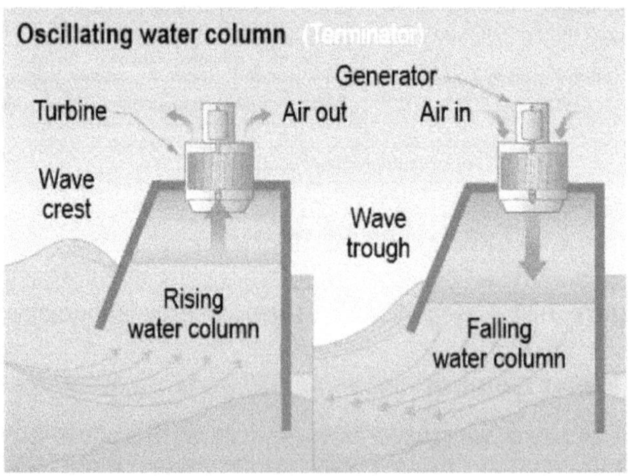

Figure 3.34 A typical OWC Power Plant schematic.

Figure 3.35 (Top Left)-Ocean Power Technologies Inc.; (Top Right)–Pelamis Wave Power; (Bottom Right)–Wave Star AS;(Bottom Left)–Aquamarine Power.

3.5.3 Overtopping Devices

Overtopping converters (or terminators) consist of a floating or bottom fixed water reservoir structure, and also usually reflecting arms, which ensure that as waves arrive, they spill over the top of a ramp structure and are restrained in the reservoir of the device. The potential energy, due to the height of collected water above the sea surface, is transformed into electricity using conventional low head hydro turbines (similar to those used in mini-hydro plants).

The main advantage of this system is the simple concept. It stores water and when there is enough, lets it pass through a turbine. Key downsides include the low head (in the order of 1-2 m) and the vast dimensions of a full scale overtopping device. Some representative devices are shown in Figure 3.36.

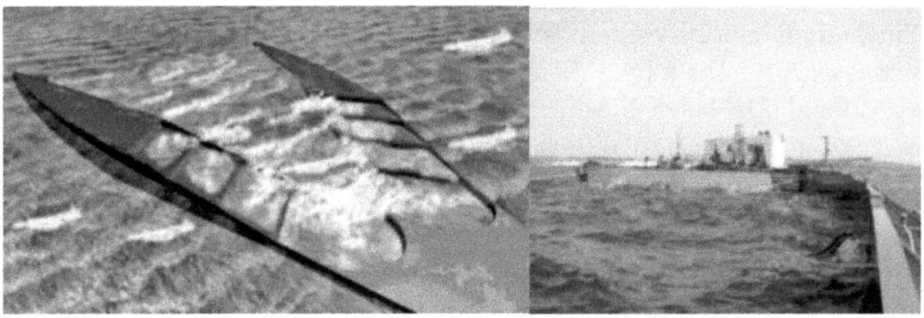

Figure 3.36 Overtopping wave power technologies: Wave Cat (Left) and Wave Dragon (Right).

Power take-off (PTO) systems: There are a number of different PTO systems that can be used to convert the wave energy into electricity: turbines, hydraulic systems, electrical linear generators as well as full mechanical systems. OWCs use air turbines (pneumatic systems) to convert the wave motion into electricity, whilst oscillating bodies and overtopping converters predominantly use a variety of hydraulic PTO systems or turbines. PTO systems have to be adapted to be used in WECs, as the energy flow provided by wave energy is random and highly variable per wave, per day, and per season. As a consequence, air turbines can only reach efficiencies of 50-60%, while hydraulic turbines can reach efficiencies from 70-90%. Furthermore, high-pressure oil hydraulic motors are being explored that include gas accumulator systems capable of storing energy over a few wave periods, smoothing out the irregularities provided by wave energy. Other technological advances in PTO systems include multistage rotor turbines, and adjustable inlet guide vanes to increase the efficiency of the systems.

3.6 Tidal Energy

Tidal range technologies harvest the potential energy created by the difference in head between ebb tide and flood tide. Such resources exist in locations where due to geological and ecological conditions, large water masses flow into compounded areas or bays and estuaries. Furthermore, tidal range energy is predictable, as the energy production is not influenced by weather conditions, but rather by the cyclical constellations, the gravity of the moon, sun and earth, providing a predictable bi-week, biannual and annual cycle.

The first tidal barrage was completed in the Rance River in north-western France (Brittany) in 1966, but due to plans for greater use of nuclear energy, the further pursuit of tidal energy was abandoned. Between 1966 and 2011, a number of small tidal plants were built in countries such as Canada, China, Iran and Russia, where tidal energy resource is abundant. The largest and newest tidal barrage in the world is the Sihwa dam in north-eastern South Korea, which was built in 2011 and became operational in 2012. The Sihwa dam, with capacity of 254 MW, is an example of a multi-functional tidal barrage, which improves the ecology of a formerly closed sea-arm by creating openings in an existing sea defence and installing hydro-turbines. This project is a relevant example for a combined tidal range solution, where in the end the priority was placed on ecological water quality improvement.

Operation: Most conventional tidal schemes use bulb turbines, which are comparable to hydropower turbines that are installed in a dam (run of rivers hydro power plant).

One way power generation at ebb tide: The reservoir is filled at flood tide through sluice gates or valves that are closed once the tide has reached its highest level. At the ebb tide, the water in the reservoir is released through the turbines and power is generated. With this single cycle, power is generated for only four hours per day. Annapolis in Canada is an ebb generation plant.

One way power generation at flood tide: At flood tide the sluice gates are kept closed to isolate the reservoir while at its lowest level. When the tide is high, the water from the sea-side flows into the reservoir via the turbines, thus generating power. The disadvantage of this cycle is that it has less capacity and generates less electricity, and it may be ecologically disadvantageous as the water level in the impoundment is kept at a low level for a long time. Sihwa is a flood generation plant.

Two way power generation: Both incoming and outgoing tides generate power through the turbines. This cycle generates power for four hours twice daily.

3.6.1 Technologies

Tidal current or tidal stream technologies convert the kinetic energy into useable energy. Technology developments are comparable to the development of wind turbines. Although most larger-scale demonstration projects use horizontal axis turbines. They can be categorised in to 3 varieties.

Horizontal-axis axial and vertical-axis cross flow turbines: Horizontal and vertical axis tidal turbines currently use blades that are positioned either in parallel (horizontal) or perpendicular (vertical) to the direction of the flow of water (Figure 3.31). The turbines are similar to designs used for wind turbines, but due to the higher density of water the blades are smaller and turn more slowly than wind turbines. Furthermore, they have to withstand greater forces and movements than wind turbines. Most designs use blades that are connected to a central rotor shaft, which through a gearbox, is connected to a generator shaft. Open-centre turbines have a different design in that the blades are mounted on an inner, open-centred shaft housed in a static tube. As the water flows through the shaft, it rotates and electricity is generated. The advantage of this design is that it eliminates the need for a gearbox. The blades of horizontal or vertical turbines can also be enclosed within a duct. The latter turbines are referred to as Enclosed, Ducted or Shrouded turbines (Figure-3.37). Due to the enclosure, the ocean current is concentrated and streamlined so that the flow and power output from the turbines increases. Based on an overview of existing tidal current projects, 76% of all turbines are horizontal axis turbines and 12% are vertical axis turbines.

Figure 3.37 Horizontal Axis Turbines (left), Vertical Axis Turbines (middle), and Enclosed Turbines (right).

Reciprocating Devices: Reciprocating devices have blades called hydrofoils - shaped like airplane wings that move up and down as the tidal stream flows on either side of the blade. The up and down movement of the hydrofoils is subsequently converted into rotation to drive a rotating shaft, or connected to pistons to support a hydraulic system for power generation (Figure 3.38). The advantage of reciprocating devices is that the length of the blade is not constrained by water depth, however it also requires complex control systems to pitch the blades correctly. Around 2% of all R&D investments in tidal current technologies went into reciprocating devices.

Figure 3.38 Hydrofoil shaped design concepts.

Other designs: There are a number of other designs that are in the research and development stage. This category includes rotating screw-like devices and tidal kites that carry turbines below their wings. Besides the conversion technology, there are a number of additional technological aspects that determine the performance and costs of tidal current technologies: 1) support structures, 2) array formation, and 3) electrical connections to shore. Some of the turbines developed by various companies are shown in Figure 3.39.

Figure 3.39 Tocardo (Top Left), ANDRITZ HYDRO GmbH, Hammerfest (Top Right), Marine Current Turbines- a Siemens Business (Bottom Left), Sabella (Bottom Right).

3.7 Geothermal Energy

The term "geothermal" means "Earth heat" or "heat of the Earth." Energy from geothermal resources has benefited humankind from its earliest origins. Prehistoric civilizations used hot springs and steam discharges (fumaroles) for cooking, heating, and therapeutic bathing; in modern terms, these uses are known as geothermal direct-use applications. In the United States, geothermal energy has provided affordable, reliable, and renewable energy since the 1890s, when the city of Boise, Idaho, began using geothermal resources for direct heating of commercial and residential buildings. Since then, use of geothermal energy in the United States has expanded to include utility-scale electricity production, distributed heating and cooling applications, and the augmentation of various industrial processes. United States of America was the global leader in both geothermal power generation and installed capacity. Geothermal heat pumps (GHPs) are another key geothermal technology considered in the GeoVision analysis. GHPs have been deployed since the 1940s, supplying reliable, quiet, efficient, and cost-competitive residential space heating and cooling. Commercial geothermal electric power production began in the United States as early as September 1960, at The Geysers geothermal field in California. The Geysers remains the world's largest geothermal field in terms of the number of operating power plants and wells, installed generation capacity, and the physical dimensions of the wellfield. As of 2017, the United States was the global leader in both geothermal power generation and installed capacity. However, the vast majority of medium- and high temperature geothermal systems, which are suitable for power generation, are located close to areas of volcanic activity- for example, situated along plate boundaries (subduction zones, such as the majority of the Pacific "Ring of Fire"), mid-oceanic ridges (such as Iceland and the Azores) and rift valleys (such as the East African Rift) or near hot spots (such as in Hawaii) as shown in Figure 3.40.

As of 2023, the global installed geothermal power capacity was 16,335 MW. The top countries with the highest geothermal capacity include: United States: 3,900 MW, Indonesia: 2,418 MW, Philippines: 1,952 MW, Turkey: 1,691 MW, New Zealand: 1,042 MW, Kenya: 985 MW, Mexico: 976 MW, Italy: 916 MW, Iceland: 754 MW, Japan: 576 MW. Figure 3.41 presents electric generation through Geothermal since 1980. However the generation in 1980 and 1985 are estimated.

The heat content of a geothermal field will define the power generation technology to be used. Power generation from geothermal resources requires resources with high to medium heat content. Geothermal power generation currently is based on the following four technology options.

3.7.1 Direct dry steam plants

In this case, the conversion device is a steam turbine designed to directly use the low pressure, high-volume fluid produced in the steam field. Dry steam plants commonly use condensing turbines. The condensate is re-injected (closed cycle) or evaporated in wet cooling towers as shown in Figure 3.42.

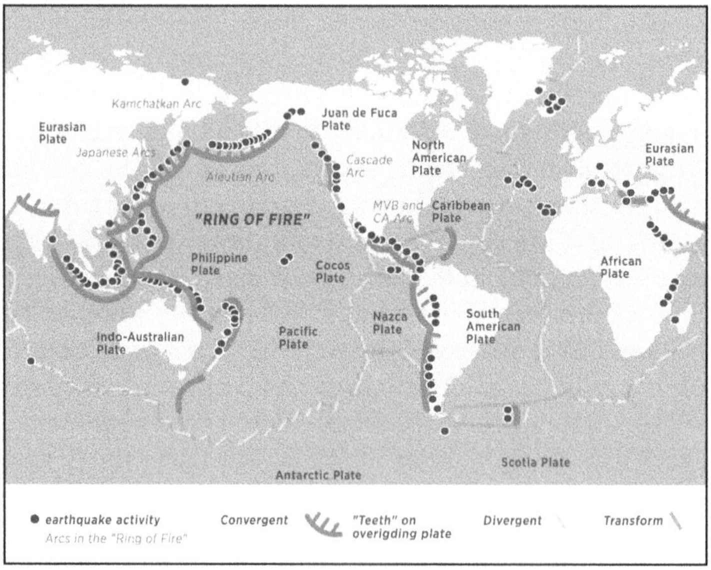

Adapted from National Park Service (U.S.), 2014

Figure 3.40 Tectonic plates and global geological activity.

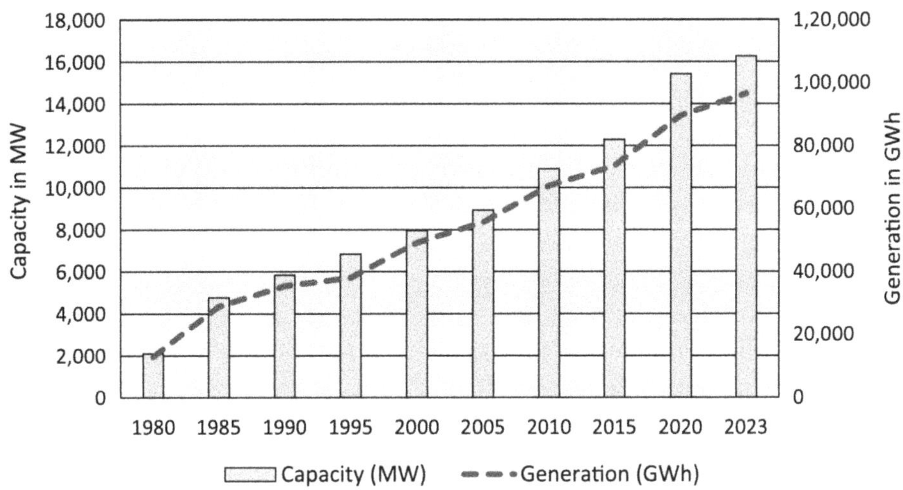

Figure 3.41 World geothermal capacity and electric generation between 1980 and 2023.

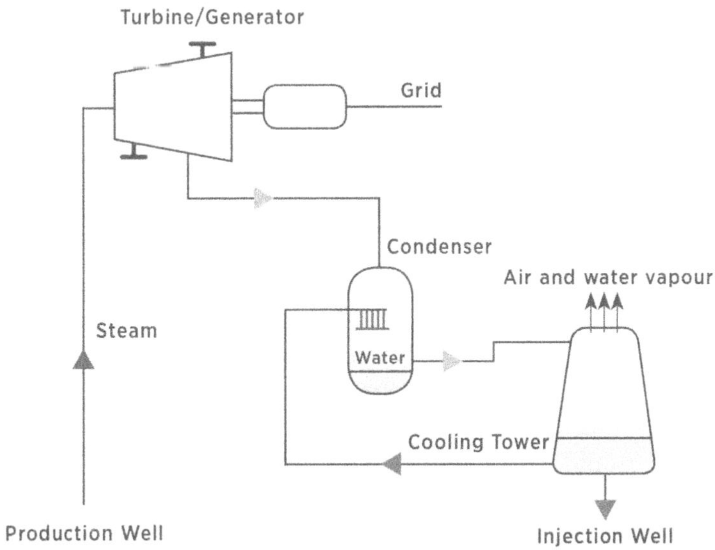

Figure 3.42 Dry Steam Plant.

This type of geothermal power plant uses steam of 150 degrees Celsius or higher, and, generally, the steam entering the turbine needs to be at least 99.995% dry to avoid scaling and/or erosion of the turbine or piping components. Direct dry steam plants range in size from 8 MW to 140 MW.

3.7.2 Flash Plants

These are the most common type of geothermal electricity plants in operation today. They are similar to dry steam plants; however, the steam is obtained from a separation process called flashing. The steam is then directed to the turbines, and the resulting condensate is sent for re-injection or further flashing at lower pressure as shown in the Figure 3.43. The temperature of the fluid drops if the pressure is lowered, so flash power plants work best with well temperatures greater than 180°C. The fluid fraction exiting the separators, as well as the steam condensate (except for condensate evaporated in a wet cooling system), are usually re-injected. Flash plants vary in size depending on whether they are single- (0.2-80 MW), double - (2-110 MW) or triple-flash (60-150 MW) plants.

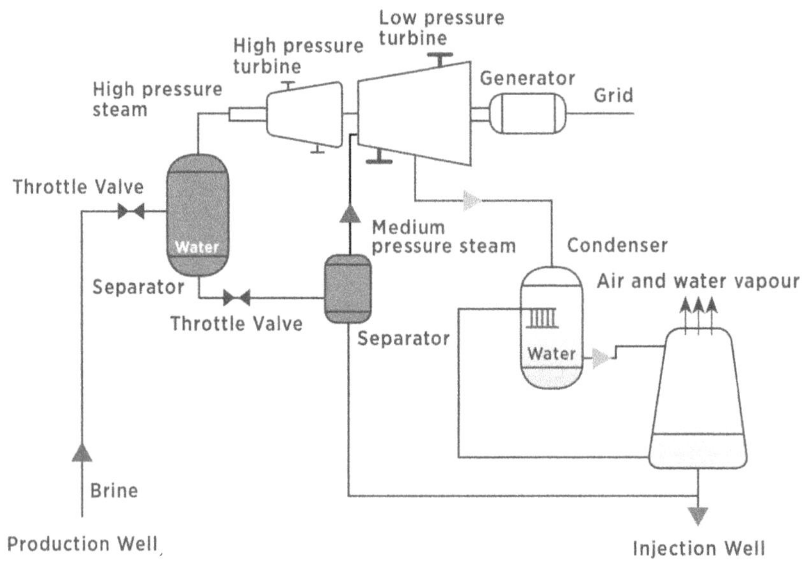

Figure 3.43 Double Flash Power Plant.

3.7.3 Binary Plants

These plants are usually applied to low/medium-enthalpy geothermal fields where the resource fluid is used, via heat exchangers, to heat a process fluid in a closed loop as shown in Figure 3.44. The process fluid (e.g., ammonia/water mixtures used in Kalina cycles or hydrocarbons in organic Rankine cycles) which have boiling and condensation points that better match the geothermal resource temperature are used. Typically, binary plants are used for resource temperatures between 100°C and 170°C. Although it is possible to work with temperatures lower than 100°C, the efficiency of the electricity output decreases. Binary plants range in size from less than 1 MW to 50 MW.

3.7.4 Combined-Cycle or Hybrid Plants

Some geothermal plants use a combined cycle which adds a traditional Rankine cycle to produce electricity from what otherwise would become waste heat from a binary cycle as presented in Figure 3.45. Using two cycles provides relatively high electric efficiency. The typical size of combined-cycle plants ranges from a few MW to 10 MW. Hybrid geothermal power plants use the same basics as a stand-alone geothermal power plant but combine a different heat source into the process; for example, heat from a concentrating solar power (CSP) plant. This heat is added to the geothermal brine, increasing the temperature and power output.

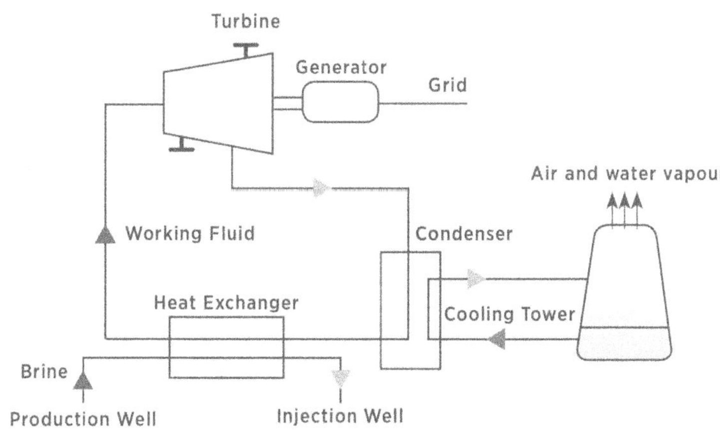

Figure 3.44 A Typical layout of Binary Plant.

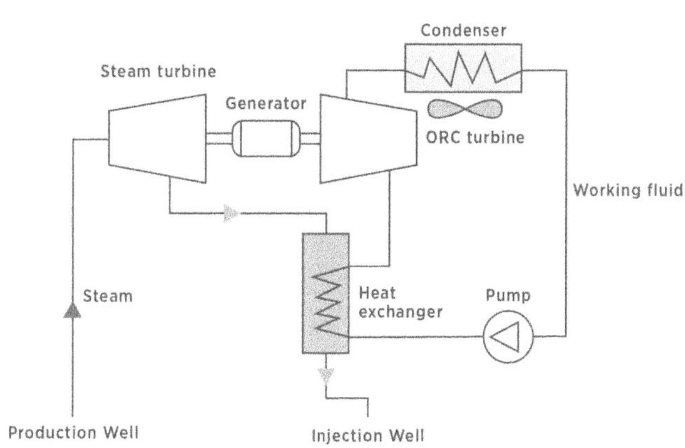

Figure 3.45 Geothermal Hybrid or Combined cycle power plant layout.

3.7.5 Enhanced Geothermal Systems

A large part of the geothermal potential is heat stored at depths greater than commonly drilled. Standard hydrothermal technologies depend on permeable aquifers, which allow the flow of water through them, to produce hot water. However, at greater depths the ground becomes less porous and water flow is restricted. Research and demonstration projects are being developed to overcome this limitation. Instead, artificial fractures are created to connect production and injection wells by hydraulic or chemical stimulation. Stimulation is accomplished by injecting water and a small amount of chemicals at high pressure to create or re-open fractures in the deep rock as shown in Figure 3.46. To prevent these fractures from closing again when the injection pressure is reduced, special materials called proppants are added. This approach, known as enhanced geothermal (EGS), uses binary plants to produce power from the hot brine. As there is no natural flow of water, all the brine has to be re-injected into the reservoir to keep the pressure and production stable. This helps prevent air emissions during the service life.

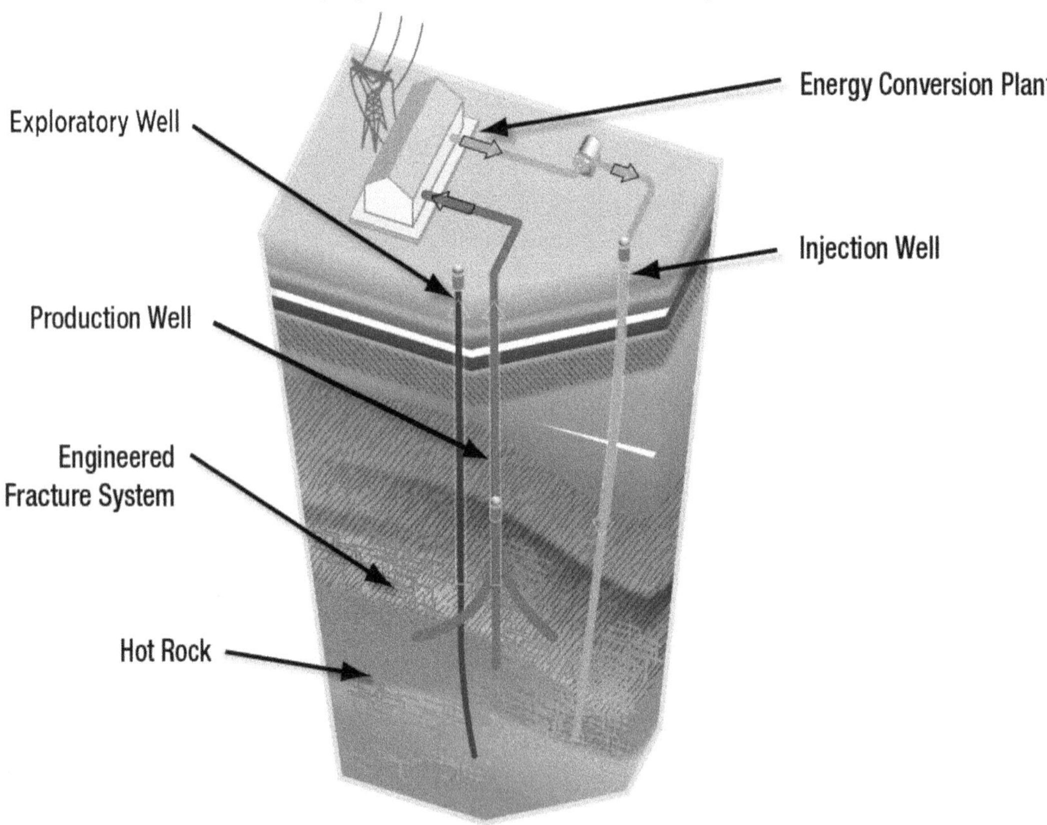

Figure 3.46 Enhanced geothermal system.

3.8 Biomass Energy

On a global scale, biomass supplies more than 1% of the electricity demand, i.e. some 257 TWh per year. Based on up-to-date combustion technologies, biomass and waste also supply approximately 4.5 EJ (105 Mtoe) of direct heat to the industrial and residential sectors, and 2 to 3 EJ (47 to 70 Mtoe) of heat from combined heat and power (CHP) plants. These estimates do not include traditional biomass combustion mostly used in developing countries. The use of combustible renewables (especially solid biomass) has a significant impact on the energy balance of countries and regions with abundant primary resources such as the European Nordic countries, Austria, and Switzerland while the use of biogas is increasing in Germany, the Netherlands, the United Kingdom, and Italy. Power generation and CHP based on biomass and waste, as well as on biomass co-firing in coal-fired power plants, are also rapidly growing. In Germany, for instance, the growth of biomass-based CHP amounted to 23% per year in the period 2004-2008 and state-of the-art plants are characterised by high-performance steam parameters and efficiency. The capacity of biomass CHP plants varies considerably. Biogas anaerobic digestors are usually associated to gas-fired engines for heat and power generation with electrical capacity from tens of kWe up to a few MWe. Biomass-fired power plants and CHP plants have capacities ranging from a few MWe up to 350 MWe. Small and medium-size CHP plants are usually sourced with locally available biomass. Large CHP plants and coal/biomass co-firing power plants require biomass sourcing from a wide region and/or imported wood or forestry residues. Biomass CHP plants are mature technologies while biomass integrated gasification combined cycles (BIGCC), which offer high technical and economic performance, are currently in the process of entering the market, following the industrial demonstration phase.

Process and Technology Status: In the IEA countries, the use of solid biomass for heat and power generation, including combined generation, started during the 1970s, in the wake of the oil crises. Currently, on a global scale, biomass supplies more than 1% of the electricity demand, i.e. approximately 257 TWh per year. Along with waste, biomass also supply approximately 4.5 EJ (105 Mtoe) of direct heat to the industrial and residential sectors, and between 2 and 3 EJ (47 to 70 Mtoe) of heat from combined heat and power plants (CHP). In Europe, the use of biomass is significant in regions with ample biomass resources, e.g. the Nordic countries, Austria, and Switzerland. The status of major biomass technologies is highlighted in Figure 3.47.

Water-cooled vibrating grate (VG) boilers (Figure 3.48): These are an established technology for power generation from wood residues. Based on natural circulation, these boilers are designed to burn low-heating-value (LHV of about13.8 MJ/kg) wood residues, with 30% humidity. The typical power plant capacity is in the order of 10 MWe.

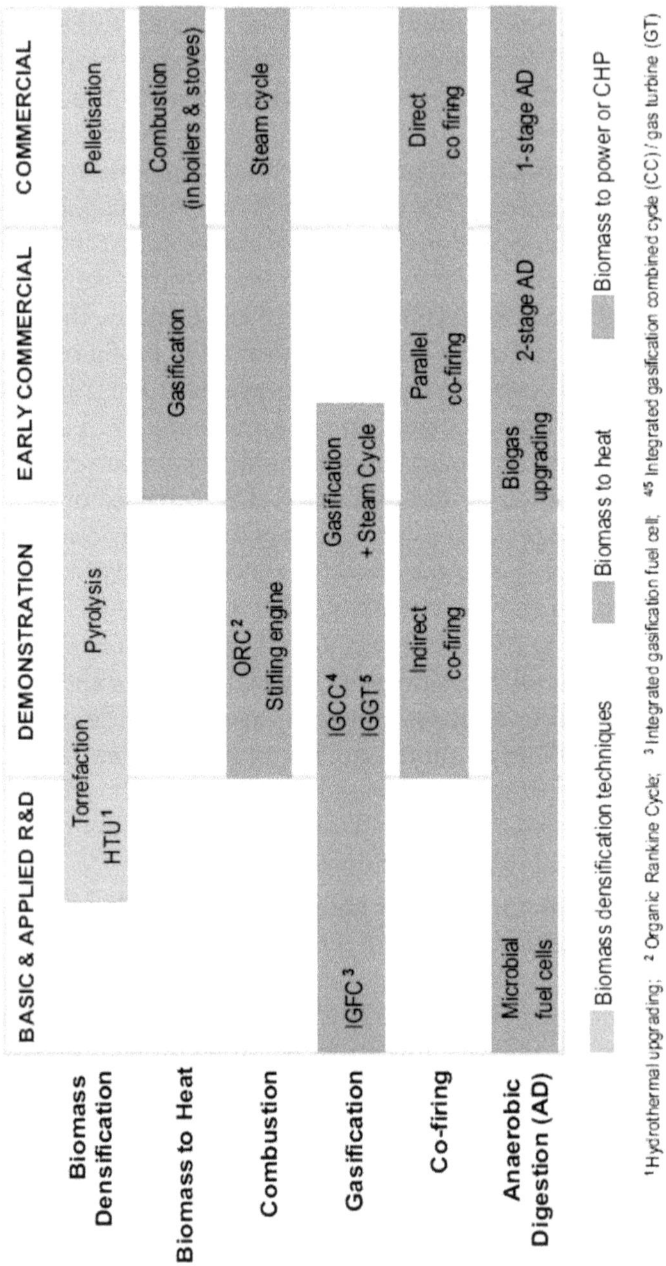

Figure 3.47 Status of major biomass technologies (IEA Bioenergy, 2009).

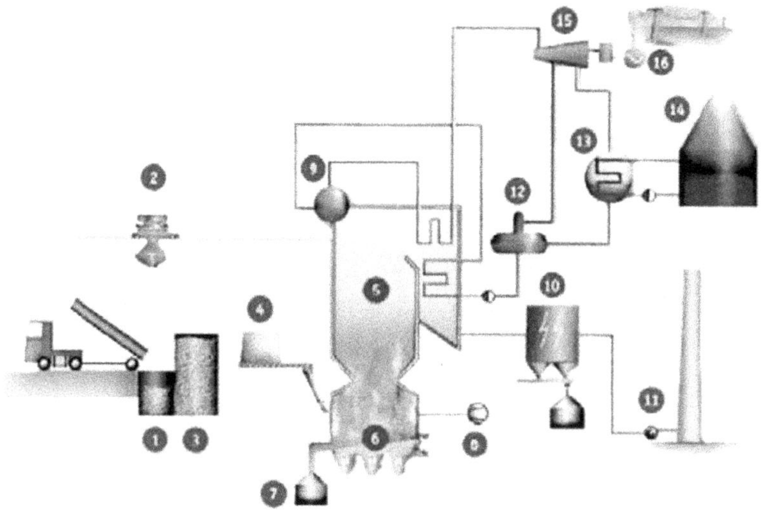

1 Pit	7 - Ashtrays	13 - Condenser
2- Crane	8 - Fans	14 – Refrigerator tower
3 - Silo	9 - Holdfast	15 - Turbine
4 - Hopper	10 – Electrostatic precipitator	16 - Generator
5 - Boiler	11 - Chimney	
6 - Grate	12- Desgasifier	

Figure 3.48 Water-cooled vibrating grate (VG) boiler base CHP & Power Plant.

Bubbling fluidised bed combustion (BFBC) boilers: Figure 3.49 shows Scheme of bubbling fluidised bed combustion boiler based CHP and Power Plant. Solid biomass and other feedstock are also a proven and commercial option, but continued improvements in CHP technology have made available a new generation of plants that offer advanced steam parameters and high efficiency. In the BFBC boilers, the ascending air speed is sufficiently high to maintain the bed in a state of fluidisation, with a high degree of mixing, but it is low enough to make most of the solid particles lifted out of the bed fall back. The result is a dense bed with uniform temperature and burning char, and rather small over-temperatures. The dense part of the fluidised bed has a void fraction that is near to minimum fluidisation requirement. Within the dense part of the bed, a bubble phase exists, with a low content of solids. The bubbles formed from air in excess rise through the dense phase. As in gas-liquid systems, the bubble flow in the fluidised bed induces solids transport and mixing in the dense region. The upward velocity of air/combustion gases is 2 - 3 m/s, and bed heights are 0.5 to 1.5 m. Solid materials mostly stay in the well stirred bed, although small particles will leave the bubbling bed and be thrown up into the freeboard region. Cyclones and other particulate removal equipment are used to collect them before the flue gas is channelled to the heat recovery systems. Coarse bed material is also withdrawn from the bottom of the bed to maintain high sulphur-capture capacity and to avoid ash contamination which might cause bed agglomeration.

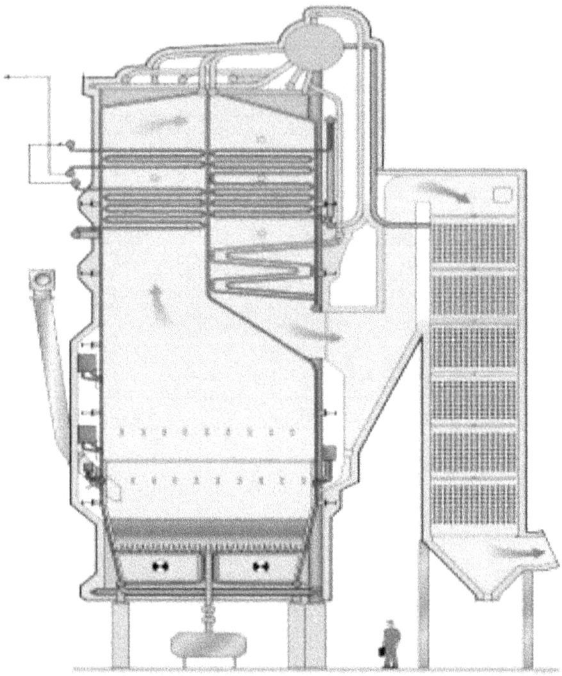

Figure 3.49 Scheme of bubbling fluidised bed combustion boiler.

Circulating fluidised bed combustion (CFBC) boilers: They offer a further option for biomass-fired CHP. In CFBC, a distinction between the bed and the freeboard area is no longer applicable. A large fraction of the particles rises up from the bed and is re-circulated by a cyclone. The circulating bed material is used for temperature control in the boiler. The choice between BFBC and CFBC depends inter alia on the fuel used. CBFC boilers are used in large CHP or power plants, with capacity of hundreds of MWe, but they may also be competitive in smaller biomass-fired plants.

Co-firing of biomass in coal-fired power plants: Biomass co-firing in coal-fired power plants offers significant advantages: it is highly efficient, approximately between 36% and 44%, depending on the efficiency of the coal-fired unit (39% - 46%); coal- fired power plants have coal access facilities, which may also facilitate biomass supply; they also have advanced flue gas cleaning equipment, which in some cases may obviate separate cleaning for biomass. Coal-Fired Power, today's maximum efficiency of a pulverised coal-fired power (PC) plants is around 46%, with potential for reaching 50% or more. Biomass co-firing in coal power plants requires significant boiler retrofitting, as well as specific equipment and space for biomass logistics, and tailoring of flue gas cleaning equipment (i.e. electrostatic precipitator, flue gas desulphurisation, and de-NOx, if applicable), especially if significant amounts of biomass are co-fired. Nox emissions in coal/biomass co-firing depend significantly on the emission reduction technology, e.g. separated overfire air, or NOx selective catalytic reduction (SCR). For a retrofitted coal-fired power plant in Poland, NOx emissions are less than 200 mg/Newtom-m cube equivalent to 500 g NOx/MWh. Biomass co-firing may reduce NOx emissions compared to coal as biomass has a lower nitrogen content.

Anaerobic digestion of wet biomass with CHP: Anaerobic digestion for biogas production from wet biomass is a small-scale biomass CHP application. Figure 3.50 shows the stages from wet biomass to bio- methane. The use of biogas is gaining importance in Germany, the Netherlands, the United Kingdom, and Italy. Biogas may also be upgraded to be mixed with natural gas and used in natural gas grids or to power vehicles as compressed natural gas (CNG). Also, anaerobic digestion of wet manure and co-digestion of wet manure along with agricultural residues may be economically viable for the generation of heat and power using internal combustion gas engines.

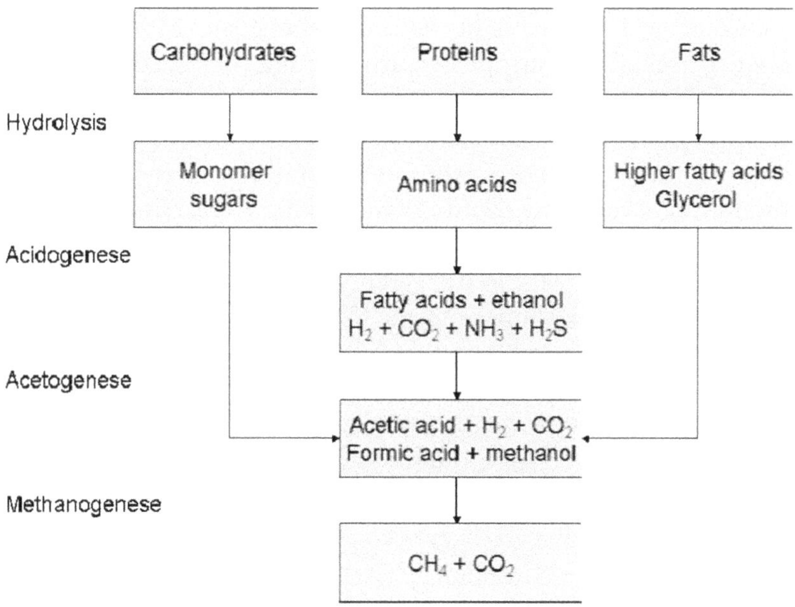

Figure 3.50 Stages from wet biomass to bio- methane.

3.9 Variable RE Integration and Overcoming Intermittence

Variable source generators are connected to the grid using power electronic converters. These generators are fundamentally different from conventional power plants in terms of capacity, inertia, etc. Intermittent sources act rigidly and their high penetration reduces the flexibility of the power system and may lead to new challenges related to energy quality, stability, and protection of the power grid. A variety of solutions are available to meet the challenges of integrating variable energy into the power grid. For example, power grid expansion and strengthening, advanced forecasts of solar and wind production, demand response, use of flexible production sources, and energy storage. The use of flexible generation sources (gas turbine, hydro, etc.), and grid expansion may require significant capital investment. Demand response and renewable generation forecasting allow the grid operator to improve the flexibility of the power system, but do not solve the power quality and stability problems caused by high VRE penetration. Traditional generators are not well-suited for providing the flexibility required by modern power grids with high renewable penetration. They are typically designed to operate at a constant output level and are not able to ramp up or down quickly in response to changes in demand or supply. Because of its ability to unlink supply and demand and to allow for momentary shifts, energy storage is considered as a solution to intermittent production. Storage can reduce voltage fluctuations, increase system flexibility and enable the storage and dispatch of renewable energy. Multiple energy storage technologies exist, including pumped hydro storage (PHS), compressed air energy storage (CAES), flywheels, supercapacitors, and batteries. Several studies have conducted comparative analyses of these various ESS technologies, considering factors such as their characteristics, limitations, and applications. ESS technologies have their own advantages and disadvantages. One example of an energy storage technology is pumped hydro storage (PHS), which offers a high energy storage capacity. However, it requires specific geological features for implementation and operation. Compressed air energy storage (CAES), another technology, exhibits a high energy density but is dependent on specific geological features as well, and it tends to have lower efficiency. Flywheels, on the other hand, are characterized by high power density but relatively lower energy density compared to other ESS technologies. Supercapacitors are known for their high power density but tend to have lower energy density compared to other energy storage system (ESS) technologies. On the other hand, battery energy storage systems (BESS) offer several advantages over other ESS technologies. They exhibit higher round-trip efficiency compared to pumped hydro storage (PHS) and compressed air energy storage (CAES).

BESS also provide flexibility in terms of installation and find applications across various sectors and scales due to their versatility and modularity. For example, in

August 25, 2018, when the Heywood interconnector between South Australia and Victoria was lost, the battery provided fast frequency response (FFR) and contingency support. As the frequency of the system dropped below the lower threshold, the battery responded by supplying active power. It continued to provide incremental response as the frequency reached its lowest point. When the Heywood interconnector tripped and the frequency rose, the battery's fast frequency response was activated to track its droop curve. Ultimately, the battery helped to stabilize the frequency and returned it to the normal operating range, deactivating its FFR.

There have been demonstrations how adding BESS to the grid increases the flexibility required to cope with power fluctuations. However, although energy storage is widely used, it faces the challenge of choosing the most suitable technology for grid-scale applications. Pumped Hydro Storage is the most commonly used method to store energy in large quantities, with a total installed capacity of approximately 160 GW as of 2021. Although batteries have a smaller capacity than hydro-pumped storage, the total installed grid-scale capacity for batteries in 2021 globally was 16 GW. Due to their mobile and scalable characteristics, storage batteries are attractive in stationary applications. However, the positioning and sizing of storage batteries in electrical systems are critical to maximizing their potential. An inadequate deployment reduces the reliability of the power grid, meanwhile, oversizing results in significant costs for the power grid operation. Therefore, significant studies are being conducted for the optimal deployment of battery energy storage systems (BESS) in the power grid.

This study investigates the impact of high variable renewable energy penetration to the grid and the role of electrochemical batteries in mitigating these effects. It aims to identify the most suitable battery technologies for grid applications. The study also proposes an incentivized approach, incorporating grid tariffs into capacity expansion planning, to encourage the integration of battery storage to the grid. The subsequent chapter deals with various energy storage systems along with their characteristics, technologies and applications.

References

1. https/www.stannescet.ac.in/cms/staff/qbank/EEE/Notes/EE8703-RENEWABLE%20ENERGY%20SYSTEMS-644988498-RES%20UNIT%203.pdf
2. https://courses.edx.org/c4x/DelftX/ET.3034TU/asset/solar_energy_v1.1.pdf
3. https://electricalnotebook.com/i-v-characteristics-of-solar-pv-cell/
4. Nacer K. M'Sirdi etal "Review of the Best MPPT Algorithms for Control of PV Sources RUCA Tracking Algorithm" cite press, 2017, pp 1-8
5. Ariyur, K. B. and Krstic, M. (2003). Real-time optimization by extremum-seeking control. John Wiley & Sons

6. Brunton SL, Rowley CW, Kulkarni SR, Clarkson C. Maximum power point tracking for photovoltaic optimization using ripple-based extremum seeking control. IEEE Transactions on Power Electronics 2010; 25(10): 2531-40.
7. E. H. Camm *et al.*, "Characteristics of wind turbine generators for wind power plants," *2009 IEEE Power & Energy Society General Meeting*, Calgary, AB, Canada, 2009, pp. 1-5, doi: 10.1109/PES.2009.5275330
8. https://www.worldscientific.com/doi/pdf/10.1142/9789811225925_0001?srsltid=AfmBOoq9kxZ5LaZqVsgtRocYwUqUGKewWEIT_t5aXQ29z2kUvN226gaE
9. https://ourworldindata.org/grapher/cumulative-installed-wind-energy-capacity-gigawatts?facet=none#explore-the-data
10. https://engineering.purdue.edu/~dionysis/Wind_Energy_Conversion.pdf
11. Kumar, A., T. Schei, A. Ahenkorah, R. Caceres Rodriguez, J.-M. Devernay, M. Freitas, D. Hall, Å. Killingtveit, Z. Liu, 2011: Hydropower. In IPCC Special Report on Renewable Energy Sources and Climate Change Mitigation [O. Edenhofer, R. Pichs-Madruga, Y. Sokona, K. Seyboth, P. Matschoss, S. Kadner, T. Zwickel, P. Eickemeier, G. Hansen, S. Schlömer, C. von Stechow (eds)], Cambridge University Press, Cambridge, United Kingdom and New York, NY, USA.
12. Paul Breeze, Hydro Power, 125 London Wall, London EC2Y 5AS, United Kingdom Academic Press is an imprint of Elsevier
13. https://www.ipcc.ch/site/assets/uploads/2018/03/Chapter-6-Ocean-Energy-1.pdf
14. https://www.irena.org/-/media/Files/IRENA/Agency/Publication/2014/Wave-Energy_V4_web.pdf
15. https://www.irena.org//media/Files/IRENA/Agency/Publication/2014/Tidal_Energy_V4_WEB.pdf
16. https://www.ipcc.ch/site/assets/uploads/2018/03/Chapter-4-Geothermal-Energy-1.pdf
17. https://www.irena.org//media/Files/IRENA/Agency/Publication/2017/Aug/IRENA_Geothermal_Power_2017.pdf
18. https://www.cedengineering.com/userfiles/Introduction%20to%20Geothermal%20Energy%20(R03-014).pdf
19. https://www.irena.org/-/media/Files/IRENA/Agency/Publication/2015/IRENA-ETSAP_Tech_Brief_E05_Biomass-for-Heat-and-Power.pdf
20. https://www.undp.org/sites/g/files/zskgke326/files/migration/lk/Feasibility-of-BM-Power-Plants.pdf
21. https://iea-etsap.org/E-TechDS/PDF/E05-BiomassforHP-GS-AD-gct.pdf
22. Lere Deguenon, Daniel Yamegueu, Sani Moussa kadri, Aboubakar Gomna, “Overcoming the challenges of integrating variable renewable energy to the grid: A comprehensive review of electrochemical battery storage systems”,Journal of Power Sources, Volume 580, 2023, 233343, ISSN 0378-7753
23. IEA, “Grid-Scale storage,” paris [Online]. Available: https://www.iea.org/reponts/grid-scale-storage, 2022.

CHAPTER

4

Electrical Energy Storage Systems

4.1 Introduction

Energy storage will play a crucial role in enabling the next phase of the energy transition. Along with boosting solar and wind power generation, it will allow sharp decarbonization in key segments of the energy market. As variable renewables grow to substantial levels, electricity systems will require greater flexibility. At very high shares of VRE, electricity will need to be stored over days, weeks or months. By providing these essential services, electricity storage can drive serious electricity decarbonization and help transform the whole energy sector.

Energy storage can be achieved by transforming into another form of storable energy and to be transformed back when needed. There are many possible techniques for energy storage, found in practically all forms of energy i.e., mechanical, chemical, and thermal. The storage technologies that answer to specific technical and economic criteria, which vary considerably as a function of the applications and needs, will obviously be of different types. The storage techniques can be divided into four categories, according to their applications:

1. Low-power application in isolated areas, essentially to feed transducers and emergency terminals.
2. Medium-power application in isolated areas (individual electrical systems, town supply).
3. Network connection application with peak shaving
4. Power-quality control applications.

The first two applications belong to small-scale systems where the energy could be stored as kinetic energy (flywheel), chemical energy, compressed air, hydrogen (fuel cells), or in supercapacitors or superconductors. The next two applications are related to large-scale systems where the energy could be stored as gravitational energy (hydraulic systems), thermal energy (sensible, latent), chemical energy (accumulators, flow batteries), or compressed air (or coupled with liquid or natural gas storage).

4.2 Characteristics of Storage Systems

The main characteristics of storage systems on which the selection criteria for a particular application is as follows.

Storage capacity: This is the quantity of available energy in the storage system after charging. Discharge is often incomplete. Hence, it is defined on the basis of total energy stored (W_{st},) which is greater than actually retrieved (W_{ut}). The usable energy, limited by the depth of discharge. Under rapid charge or discharge, the efficiency deteriorates, and the retrievable energy becomes much lower than storage capacity.

Available Power: This parameter determines the constitution and size of the motor-generator in the stored energy conversion chain. It is generally expressed as an average value, as well as a peak value often used to represent maximum power (P_{max}) of charge or discharge.

Depth of discharge or power transmission rate: Energy storage is a slow process whereas discharge is quick on demand. The power output, or discharge, can be a limiting factor called the power transmission rate. This delivery rate determines the time needed to extract the stored energy.

Discharge time: This is the maximum-power discharge duration, $\tau(s)=W_{st}/P_{max}$. It depends on the depth of discharge and operational conditions of the system, constant power or not. It is a characteristic of system adequacy for certain applications.

For a pumped storage system, the storage capacity is dependent on the water mass and the height of the waterfall, while the maximum power is determined by the size of the conduits and the power of the reversible turbine–electrical-machine units.

Efficiency: This is the ratio between released energy and stored energy, $\eta=W_{ut}/W_{st}$. This definition is often oversimplified because it is based on a single operation point. Yet, systems have charging, no-load, and self-discharge losses. The definition of efficiency must therefore be based on one or more realistic cycles for a specific application. Instantaneous power is a defining factor of efficiency. For the storage system to be competitive, it must have good overall efficiency. This means that, for optimum operation, the power-transfer chain must have limited losses in terms of energy transfer and self-discharge.

Durability (cycling capacity): This is the number of times the storage unit can release the energy level (it was designed) for each recharge, expressed as the maximum number of cycles N (one cycle corresponds to one charge and one discharge). All storage systems are subject to wear by usage due to aging, ahead of thermal degradation. The design of a storage system that considers the endurance of the unit in terms of cycles should be a primary importance when choosing a system. In all cases, it is strongly linked to the amplitude of the cycles and/or the average state of charge.

Durability (Cycling Capacity): This refers to the number of times the storage unit can release the energy level it was designed for after each recharge, expressed as the maximum number of cycles N (one cycle corresponds to one charge and one discharge). All storage systems are subject to wear by usage due to aging, ahead of thermal degradation. However, real fatigue processes are often complex, and the cycling capacity is not always well defined. In all cases, it is strongly linked to the average state of charge.

Autonomy: This is the maximum amount of time the system can continuously discharge. It is defined by the ratio between the energy capacity and maximum discharge power, $a=W_{ut}/P_d$. The autonomy of a system depends on the type of storage and thc type of application.

Costs: Like any other investment, a storage system is an interesting venture when total gains exceed total investment. The capital invested and operational costs (maintenance, energy lost during cycling, aging) are the most important factors to consider for the entire life of the system. Low-efficiency systems with low cycling capacity generally require the lowest initial investment. It is therefore crucial to carry out an analysis of the estimated durability of the entire system, including the storage unit. In terms of sustainable development, the overall costs over the entire life of the system, including materials, energy, and other environmental costs from fabrication to recycling, must be considered. The investment costs of storage is factored out using the following formula:

$$C = C_1 W_{ut} += C_2 P_d \quad .....(4.1)$$

The operational costs, spread over the lifespan of the system, are supposed to be proportional to the investment, for a total cost of $C_t = (aC_1 + C_2)P_d$. Here a is the autonomy.

Feasibility and adaptation to the generating source: A highly efficient storage system is closely adaptable to the type of application (small to medium power in isolated areas, network connection, etc.) and to the type of production (permanent, portable, renewable, etc.). It needs to be harmonized with the network.

Self-discharge: This is the amount of stored energy that which has dissipated over a given amount of non-use time (in idle condition).

Mass and volume densities of energy: These represent the maximum amounts of energy accumulated per unit of mass or volume of the storage unit and demonstrate the importance of mass and volume for certain applications (especially for mass density of energy in portable applications, but less so for permanent applications).

Monitoring and control equipment: This equipment, on both the quality and safety of storage levels, has repercussions on the accessibility and availability of the stored energy to control and monitoring equipment.

Operational constraints: These are related to the safety (explosions, waste, bursting of a flywheel, etc.) or other operational conditions (temperature, pressure, etc.). They can influence the choice of a storage technology as a function of energy needs.

Reliability: Storage-system reliability is always an important factor because it is a guarantee of on-demand service.

Environmental aspect: While this parameter is not a criterium of storage-system capacity, the environmental aspect of the product (recyclable materials) is a strong sales pitch. For example, in Nordic countries (Sweden, Norway), a definite margin of the population prefers to pay more for energy than to continue polluting the country. This is a feature must not be ignored.

Other characteristics: The ease of maintenance, simple design, operational flexibility, fast response time for the release of stored energy, etc., It is important to note that these characteristics apply to the complete storage system.

4.3 Energy Storage Technologies

Electrical energy storage systems are today, very vital to the energy generation industry. This is because their performance determines the efficiency of the system, as well as its cost of operation. Today, several types of storage applications are being explored, and most of them are being carried out with the goal to reduce cost as well as ensuring longevity. Currently, the main issue affecting the utilization of electrical energy storage systems has to do with the initial capital required and the cost of operating the system. It is also another goal of researchers to ensure these storage devices do not impact the environment negatively. Electrical storage systems can be largely classified as mechanical storage system, electrochemical systems, chemical storage and thermal storage systems.

An electrical energy storage system is made up of a storage unit, as well as a power-converting unit. The dc voltages are converted into ac with help of converter. With the aid of an alternating current transmission system, the alternating power can be added to the grid. For some electrical energy storage systems, a rectifier transforms the alternating current to a direct current for the storage systems. The efficiency of the grid can be improved based on the performance of the energy storage system. The energy storage device can ensure a baseload power is utilized efficiently, especially during off-peak times. This can significantly reduce the cost of generation.

The IEA have concluded that an effective installed energy storage capacity will reduce global warming by 2 °C, provided the installed capacity increases by 450 GW in 2050 as compared to 140 GW in 2014. Information gathered indicated that the installed grid connected capacity for Energy Storage System was 140976 MW as of 2014. Nearly 99.3% of the capacity that was stored was in the form of pumped hydro storage. The rest were all obtained from other types of storage techniques. As of 2018, the energy storage system is still gradually increasing, with a total installed grid capacity of 175 823 MW. The pumped hydro storage systems were 169557 GW, and this was nearly 96% of the installed energy storage capacity worldwide. All others combined increased approximately by 4%. Figure 4.1 depicts the total installed capacity for energy storage systems.

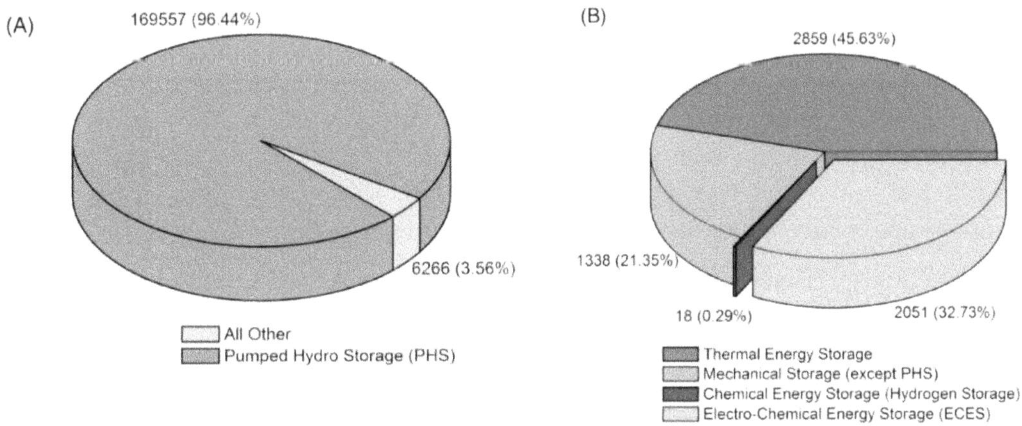

Figure 4.1 (A) Operational capacity for installed energy storage systems (B) Grid connected operational capacity for other types of energy storage systems.

Figure 4.2 presents detailed classifications of the various types of storage systems. The form of energy being stored determines the type of the storage device that is suitable for the application.

Mechanical Energy Storage Systems	Thermal Energy Storage Systems	Electrochemical Energy Storage Systems	Battery Energy Storage Systems	Chemical Storage Systems
• Flywheel Energy Storage (FES) • Pumped Hydro Storage(PHS) • Compressed Air Energy Storage(CAES)	• Sensible Heat Storage(SHS) • Latent Heat Storage(LHS)	• Capacitor Storage(CS) • Super Capacitors Storage(SCS) • Superconducting Magnetic Energy Storage(SMES)	• Lead-Acid Batteries • Li-ion Batteries • Nickel-Cadmium Batteries • Nickel Metal Hydride Batteries • Sodium Sulphur Batteries • Vanadium-Redox Flow Batteries	• Hydrogen Fuel Cells • Direct ethanol Fuel Cells • Molten Carbonate Fuel Cells • Solid Oxide Fuel Cells • Metal-Air Batteries

Figure 4.2 Classifications of the various types of storage systems.

4.3.1 Mechanical Energy Storage Systems

Mechanical energy storage systems store the electrical energy into mechanical and delivers as electrical energy. During off-peak when demand is low, the electrical energy is converted to mechanical energy via the principle of potential, kinetic or even pressurized gas. When demand is high, the energy is converted to electrical energy. The common types of mechanical energy storage systems are pumped hydro storage (PHS), flywheel energy storage (FES), compressed air energy storage (CAES), and gravity energy storage systems (GES). The next sections discussed the various types of mechanical energy storage systems.

Flywheel Energy Storage Systems (FESS): This system is made up of a large cylinder (i.e. a rim attached to a shaft) fixed on a stator by magnetic glide bearings. These bearings serve as the major component for high-speed flywheel energy storage systems, as shown in Figure 4.3. The application takes place in a vacuum, to reduce wind shear. Mechanical Energy is stored as the kinetic energy of the rotor mass spinning at very high speeds typically 30,000 – 50,000 rpm. Energy stored in the flywheel can be reused by reducing the speed of the flywheel by a torque. The kinetic energy is returned to the electrical motor, which functions as an electric generator. The speed of the flywheel determines the energy produced, which operates at both low speed and high speeds of 600 rpm and 50,000 rpm respectively, with a specific energy of 5 J/kg. Also, the energy capacity of FESS is about 19 MWs and can deliver 10 s of ride-through at 1.65 MW load and proportionately a longer ride-through at lesser loads. Table-4.1 presents comparison between low-speed flywheel energy storage and high-speed flywheel energy storage.

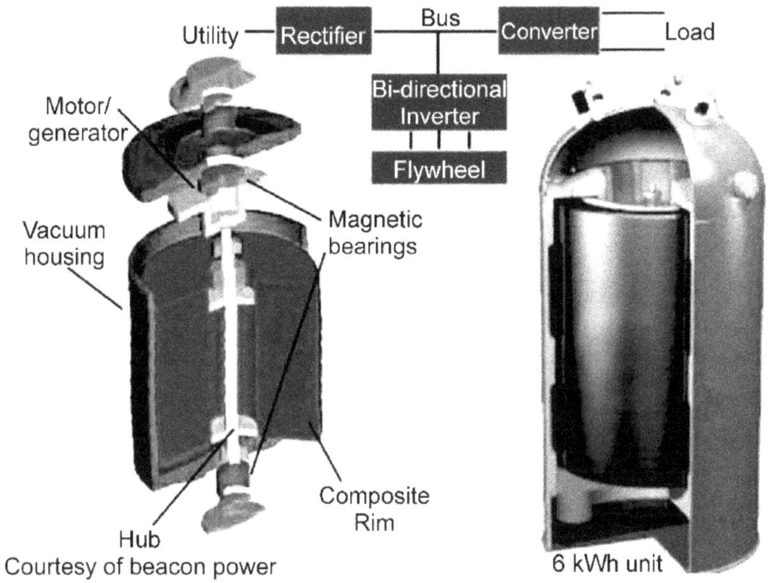

Figure 4.3 Typical Layout of a flywheel storage system.

Table 4.1 Comparison between low and high-speed flywheel energy storage.

Characteristics	Low speed flywheel energy storage	High speed flywheel energy storage
Material	Steel	Composite
Electrical machine	Induction, permanent magnet synchronous as well as reluctance machine	permanent magnet synchronous as well as reluctance machine
Atmospheric conditions	Partial vacuum and partial gas	Absolute vacuum
Weight for the enclosure	Double of flywheel weight	Half of flywheel weight
Applications	Enhancement of power quality	Aerospace and traction
Cost	Cheap and commercialized	Expensive and suitable for specific applications

Energy efficiency in flywheels is about 90% at rated power, self-discharge rates for complete flywheel systems are high, with a minimum rate of 20% of stored capacity per hour. These high rates have a deteriorating impact on efficiency when cycling is not continuous, for example, when energy is stored for a period between charge and discharge. These high discharge rates support the concept that flywheels are not suitable for the storage of energy for longer period but only ideal for standby power.

Pumped Hydro Storage (PHS) Systems: PHS is a large-scale energy storage system, energy storage technology. It provides important capacity for grid reliability and ancillary services, while also utilizing attitude intercepts to store electrical energy. The overall performance of pumped hydro storage system depends on the management of gravitational potential energy of water by pumping it from a lower reservoir to an upper reservoir during periods of low power demand. When power demand is high, water flows from the upper reservoir to the lower reservoir, activating the turbines to generate electricity. The volume and height of the water determines the amount of energy that can be stored. Currently, this application is a commercially available technology and holds more than 99% installed ESS capacity.

Figure 4.4 depicts a typical layout of a pumped hydro energy storage system. The components of PHES include pump turbine, motor, generator, penstock, inlet valve, penstock valve, upper reservoir, and lower reservoir. The water in the lower reservoir is pumped to the upper reservoir when the demand for electricity is low. The water from the upper reservoir flows to the lower reservoir when the demand for electricity is high, accelerating the turbine to produce electricity. The stored water can be later used to generate electricity to cover temporary peaks in demand from consumers or unplanned outages at other power plants. The application process in PHS is subdivided in a 24 h time scale, which prolongs energy storage in time, as well several days following.

The advantages of PHS capacities have been recognized worldwide. In type-one applications, the potential is around 1675 GW, and for type-two, 1454 GW. This technology is the most used for high-power applications. PHS can inject 1728 MW for 5 h, including high-power ramp rates, as the entire system can move from 0 to 1320 MW power injections in 12 s by means of managing 6 motor-generators of 330 MW. These motor-generators are activated by reversible Francis water turbines, installed in Europe's largest man-made cavern. In general, the lifetime of PHS installations is around 30–50 years, with an acceptable round-trip efficiency of 65–75% and power capital costs of 500–1500 D/kW and 10–20 D/kW h. In addition to the large power volumes and energy management in PHS installations, it possesses fast response times, (less than 1 min), which enables the PHS systems to become important components that the frequency in the event of reserve generation. The third category of PHS is micro, potentially has a capacity of up to 100 kW. This plant has the capability of providing power in remote areas, as well as connecting to the grids where wind and other renewable sources of energy are integrated. PHS of Pico size type has a capacity of 5 Kw. Regardless, the system is responsible for storing energy produced from wind or solar photovoltaic systems. Pumped storage has been further classified into three major types of storage:

- **Closed loop:** It consists of two reservoirs split by a vertical span, none of which is connected to another body of water.
- **Semi-open:** Consists of one artificial or modified reservoir and one modified lake or river impoundment with continuous through flow.
- Open-system (pump-back) is a system where there is continuous flow of water through both the upper and lower reservoir.

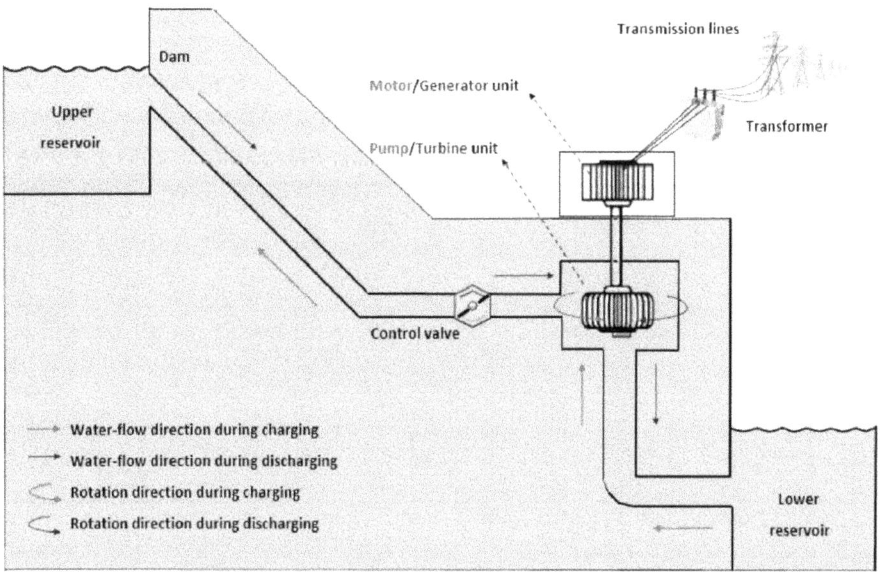

Figure 4.4 Pumped hydro storage systems.

The following are the important applications of PHS systems

- Flexible PHS generation will provide up and down control, which could stabilize the unstable production of the power system's renewable energy resources.
- PHS 's quick starting ability makes it ideal for black stars and provision of spinning and standing reserve.
- Act as fast response peaking plant to complement high inertia nuclear power and as an integrator for variable wind/solar power.

Compressed Air Energy Storage (CAES) systems: These systems work on the principles of gas turbines which have a storage capacity of 50–300 MW. Storage of energy is carried out via the compression of air kept in cavern underground. The compressed air is released from the cavern during the injection of energy into the grid (on demand). The released compressed air is then heated and made to turn turbines leading to the generation of kinetic energy and then lengthened in a set of high-and low-pressure turbines, which convert this kinetic energy of the compressed air into rotational energy which drives electric generator. To improve the efficiency of the system, heat generated from the compression stage of the cycle can be stored and utilized during the expansion process. This process improves in the overall performance of the system by 10% - 15%. In the case of isothermal compressed air energy storage, the compressed gas remains at a constant temperature throughout the compression or expansion process (i.e. internal heat is removed and added at the same rate). The storage of air underground has been reported as being one of the pragmatic approaches that can be adopted in the storage of energy in an efficient and safe manner. These characteristics are usually specifically attributed to underground storage in salt caverns. The additional air mixed with natural gas combusts, while the turbines are connected to electrical generators to obtain electrical energy. The turbine exhaust is used to heat the cavern air. Figure-4.5 illustrates typical layout for compressed air energy storage system.

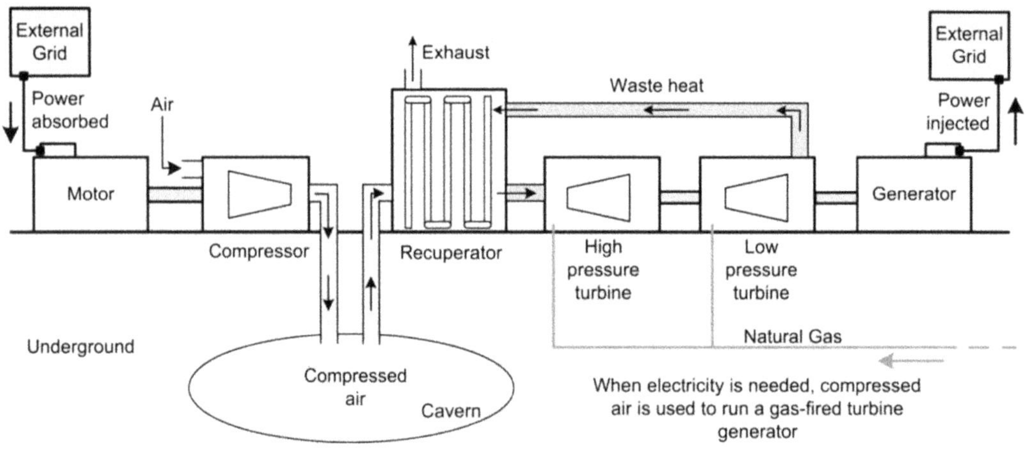

Figure 4.5 Schematic diagram for compressed air energy storage system.

Two plants of capacities 290 MW and 110 MW are built in Germany and the USA, respectively. Despite this, the concept is gaining much attention from the research community, as one of the biggest projects is being constructed in Iowa Stored Energy Park, with 2700 MW of turbine power. This development is occurring on a large wind farm to store excess wind energy generated. Modern CAES systems overlap with the energy density offered by chemical batteries, and can offer higher power densities if need, and operating at higher efficiencies.

Recent development of this application has led to the development of "Advanced-Adiabatic CAES (AA-CAES), in which the air is adiabatically compressed and then pumped into an underground cavern. The important part of this application is the heat exchangers, which are unfortunately very expensive. The effectiveness and the economics of these heat exchangers, and the compressor and expander trains are the main concerns for the success of AA-CAES. The lifetime of CAES installations is approximately 40 years, with an energy efficiency of 71%. Due to low self-discharge properties, they are being considered as long-term scale storage installations, CASES, has additional advantages of not requiring large, costly installations, having lower emissions of greenhouse gases than normal gas plants, and providing a viable option for storing energy in large quantities. A schematic diagram for diabatic and adiabatic compressed air energy storage systems is represented in Figure-4.6.

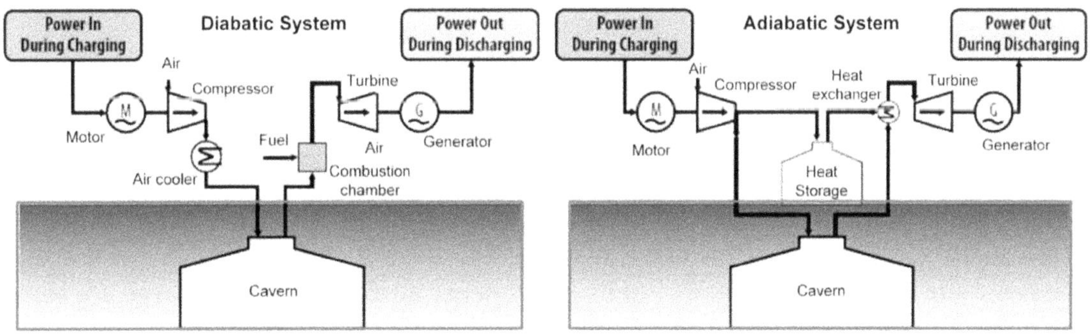

Figure 4.6 Layout of diabatic and adiabatic compressed air energy storage system.

The main challenge associated to compressed air energy storage has to do with the suitability of geological formations suitable for such projects. The other advantages include scalability, economical (20–25% the cost of batteries per kWh of storage), minimal environmental effects, long life and wide energy storage periods (hours to several days).

4.3.2 Thermal Energy Storage (TES) Systems

They are mostly applied in various industrial as well as residential buildings, e.g., space heating or cooling, process heating and cooling, producing water with varying temperatures as well as the generation of electric power. These techniques can be classified in to (a)sensible heat, (b)latent heat, (c)absorption and (d)adsorption system. The thermal energy storage installed capacity is nearly 3.3 GW, which is nearly 1.9% of total world energy storage till 2017. The TES systems can significantly support domestic heating, as well as cooling. They can also be utilized in the industrial sector. Thermal energy storage is usually made up of a thermal storage tank, a medium for transferring the heat and a containment control system. The stored heat is kept in an insulated reservoir with the aid of distinctive technology. Figure 4.7 depicts a diagrammatic representation for a thermal energy storage system.

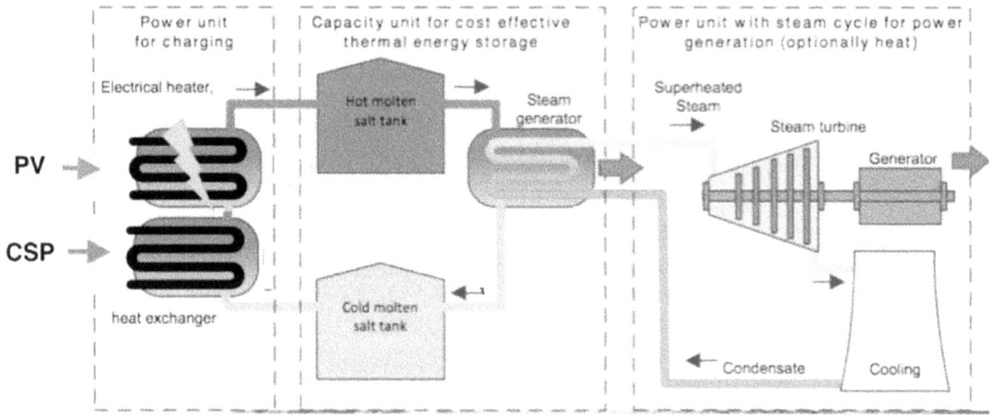

Figure 4.7 Schematic diagram of thermal energy storage system.

There is initially transfer of stored heat energy which is then transformed to electricity with the aid of heat engine cycle. The overall cycle efficiency for thermal energy storage is low (30–50%), but its high energy and daily self-discharge are some notable advantages of this useful technology. They are environmentally friendly, and the initial capital cost required for starting the project is low. This makes them suitable for large energy storage systems. Thermal energy storage systems are classified into low temperature and high temperature thermal energy storages. The low temperature thermal energy storage is made up of auriferous low temperature storages and cryogenic energy storage systems. Water cooling and reheating is predominant in low temperature thermal energy storages. Liquid air expansion is used for cryogenic energy storage, an example of this being liquid air energy storage. For load shavings, industrial cooling, and power management, the low temperature thermal energy storage system is often used. Sensible as well as latent heat are useful in high temperature thermal energy systems.

The specific heat, as well as mass of the storage medium determines the capacity of the sensible heat being stored. Some storage mediums include molten salt or concrete. The space requirement for sensible heat storage is large, which is one of the disadvantages of this technology. In latent heat storage system, energy swaps during phase changes of materials (PCM) via metallic fillers, and metal matrix structures.

Based on combinations of these phase change materials the storage systems are categorized into solid-solid, solid-liquid as and liquid gas phase change materials as presented in Figure-4.8. In thermal energy systems the solid-liquid phase change materials are usually preferred. The solid-liquid phase change materials can further be sub-classified into organic and inorganic. Table 4.2 summarizes the differences between the types of phase change materials. The subsequent sections discussed the various types of phase change materials.

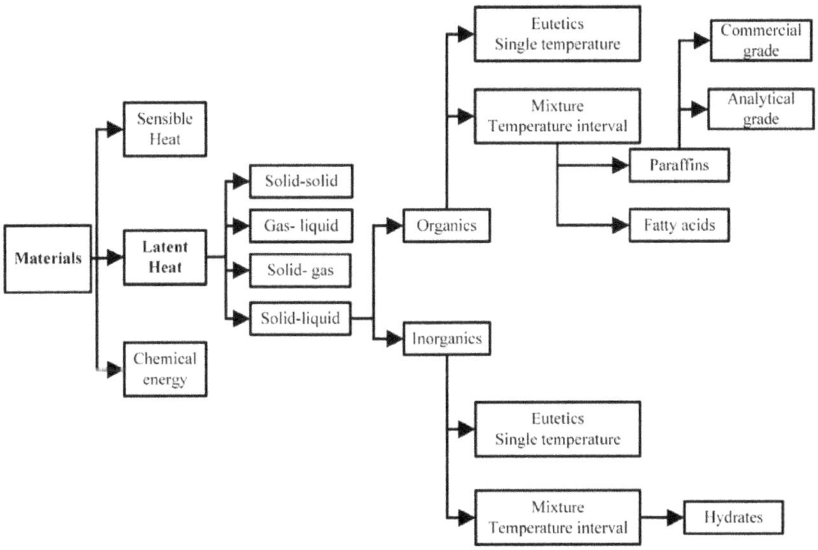

Figure 4.8 Classification of latent heat storage system based on PCMs.

Table 4.2 Performance comparison of various types of PCMs.

Category of phase change materials	Merits	Demerits
Organic	• Available in larger temperature range. • The heat of fusion required is high. • Absence of supercooling. • Stable chemically. • Can easily be recycles.	• Lower conductivity thermally. • Large volume change. • Flammable
Inorganic	• The heat of fusion required is high • Higher thermal conductivity. • Low volume change. • Cheap.	• Supercooling. • Susceptible to corrosion.
Eutectics	• Sharp melting characteristics. • High thermal storage density.	• Unavailability of current physical data of thermo-physical properties.

4.3.3 Electrochemical Energy Storage (EES) Systems

Electrochemical energy storages come under the category of electrostatics, which include capacitors and supercapacitors, magnetic or current energy storage systems. Capacitors in EESS are used for high currents but are only used for short periods due to their relatively low capacities. Supercapacitors are being used to substitute capacitors, only when they offer very high capacitance in small sizes. Superconducting magnetic energy storage (SMES) systems are mainly used in power plants to stabilize output or on industrial sites. They can also be used to during peak hours of energy consumption (e.g. steel plants or rapid transit railway).

Capacitors: Electric capacitors are mainly used for electricity storage, which consists of two metal plates separated by a nonconducting material called a dielectric. During charging, one plate is charged with from a direct current source while the other plate will be induced by a charge of the opposite sign. Capacitors store energy on the surfaces of metalized plastic films or metal electrodes. Since the energy density of capacitors is very low, they are employed to deliver and charge high currents, but only for extremely short periods. Figure 4.9 presents is a schematic diagram of a capacitor application.

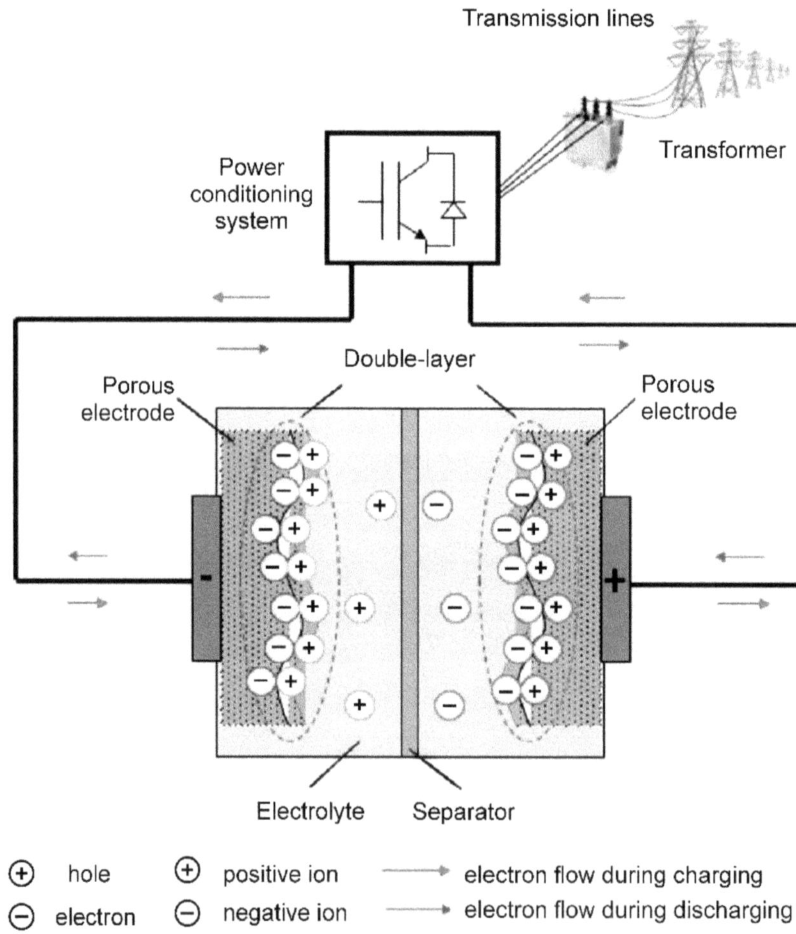

Figure 4.9 A schematic diagram of a capacitor application.

Super Capacitors: Recent progress in the supercapacitors could lead to much greater capacitance and energy density than conventional capacitors, thus enabling compact designs. The supercapacitors store energy by means of an electrolyte solution between two solid conductors rather than the more common arrangement of a solid dielectric between the electrodes. The electrodes are often made from porous carbon or another high surface area material as the conductor with an aqueous or non-aqueous electrolyte. Since the surface area of activated carbons is very high, i.e. up to 2000 m^2 per gram, and since the distance between the plates is very small (less than 1 nm), very large capacitances and stored energy are possible using supercapacitors. The energy storage capabilities of supercapacitors are substantially greater than that of conventional capacitors, by approximately two orders of magnitude (10- 100 s kW).

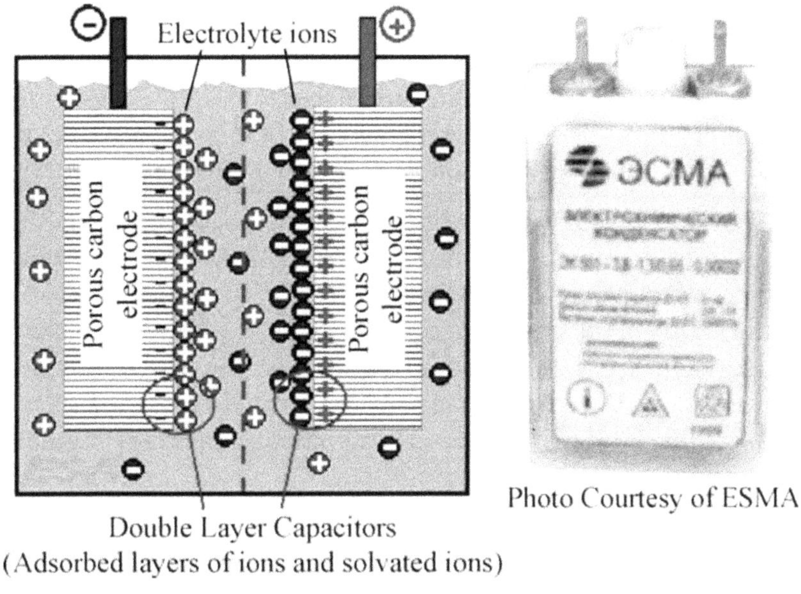

Figure 4.10 Doubled layered super capacitor.

The limitations associated with these capacitors are the short durations and high energy dissipations due to self-discharge loss. Therefore, like flywheel, capacitors are mainly used in power quality applications such as ride-through and bridging, as well as for energy recovery in mass transit systems. On the other hand, although the small electrochemical capacitors are well developed, large units with energy densities over 20 kWh/meter cube are still in the development stage. There are many developers of supercapacitors. The leading companies include SAFT (France), NESS (Korea),

ESMA (Russia), PowerCache (Maxwell, USA), ELIT (Russia), PowerSystem Co. (Japan) and Chubu Electric Power (Japan), etc.

Superconducting Magnetic Energy Storage Systems (SMES): SMES systems store energy in the magnetic field generated by superconducting magnets working at cryogenic temperature. An SMES system consists of a superconducting coil, a DC/AC converter, a quench protection system and a magnet cooling system as shown Figure-4.11. The AC/DC converter rectifies the grid alternate current (AC) to generate the magnetic field where energy is stored. After charging, the current does not decay, and the energy can be stored indefinitely with negligible or no losses. The stored energy can be released back to the grid by discharging the coil through the AC/DC converter. The quenching protection system is needed to dissipate the magnetic energy in case of a superconductor quench. The cooling system is needed to keep the superconducting coil below the critical temperature (i.e. about 4K in the case of NbTi and Nb_3Sn superconductors). SMES systems offer high storage efficiency (>90%) ~~[20]~~ and high energy density with very short response time (close to zero). Power is available almost instantaneously. High power output can be provided for short time periods. Energy losses are associated to the AC/DC converter (2–3%) and to the superconductor refrigeration system.

The basic SMES technology currently available are small-scale superconducting magnets ~~are~~ used in commercial devices (e.g. magnetic resonance) while large-scale superconducting magnets are being developed for fusion power experiments and particle detectors. Due to the energy needed for refrigeration, SMES is well-suited to short-term energy storage. Several demonstration SMES with power between 200 and 800 kW have been tested in Japan and the United States for distributed grid regulation and stabilization and power-quality industrial voltage regulators.

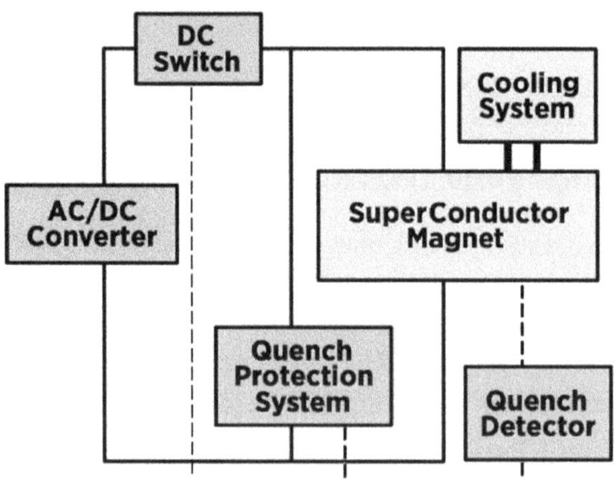

Figure 4.11 Schematic diagram of superconducting magnetic energy storage.

4.3.4 Battery Energy Storage Systems (BESS)

These systems have been predominantly used, and it is one of the common energy storage devices available worldwide. Energy storage is done in an electrochemical form, in multiple cells which are connected in series or parallel to generate a desired voltage and capacity. BESS cells are made up of two conductor electrodes and an electrolyte, tied together in a sealed container. Chemical reaction involves transfer of electrons between two electrodes from one another, with all electrons moving via an external electrical circuit. Despite the significant progress made in this storage systems recent research activities are being geared towards Lead-Acid batteries. Nonetheless, a substantial exertion has gone into novel BESS systems, such as nickel–cadmium and lithium-ion batteries, flow batteries, metal air batteries etc.

Lead-acid Batteries: This technology has been in use for a long time, it is easy and cheap for installation and maintenance, so this is the main reason for wide use of this technology and this technology is also one of the most common for stationary applications worldwide. Lead-acid batteries have ability to perform a deep discharge when it is required and the main problem with lead-acid batteries is that battery performance largely depends on temperature. Nominal voltage of this technology cell is around 2 volts. Lead-acid battery technology is based on positive and negative electrodes submerged into electrolyte, which is a combination of sulfuric acid and water, lead dioxide is used as a positive electrode and lead is used as a negative electrode. Lead-acid battery technology and its cell characteristic are shown in Figure 4.12

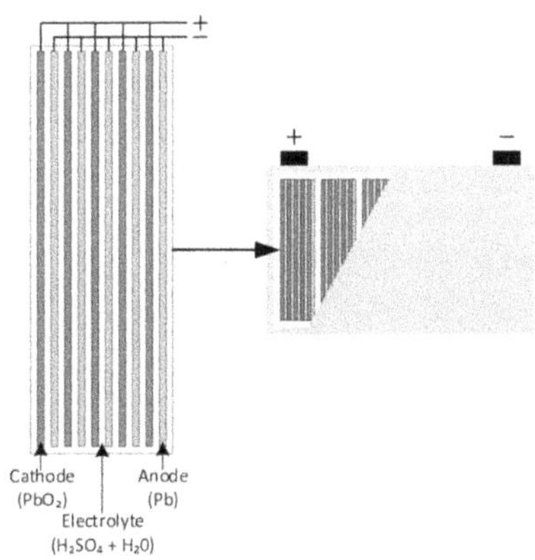

Characteristics	Value
Cell voltage	2 - 2.1 V
Specific energy	25 – 50 Wh/kg
Specific power	150 - 400 W/kg
Energy density	25 - 90 kWh/m³
Power density	10 - 400 kW/m³
Efficiency	63 - 90 %
Working temperature	18 - 45 °C
Lifetime cycles	250 – 2000
Lifetime	2 - 15 years
Max. depth of discharge	80 %
Self-discharge rate	0.1 – 0.3 % per day
Power rating	0 – 20 MW
Energy cost	40 – 170 €/kWh
Power cost	250 - 500 €/kW

Figure 4.12 Lead-acid battery technology and its cell characteristic.

Some advantages of the lead-acid technology are low cost, high cell voltage, suitability for intermittent charge applications and good ability of recycling.

Disadvantages are limited energy density and the number of lifetime cycles which is lower in comparison to other technologies.

Lithium-ion (Li-ion) Batteries: This technology is one of the most advanced battery technologies widely used today. Cellphones, smartphones, tablets, laptops, all gadgets are powered with the Li-ion battery. There are many pros of the lithium-ion technology: high power, energy capacity, long battery lifetime and relatively low weight and because of these pros, Li-ion technology is being used to power hybrid and electric vehicles.

Each battery consists of two electrodes (positive and negative) with a separation region between them. The anode is considered a negative electrode, whereas the cathode is considered a positive electrode, and the separator between them is fabricated using porous material. Once the electrodes are placed, the remaining distance between the separator and the electrode is filled with liquid electrolyte.

During the charging process of a battery, the power source accelerates the lithium ions to move from the cathode toward the anode. Then, the electrons migrate from the anode (negative electrode) to the cathode (positive electrode). Consequently, as lithium ions move between electrodes, they emit continuous ions of electricity. Once the battery reaches its maximum storage capacity, the oxidation process is completed.

The discharging process involves moving the positively charged lithium ions (Li+) from a negative electrode (anode) to a positive electrode (cathode). Therefore, electrons will be transferred from the cathode to the anode.

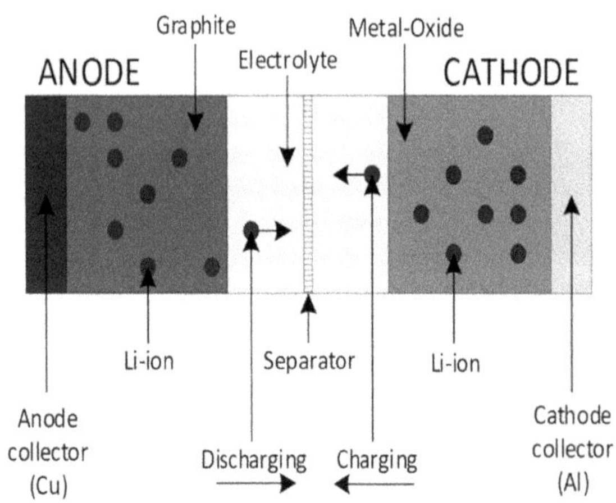

Characteristics	Value
Cell voltage	2.5 - 5 V
Specific energy	80 - 250 Wh/kg
Specific power	200 - 2000 W/kg
Energy density	95 - 500 kWh/m^3
Power density	50 - 800 kW/m^3
Efficiency	75 - 97 %
Working temperature	20 - 65 °C
Lifetime cycles	100 - 10000
Lifetime	5 - 15 years
Max. depth of discharge	100 %
Self-discharge rate	0.1 - 0.3 % per day
Power rating	0 - 0.1 MW
Energy cost	500 - 2100 €/kWh
Power cost	1000 - 3400 €/kW

Figure 4.13 Working principle of Li-ion battery and its characteristics.

Advantages of the lithium-ion technology are long battery lifetime, the number of lifetime cycles, high energy density, low maintenance cost and there is no memory effect. Disadvantages are high cost in comparison to other technologies and poor performance at high temperature. Lithium-ion battery working principle and its characteristics are presented in Figure-4.13.

Nickel-Cadmium Batteries: Nickel-cadmium (Ni-Cd) technology is in use for a long time in applications that require a long battery life and in difficult environmental conditions due to its low cost and robustness. Nickel-cadmium technology is based on cathode made from nickel oxide hydroxide and anode made from metallic cadmium while electrolyte used for Ni-Cd batteries is potassium hydroxide. Cross section of the nickel-cadmium battery is shown in Figure-4.14(a) and its characteristic in Figure-4.14(b). Ni-Cd batteries can be charged with a high charge rate, but in this case the charging process must be stopped when the battery is full, otherwise the battery will heat very fast which leads to damage.

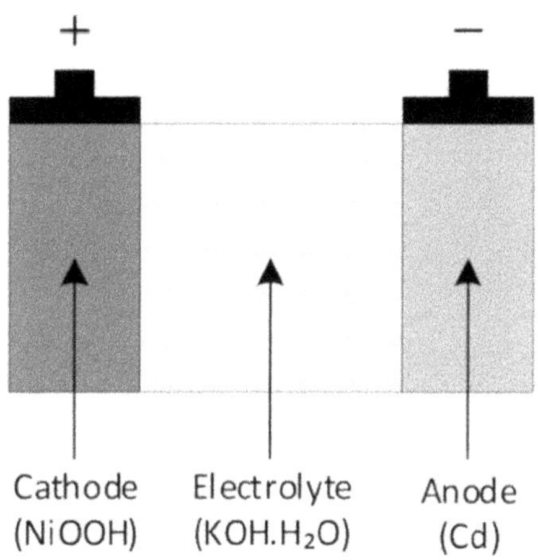

Characteristics	Value
Cell voltage	1.2 - 1.3 V
Specific energy	30 80 Wh/kg
Specific power	80 - 300 W/kg
Energy density	15 - 150 kWh/m^3
Power density	40 - 140 kW/m^3
Efficiency	60 - 90 %
Working temperature	- 40 - 50 °C
Lifetime cycles	1000 - 5000
Lifetime	10 - 20 years
Max. depth of discharge	80 %
Self-discharge rate	0.2 - 0.6 % per day
Power rating	0 - 40 MW
Energy cost	680 - 1300 €/kWh
Power cost	420 - 1300 €/kW

Figure 4.14 Ni-Cd Battery construction and its Characteristics.

The main problem with Ni-Cd batteries is a memory effect which means that battery loses full capacity if it is slightly discharging and re-charging every time during a certain period. Advantages of Ni-Cd technology are low maintenance cost, the number of lifetime cycles, suitability for long- term storage and ability to resist electrical and physical stress. Disadvantages are high cost in comparison to the lead-acid technology, limited energy density, toxic and caustic elements in batteries and the memory effect.

Nickel-Metal Hydride (Ni-MH) **Batteries**: This technology has been deployed in several applications such as energy storage for smart energy systems, robust battery systems which work at high temperatures, hybrid electric cars and public transport. Ni-MH battery cell cross section with the main parts is shown in Figure-4.15 along with its characteristics. Nickel-metal hydride technology is based on the negative electrode made from hydrogen-absorbing alloys which have the possibility to absorb releasing hydrogen, and the positive electrode made from nickel oxy-hydroxide. A separator provided which separates positive and negative electrodes to avoid shorting of electrodes. Electrolyte used in this technology is potassium hydroxide (KOH). There is a current collector made of metal which minimizes the internal battery resistance. A vent is provided to release gases produced during the overcharging or shorting.

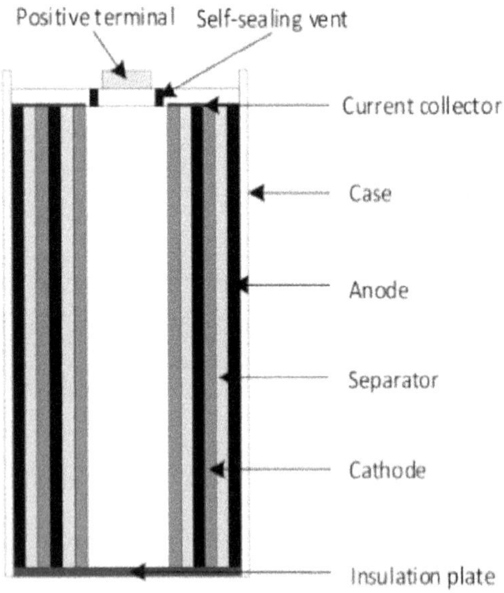

(a): Nickel Metal Hydride Battery.

Characteristics	Value
Cell voltage	1.2 - 1.35 V
Specific energy	40 - 110 Wh/kg
Specific power	200 - 300 W/kg
Energy density	40 - 300 kWh/m^3
Power density	10 - 600 kW/m^3
Efficiency	50 - 80 %
Working temperature	- 30 - 70 °C
Lifetime cycles	300 - 1800
Lifetime	2 - 15 years
Max. depth of discharge	100 %
Self-discharge rate	5 - 20 % per day
Power rating	0.01 - 3 MW
Energy cost	170 - 640 €/kWh
Power cost	200 - 470 €/kW

(b): NI-MH Battery Characteristics

Figure 4.15 Ni-MH battery cell cross section with the main parts and its characteristic.

Some advantages of NiMH technology are long battery lifetime, high number of lifetime cycles, good performance at the high temperatures, high energy density, good ability of recycling and the high tolerance to battery overcharging and over discharging. Disadvantages are high cost in comparison with lead-acid technology and poor performance at the low working temperatures.

Sodium-sulfur (NaS) Batteries: This battery technology is one of the most suitable for use in energy storage systems due to the high energy density. The cross-sectional view of NaS battery is shown in Figure-4.16(a). This technology is based on the use of sodium as anode and sulfur as cathode, electrolyte is beta alumina ceramics. For this technology is interesting that electrolyte is also a separator. NaS technology has a low internal cell resistance because of the use a ceramic electrolyte and this is good for two reasons. With the low resistance, power to weight ratio is being increased and heat produced during the charging process is being decreased. Typically working temperature of the sodium-sulfur batteries is between 300°C and 350°C. Reason for high temperature is to keep electrodes in a liquid state. High temperature decreases efficiency of the cycle which increases the number of operation cycles.

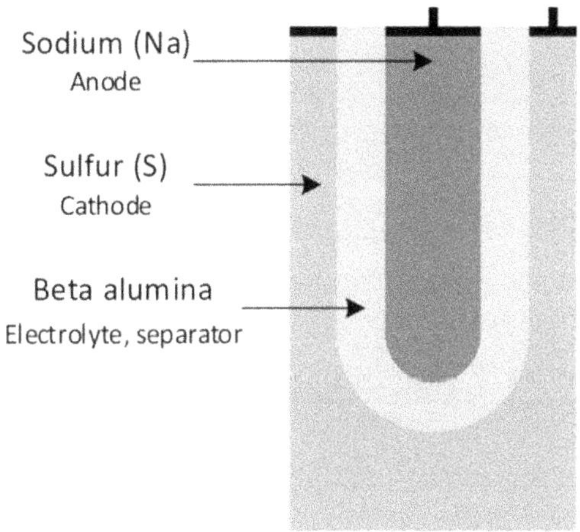

(a): Croce sectional view of NaS battery.

Characteristics	Value
Cell voltage	1.8 - 2.71 V
Specific energy	150 - 240 Wh/kg
Specific power	90 - 230 W/kg
Energy density	150 - 350 kWh/m^3
Power density	1.2 - 50 kW/m^3
Efficiency	75 - 90 %
Working temperature	300 - 350 °C
Lifetime cycles	2500 - 40000
Lifetime	10 - 15 years
Max. depth of discharge	100 %
Self-discharge rate	0 % per day
Power rating	0.05 - 34 MW
Energy cost	250 - 420 €/kWh
Power cost	850 - 2500 €/kW

(b): Characteristics of NaS Battery

Figure 4.16 NaS battery cell cross section with the main parts and its characteristic.

Advantages of sodium-sulfur technology are high energy density, high number of lifetime cycles, capability of pulse power and low self-discharging. Disadvantages are high cost and high temperature required for battery operation. Detailed characteristics of sodium-sulfur battery technology are shown in Figure-4.16(b).

Vanadium-Redox Flow Batteries (VRFB): VRFB is a modern technology with a lot of the very good characteristics such as long lifetime, very fast response time and long storage time which is ideal for long-term energy storage. Power and energy of VRFB are independent, power depends on the number and size of the cells and energy depends on the available electrolyte, respectively tank size. Working principle of the VRFB technology is shown in Figure-4.17(a). VRFB technology is based on two tanks in which vanadium ions electrolytes are stored, one electrolyte is positive, and the other is negative. Flow of electrons is caused by oxidation and reduction processes in the ion selective membrane through which electrolytes are being pumped. Response time of this technology is fast because electrolyte flow does not change, regardless of whether battery is charging or discharging. Flow batteries can release energy continuously at a high rate of discharge for up to 10 h. Flow batteries are also distinguished from fuel cells by the fact that the chemical reaction involved is often reversible, i.e. they are generally of the secondary battery type and so they can be recharged without replacing the electroactive material.

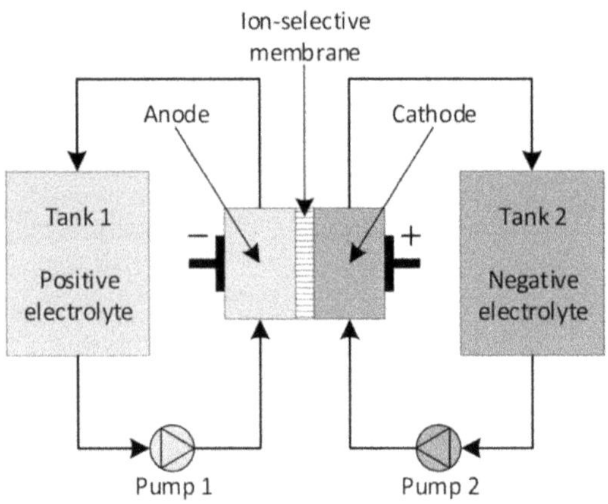

(a): Working Principle of VRFB

Characteristics	Value
Cell voltage	1.2 - 1.4 V
Specific energy	10 - 130 Wh/kg
Specific power	50 - 150 W/kg
Energy density	10 - 33 kWh/m³
Power density	2.5 - 33 kW/m³
Efficiency	75 - 90 %
Working temperature	5 - 45 °C
Lifetime cycles	10000 - 16000
Lifetime	5 - 15 years
Max. depth of discharge	100 %
Self-discharge rate	0 % per day
Power rating	0.03 - 3 MW
Energy cost	130 - 850 €/kWh
Power cost	500 - 1300 €/kW

(b): VRFB characteristics

Figure 4.17 Working principle of VRFB and its characteristics.

4.3.5 Chemical Storage Systems

A fuel cell is an electrochemical energy conversion device. It produces electricity from external supplies of fuel (anode side) and oxidant (cathode side). These react in the presence of an electrolyte. Generally, the reactants flow in and reaction products flow out, while the electrolyte remains in the cell. Fuel cells can operate virtually continuously if the necessary flows are maintained. Fuel cells differ from batteries in that they consume reactants, which must be replenished, while batteries store electrical energy chemically in a closed system. Additionally, the electrodes within a battery react and change when a battery is charged or discharged, while a fuel cell's electrodes are catalytic and relatively stable. Many combinations of fuel and oxidant are possible. A hydrogen cell uses hydrogen as fuel and oxygen as oxidant. Other fuels include hydrocarbons, alcohols and even metal. Other oxidants include air, chlorine and chlorine dioxide.

Hydrogen Fuel Cell (HFC): A hydrogen fuel cell uses hydrogen and oxygen to produce electricity and water, as shown in Figure-4.18, and a reversible hydrogen fuel cell could also use electricity and water to produce hydrogen and oxygen. Hydrogen-based energy storage systems are receiving increasing attention at the present time, particularly in relation to their integration with renewable power sources. The essential elements comprise an electrolyser unit, to convert the electrical energy input into hydrogen, the hydrogen storage system itself and a hydrogen energy conversion system, to convert the stored chemical energy in the hydrogen back to electrical energy. There is a strong technical preference for the generator element to be a fuel cell system, to capitalize on its high energy conversion efficiency, although this does not preclude the use of hydrogen burning internal combustion engines. There are several concepts of hydrogen fuel cells, the typical examples include Proton Exchange Membrane Fuel Cells (PEM- FCs), Alkaline Fuel Cells (AFCs), Phosphoric Acid Fuel Cells (PAFCs) and Regenerative Fuel Cells (RFCs).

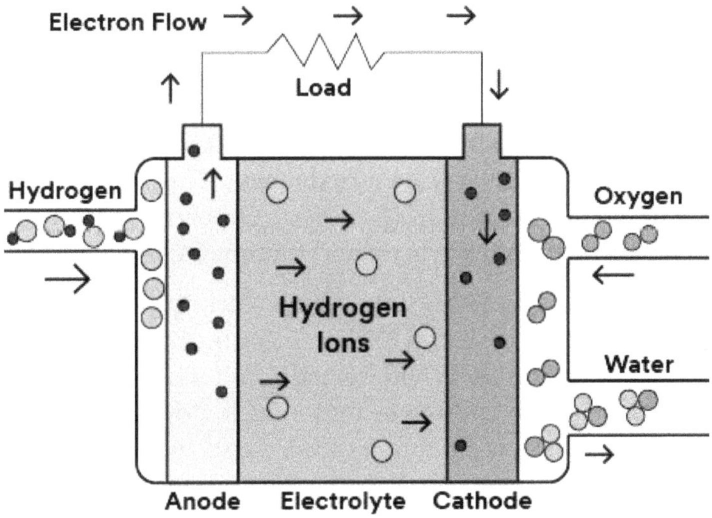

Figure 4.18 Operation principle of Hydrogen Fuel Cell.

Hydrogen fuel cells possess several inherent advantages, which include high energy density (0.6–1.2 kWh/kg), ability to implement systems over a wide range of scales, from kW scale to multi-MW capacity, independent system charge rate, discharge rate and storage capacity, modular construction with ability to add further modules and/or re-configuration later etc. However, hydrogen fuel cells are generally somewhat at a cost disadvantage ($6–20/kWh) at the present time and suffer from a relatively low round-trip efficiency (20–50%). This latter aspect may, however, not be a totally limiting aspect, due to low-cost input resource.

Direct-Methanol Fuel Cell (DMFC): This is a subcategory of Proton Exchange Membrane Fuel Cell (PEMFC), where, the fuel, methanol (CH_3OH), is not reformed, but fed directly to the fuel cell. Because methanol is fed directly into the fuel cell, complicated catalytic reforming is unnecessary, and storage of methanol is much easier than that of hydrogen, as high pressure or low temperature storage is not needed (methanol is a liquid at -97.0–64.7 °C). The DMFC depends on the oxidation of methanol on a catalyst layer to form carbon dioxide. Water is consumed at the anode and is produced at the cathode. Positive ions (H+) are transported across the proton exchange membrane to the cathode where they react with oxygen to produce water. Electrons are transported via an external circuit from anode to cathode providing power to external devices.

Because water is consumed at the anode in the reaction, pure methanol cannot be used without provision of water via either passive transport such as back diffusion (osmosis), or via active transport such as pumping. Although the need for water limits the energy density of the fuel cell, the energy density of methanol is orders of magnitude higher than highly compressed hydrogen. However, the efficiency of DMFC is low due to the high permeation of methanol through the membrane (methanol crossover), and the dynamic behavior is sluggish. Other problems include the management of CO_2 emission at the anode. Currently, DMFC is limited by the power it can produce, but it has large energy density. This means that it can produce a small amount of power over a long period. Other issues associated with the chemical properties of methanol are toxic and flammable.

Molten Carbonate Fuel Cells (MCFCs): They work quite differently from those discussed above. These cells use either molten lithium potassium or lithium sodium carbonate salts as the electrolyte. When heated to a temperature of around 650 °C, these salts melt and generate carbonate ions which flow from the cathode to the anode where they combine with hydrogen to give water, carbon dioxide and electrons. These electrons are routed through an external circuit back to the cathode, generating power.

Due to high temperature at which these cells operate, they internally reform hydrocarbons, such as natural gas and petroleum, to generate hydrogen within the fuel cell structure. Carbon monoxide poisonings not an issue at elevated temperatures. However, Sulphur remains a problem but can be tackled by replacing less expensive nickel species by platinum catalysts. The excess heat generated can also be harnessed and used in combined heat and power plants. The high temperatures however, present some issues. Since these cells take a considerable time to reach the operating temperature, they do not find applications in transport. The temperature and corrosive nature of the electrolyte is unsafe for home power generation. Due to high power generating efficiencies they can be best used in large-scale industrial processes and electricity generating turbines. Current demonstration cells have produced up to 2 MW but designs up to 50 and 100 MW capacities are still on the drawing board.

Solid Oxide Fuel Cells (SOFCs): SOFCs work at higher temperatures than that of the MCFCs. They use a solid ceramic electrolyte, such as zirconium oxide stabilized with yttrium oxide, instead of a liquid, and operate at 800–1000°C. In these fuel cells, the energy is generated by the migration of oxygen anions from the cathode to the anode to oxidize the gaseous fuel, which is typically a mixture of hydrogen and carbon monoxide. The electrons generated at the anode move via an external circuit back to the cathode where they reduce the incoming oxygen, thereby completing the cycle.

Carbon monoxide poisonings not an issue due high temperature of operation. This removes the need for external reforming to extract hydrogen from the fuel and these cells can again use petroleum or natural gas directly. SOFCs also exhibit the highest tolerance to Sulphur contamination among all the fuel cell technologies so far. These cells are more stable than MCFCs due to the solid electrolyte, but the construction materials needed to contain the high temperatures are more expensive. The SOFCs operate at an efficiency of around 60% and are expected to be used for generating electricity and heat in industry. They find potential applications in auxiliary power for vehicles.

Metal-Air Batteries (MABs): MABs can be considered as s a special type of fuel cell using metal as the fuel and air as the oxidant as is shown in Figure 4.19. The anodes in these batteries are commonly available metals with a high energy density such as Aluminum or Zinc that release electrons when oxidized. The cathodes or air electrodes are often made of a porous carbon structure or of a metal mesh covered with proper catalysts. The electrolytes are often a good OH-ion conductor such as KOH. The electrolyte may be in the liquid form, or a solid polymer membrane saturated with KOH.

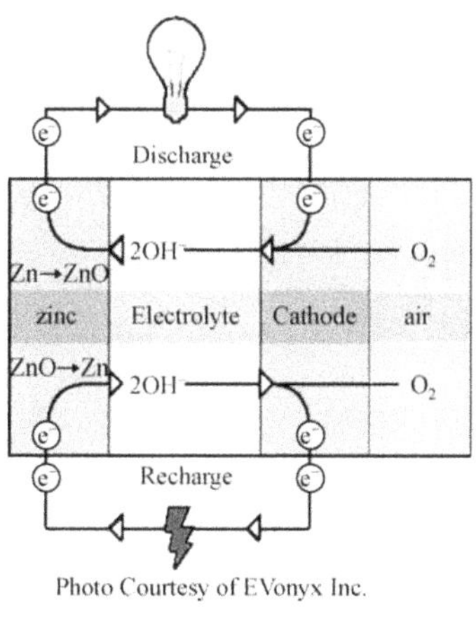

Figure 4.19 Working principle of Metal-Air Battery.

Metal-Air batteries are the most compact and, potentially, the least expensive batteries. They are also environmentally benign. However, the main disadvantage is that electrical recharging of these batteries is very difficult and inefficient. Although many manufacturers offer refuel units, where the consumed metal is mechanically replaced and processed separately. Few developers offer an electrically rechargeable battery. Rechargeable metal-air batteries that are under development have a life of only a few hundred cycles and the efficiency is below 50%. While the high energy density and low cost of Metal-Air batteries may make them ideal for many primary battery applications, the electrical rechargeability of the batteries needs to be developed further.

4.4 Applications and Performance Comparision of Various Storage Systems

4.4.1 Applications of Energy Storage Systems

Energy utilities: The use of energy storage systems in utility networks has become increasingly important and focused on as more storage options become available. Energy storage is generally deployed at five major subsystems in the electric power systems, i.e., generation, transmission, substations, distribution, and end consumers. This can help balance customer demand and generation. Intermittent power generation, such as by many renewable energy sources, results in power instability which can damage grid equipment such as generators and motors. By combining renewable energy systems with energy storage technology, renewable energy penetration levels can be increased and hence the overall system performance. It also improves, flexibility which ease grid control and maintenance. Some of the applications of energy storage systems include:

- Reduction of congestion in the transmission system
- Energy storage during periods of low demand for use during periods of high demand
- Maintaining voltage and frequency within normal operating ranges
- Compensating for contingencies such as failure of a generating unit
- Maintaining a real-time balance between generation and load

Renewable energy utilization: The integration of renewable energy use is growing rapidly, helping provide electricity to satisfy the world's demand and mitigate environmental impacts, especially power industry. However, the variability of these resources creates technical and economic challenges for their operation and use when integrated on a large scale. An important means of addressing the intermittency of renewable energy sources is energy storage. A higher penetration of renewable energy generation can be achieved with storage, as it permits excess energy produced from renewable energy sources to be stored and dispatched later when needed. In addition to intermittency of renewable energy sources, storages are also distributed. Moreover, energy storage technologies are scalable, modular, durable, and low maintenance may be suitable options for distributed renewable energy harvesting. Such technologies include flywheels, capacitors, SMES and most batteries excluding lead-acid batteries.

Buildings and communities: The current research is aimed to achieving net-zero energy buildings and communities motivates, the governments for promoting the adoption of renewable energy sources in buildings in the commercial, institutional, industrial and residential sectors. Energy storage is recognized as an important way to facilitate the integration of renewable energy into buildings (on the generation side), and as a buffer that permits the user-demand variability in buildings to be satisfied. A number of key performance indicators those facilitate the comparison of various storage technologies in the decision-making/design phase and the assessment of technical solutions. The indicators include storage capacity, maximum charge and discharge power, depth of charge, durability, specific cost of storage, maximum self-discharge rate, storage weight, and generated energy/cost savings.

Transportation: The evolution of ground, water and air transportation technologies has resulted in the need for advanced energy storage systems. Compared to conventional transportation technologies that are driven by internal combustion engines and utilize gasoline tanks for energy storage, hybrid electric vehicles use onboard energy-storage systems such as flywheels, ultra-capacitors, batteries and hydrogen storage tanks for fuel cells. The requirements for the energy storage devices used in vehicles include high power density, fast discharge, especially during acceleration, large cycling capability, high efficiency, easy of control and regenerative braking capacity.

The primary energy-storage devices used in electric ground vehicles are batteries. Electrochemical capacitors, which have higher power densities than batteries, are options for use in electric and fuel cell vehicles. In these applications, the electrochemical capacitor serves as a short-term energy storage with high power capability and can store energy from regenerative braking. A combination of a battery and an electrochemical capacitor can enhance the characteristics desired in land-based vehicles, aircraft and ships, including engine starting, high current for fast preheating of catalysts, electric power steering, and local power for actuators and distributed power systems.

However, a system consisting reduced battery size and an electrochemical capacitor must be commercially competitive with sole-battery systems to compete in the market. Flywheels can also be used for a long-range transportation system. They can be used for regenerative braking and load averaging or as the prime energy source for propulsion; the latter option is only a theoretical probability due to flywheel low energy density. To improve energy storage energy density, hybrid systems using flywheels and batteries can also be attractive options in which flywheels, with their high-power densities, can cope well with the fluctuating power consumption and the batteries, with their high energy densities, serve as the main source of energy for propulsion. However, for large vehicles such as trains, a larger flywheel needs to be used to serve such a purpose, and its weight becomes a disadvantage. The need for a storage unit to recapture vehicular braking energy can be achieved in railway systems by installing an energy storage device at the supply substations, along the railway track or on board the train. Designing and optimizing a train timetable to allow the interchange of energy among accelerating and decelerating trains with energy storage options is another approach that is proving to be effective for high traffic conditions.

Wayside energy recovery systems store energy along the railway tracks from decelerating vehicles and discharge it to accelerating ones. This increases overall system efficiency and voltage stability within the grid, and lowers peak power demands, costs and potentially CO_2 emissions depending on the energy mix. Flywheels, batteries and supercapacitors are suitable options for wayside energy storage. Pneumatic accumulators are also available options for regenerative braking energy storage, but often not considered due to their low energy density and efficiency. Some additional benefits of such installations are load levelling and support of the mains voltage, lower energy costs, reduced investment costs since fewer substations are needed, and emergency supply in case of power failures.

4.4.2 Performance Comparision of Various Energy Storage Systems

The performance of the different energy storage systems are compared base on different criteria such as costs, density of energy, specific power, recyclability, durability, energy efficiency, etc. These criteria together allow to define a "performance index" for the four categories of application:

- Low-power application in isolated areas, essentially to feed transducers and emergency terminals.
- Medium-power application in isolated areas (individual electrical systems, town supply).
- Network connection application with peak levelling
- Power-quality control applications.

To compare storage systems, Ragone's diagram is generally used to represent performance in terms of the ratio of mass to energy and power. This type of comparison is particularly interesting for portable units, for which mass is a critical aspect, but for permanent units, in a context of electrical-energy processing, life expectancy and total costs (investment, energy losses, and cycling fatigue) are the most important criteria.

The performance parameters of storage systems are often expressed for a wide variety of types and applications. The diagrams that follow show a comprehensible comparison of the different energy storage systems. The comparison is based on the "performance index" defined as the applicability of the storage technique to the specified application considering the characteristics presented in section 4.2. The data represents the typical systems in each category in Figure-4.20.

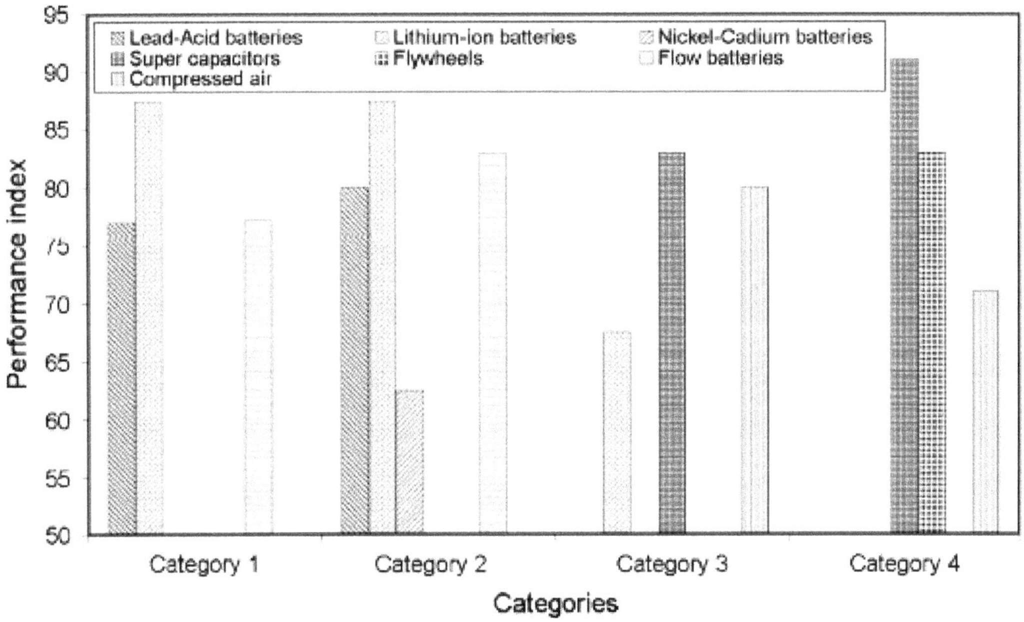

Figure 4.20 Performance index various storage systems under four categories of application.

Comparison based on Power as a function of field of application: Large-scale, permanent energy storage applications can be classified into three main operational categories as depicted in Figure 4.21.

Power quality: Stored energy is only used for a few seconds or less to ensure the quality of power delivered.

Buffer and emergency storage: In these applications, the stored energy is used for seconds to minutes to ensure service continuity when switching from one source of electricity to another.

Network management: Energy stored is used to decouple synchronization between power generation and consumption. A typical application is load levelling, which implies storing up energy during off-peak hours (low energy cost) and the use of stored energy during peak hours (high energy cost).

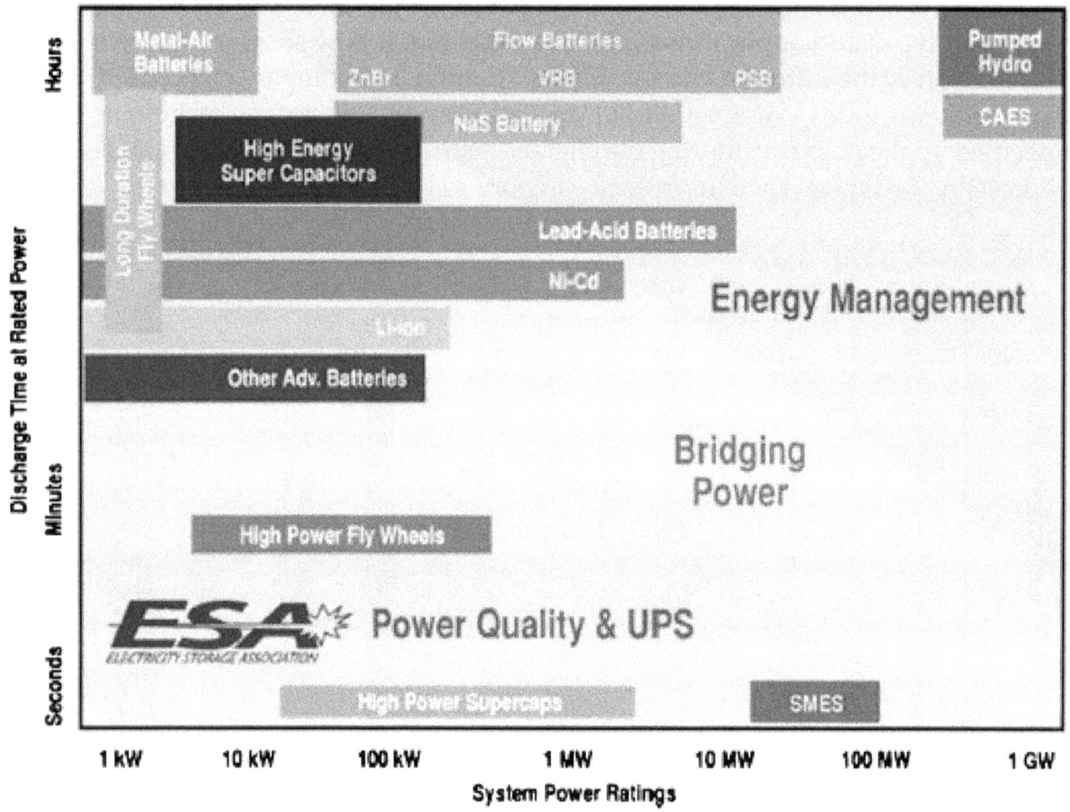

Figure 4.21 Distribution of storage techniques as a function of their application.

Comparison based on the energy efficiency and life expectancy: Energy efficiency and life expectancy (maximum number of cycles) are the two important parameters to be considered, before choosing a storage technology, as they affect the overall costs. Low efficiency increases the effective energy costs since only a fraction of the stored energy can be used. A short lifespan also increases long-term costs as the storage unit needs to be replaced more often. Actual expenses need to include the financial fees and operating expenses in order to develop a better idea of all the financial characteristics of a storage technique. Figure 4.22 illustrates the characteristics of different storage techniques with respect to efficiency and lifetime.

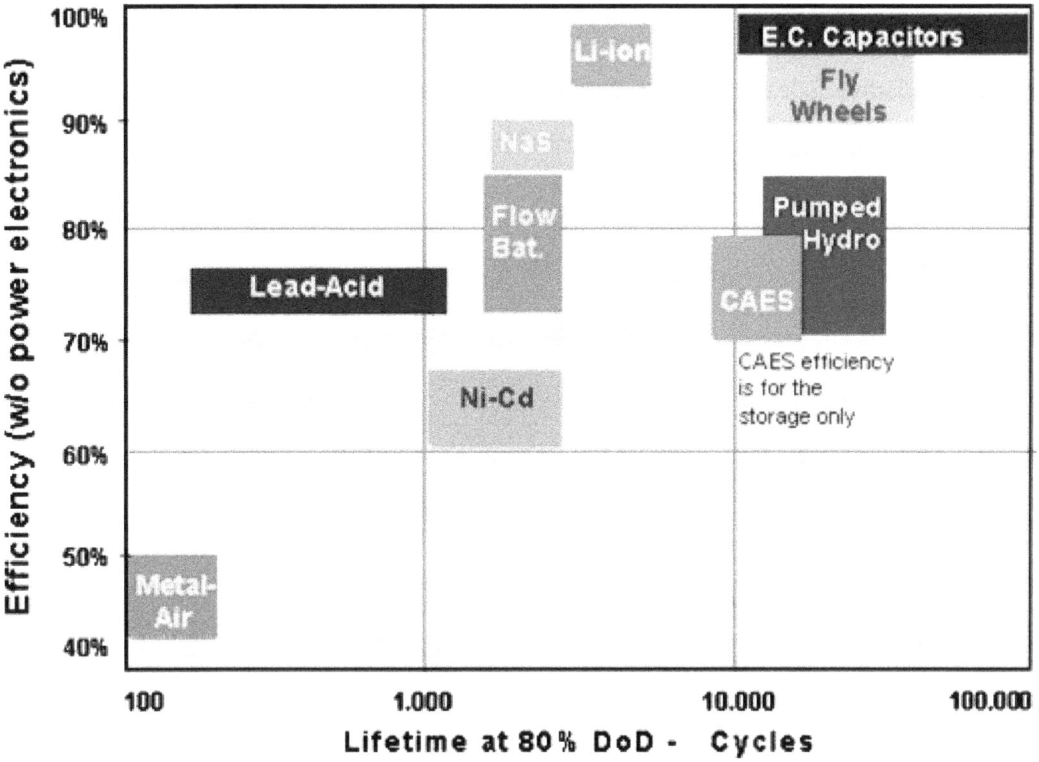

Figure 4.22 Characteristics of different storage systems with respect to efficiency and lifetime.

Comparison based on investment cost: The investment cost associated to a type of storage is an important economical parameter and affect the total cost of energy production. Thus, some types of storage systems can only become profitable if a certain minimum of energy is supplied. The overall cost of the system then needs to be considered in order to achieve a complete cost analysis. For example, despite the fact that lead-acid batteries are relatively inexpensive, they are not necessarily the least expensive option for energy management, due to their relatively low durability. The cost of batteries (Figure 4.23) was adjusted to exclude the cost of power electronics conversion. The cost of energy was also divided by the storage efficiency to obtain the cost per unit of useful energy. Installation costs also vary according to the type and size of the system.

Comparison based on investment cost per (charge-discharge) cycle: The cost per cycle could be the best way to evaluate the cost of an energy storage system designed for frequent charge-discharge applications. Figure 4.24 shows the principal component of this cost, considering durability and efficiency. For a more complete cost per cycle, research expenses as well as replacement and other costs, that remain unknown for emerging technologies, must be considered. It should be noted that the cost per cycle is not an appropriate criterion for load levelling or energy arbitration, where the storage is less frequent, and the energy costs are high and unstable.

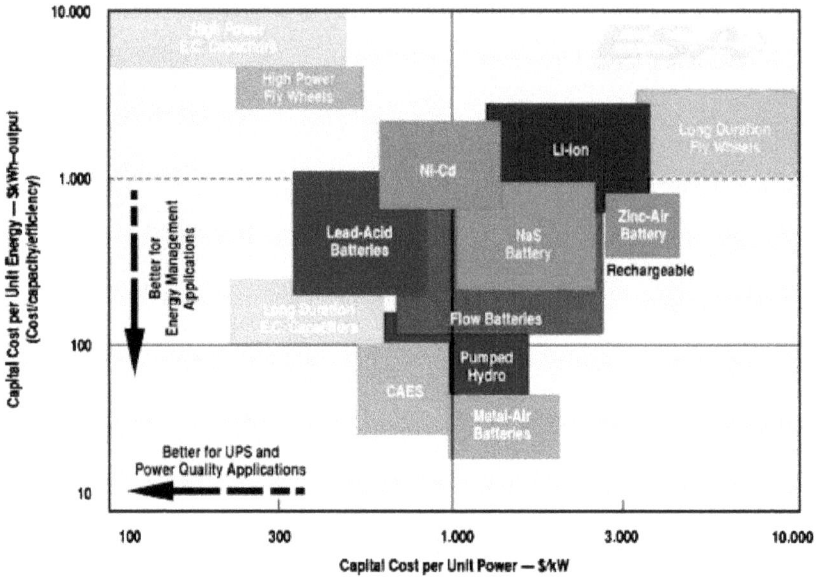

Figure 4.23 Distribution of storage systems as a function of investment costs per unit cost of Energy.

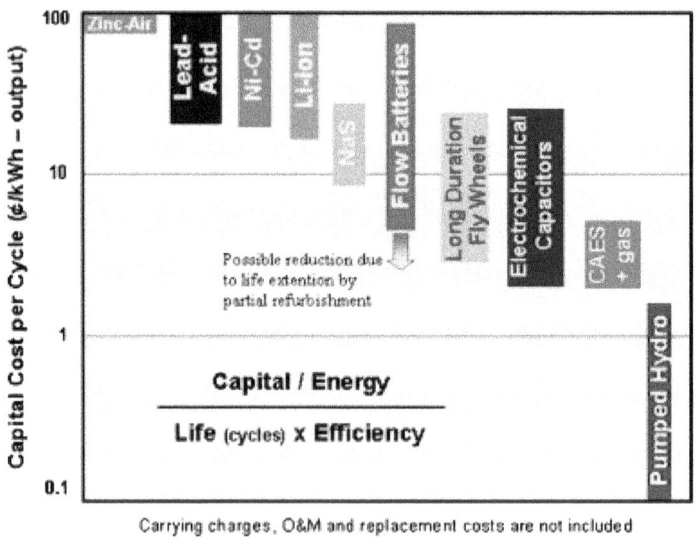

Figure 4.24 Distribution of storage systems as a function of investment costs calculated cycle.

Comparison based on mass or volume density: Various storage devices can be classified as a function of the availability of energy and power per litre (volume density) or per kilogram (mass density). This comparison is particularly important for the transportation industry, portable applications, and isolated sites. Figure 4.25 illustrates the distribution of storage devices as a function of their mass and volume densities

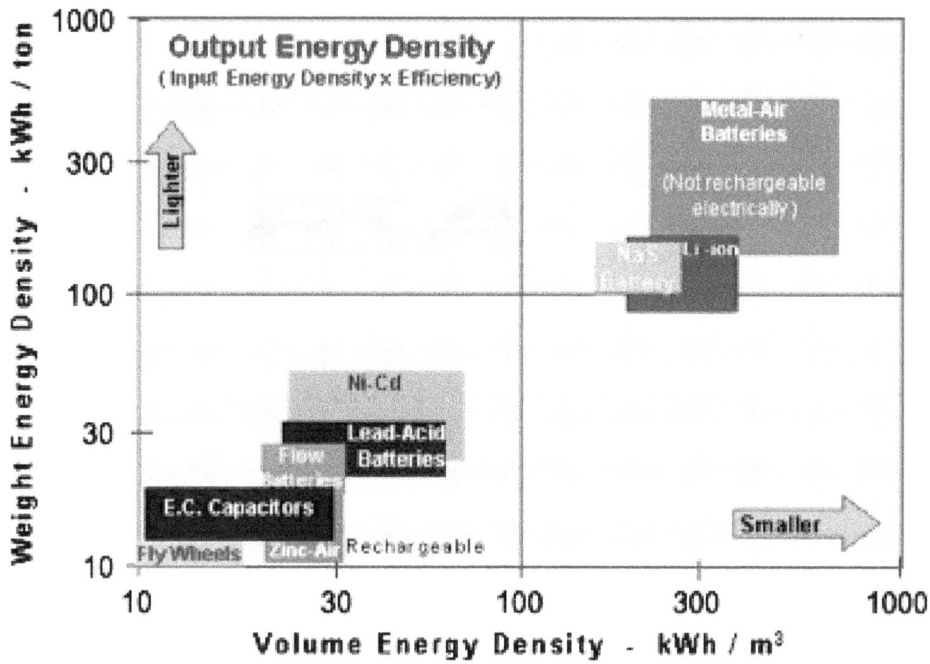

Figure 4.25 Distribution of storage devices as a function of their mass and volume densities.

Mass is an important parameter for permanent applications in terms of material costs. One could therefore choose, for an electrical installation, a material with specific energy, but far from the maximum available, at a unit cost that reduces the overall cost of the storage system. In this case, it is the price per kWh that is important, and not the kgs/kWh.

The volume of a storage system is important, if it must be installed in a restricted or costly space, for example in urban areas. Another aspect of volume is its incidence on the means and costs of containing and maintaining vacuum (flywheels for example) or cooling and thermal insulation systems. For large volume, more material and a larger construction site are needed, which leads to increased cost of the overall system.

References

1. https://www.irena.org//media/Files/IRENA/Agency/Publication/2017/Oct/IRENA_Electricity_Storage_Costs_2017.pdf
2. H. Ibrahim, A. Ilinca, J. Perron, Energy storage systems—Characteristics and comparisons, Renewable and Sustainable Energy Reviews, Volume 12, Issue 5, 2008, Pages 1221-1250, ISSN 1364-0321, https://doi.org/10.1016/j.rser.2007.01.023.
3. Couffin P, Perrin M. Le stockage des énergies intermittentes. Commissariat à 1'Énergie Atomique (CEA). Clefs CEA (50–51), hiver 2004–2005. p. 136–8.
4. Multon B, Robin G, Erambert E, Ben Ahmed H. Stockage de l'énergie dans les applications stationnaires. Colloque Énergie électrique: besoins, enjeux, technologies et applications. Belfort, France, 18 juin 2004. pp. 64–77.
5. Bonneville JM. Stockage cinétique. Institut de Recherche de l'Hydro-Québec, Varennes, Québec, Canada. Rapport interne; février 1975.
6. Faure F. Suspension magnétique pour volant d'inertie. Thèse de doctorat. Institut National Polytechnique de Grenoble, France; Juin 2003.
7. S. Koohi-Fayegh, M.A. Rosen, A review of energy storage types, applications and recent developments, Journal of Energy Storage, Volume 27, 2020, 101047, ISSN 2352-152X, https://doi.org/10.1016/j.est.2019.10104
8. A. Sharma, V.V. Tyagi, C.R. Chen, D. Buddhi, Review on thermal energy storage with phase change materials and applications, Renew. Sust. Energy Rev., 13 (2009), pp. 318-345
9. Energy Policy Council: Joint legislative commission on energy policy. Energy Storage Options for North Carolina. 2018.
10. A.s.t.e.N. Del Pero I, H. Paksoy, F. Haghighat, S. Grillo, F. Leonforte Energy storage key performance indicators for building application Sustain. Cities Soc., 40 (2018), pp. 54-65
11. S.M. Mousavi, F. Faraji, A. Majazi, K. Al-Haddad, A comprehensive review of Flywheel Energy Storage System technology, Renew. Sust. Energy Rev., 67 (2017), pp. 477-490
12. F. Nadeem, S. M. S. Hussain, P. K. Tiwari, A. K. Goswami and T. S. Ustun, "Comparative Review of Energy Storage Systems, Their Roles, and Impacts on Future Power Systems," in *IEEE Access*, vol. 7, pp. 4555-4585, 2019, doi: 10.1109/ACCESS.2018.2888497
13. https://www.energystorageexchange.org/projects
14. V.A. Boicea, Energy storage technologies: the past and the present, Proc IEEE, 102 (11) (2014), pp. 1777-1794
15. H. Chen, T. Ngoc, W. Yang, C. Tan, Y. Li, Progress in electrical energy storage system : a critical review, Prog Nat Sci, 19 (2009), pp. 291-312, 10.1016/j.pnsc.2008.07.014
16. D. Beevers, L. Branchini, V. Orlandini, A De Pascale, H. Perez-blanco, Pumped hydro storage plants with improved operational flexibility using constant speed Francis runners q137 (2015), pp. 629-637, 10.1016/j.apenergy.2014.09.065

17. D. Al, D.G. Christakis, K. Pavlopoylos, S. Stamataki, I. Dimitrelou, I. Stefanakis, *etal.,* Introduction of a wind powered pumped storage system in the isolated insular power system of Karpathos – Kasos 97 (2012), pp. 38-48, 10.1016/j.apenergy.2011.11.069
18. Hadjipaschalis, A. Poullikkas, V. Efthimiou, Overview of current and future energy storage technologies for electric power applications,13 (2009), pp. 1513-1522, 10.1016/j.rser.2008.09.028
19. Rouindej, E. Samadani, R.A. Fraser, CAES by design: a user-centered approach to designing Compressed Air Energy Storage (CAES) systems for future electrical grid: a case study for Ontario, Sustain Energy Technol Assessments, 35 (2019), pp. 58-72, 10.1016/j.seta.2019.05.008
20. P. Denholm, G.L. Kulcinski, Life cycle energy requirements and greenhouse gas emissions from large scale energy storage systems. 45 (2004), pp. 2153-2172, 10.1016/j.enconman.2003.10.014
21. L.F. Cabeza, I. Martorell, L. Miró, A.I. Fernández, C. Barreneche, Introduction to thermal energy storage (TES) systems, Woodhead Publishing Limited (2015), 10.1533/9781782420965.1
22. M. Fatih Demirbas Thermal energy storage and phase change materials: an overview, Energy Sources B Energy Econ Plann, 1 (1) (2006), pp. 85-95
23. M.S. Guney, Y. Tepe, Classification and assessment of energy storage systems 75 (2017), pp. 1187-1197, 10.1016/j.rser.2016.11.102
24. Wagner, Overview of Energy Storage Methods (2007)
25. https://www.irena.org/-/media/Files/IRENA/Agency/Publication/2012/IRENA-ETSAP-Tech-Brief-E18-Electricity-Storage.pdf
26. https://www.3mindia.in/3M/en_IN/hydrogen-technology-in/applications/fuel-cells/
27. Šimić, Zvonimir & Topić, Danijel & Knežević, Goran & Pelin, Denis. (2021). Battery energy storage technologies overview. International journal of electrical and computer engineering systems. 12. 53-65. 10.32985/ijeces.12.1.6.
28. F. A. Silva, "Lithium-Ion Batteries: Fundamentals and Applications [Book News]," in *IEEE Industrial Electronics Magazine*, vol. 10, no. 1, pp. 58-59, March 2016, doi: 10.1109/MIE.2016.2515040.
29. O. Adeleke, A. Ukil, X. Zhang, "Vanadium Redox & Lithium Ion Based Multi-Battery Hybrid Energy Storage System for Microgrid", Proceedings of the IEEE PES Asia-Pacific Power and Energy Engineering Conference, Macao, 1-4 December 2019.
30. http://www.electricitystorage.org/tech/technologies_technologies.htm [2007-03-20].
31. +Electric Mobility Canada. Preliminary analysis of eVionyx Ni–Zn battery and Zn–Air fuel cell. Available on http://www.mec.ceveq.qc.-ca/ [2007-03-20].

CHAPTER

5

Plugin Electric Vehicles

5.1 History

It is difficult to attribute the invention of the electric car to an individual or a country. However, it was a series of breakthroughs in the battery and electric motor developments in the 18th century which led to the first electric vehicle on the road. In the early part of 18th century, innovators in Hungary, the Netherlands and the United States including a blacksmith from Vermont began the concept of a battery-powered vehicle and created the small-scale electric cars. And while Robert Anderson, a British inventor, developed the first crude electric carriage. In 19th century French and English inventors built some practical electric cars. In U.S., the first successful electric car was built in 1890 by William Morrison, a chemist who lived in Des Moines, Iowa. His six-passenger vehicle capable of a top speed of 14 miles per hour which was little more than an electrified wagon, sparked the interest in electric vehicles. Over the next few years, electric vehicles from different automakers began popping up across the U.S. By 1900, electric cars were at their heyday, accounting for around a third of all vehicles on the road.

In 1800s, the gasoline-powered car with internal combustion engine started entering the market along with the electric vehicles. These gasoline cars found to be reliable having limitations such as manual effort at the time of starting (hand crank) and during the changing of gears. They were also noisy, with unpleasant exhaust. Electric cars on the other hand didn't have any of these issues. They were quiet, easy to drive and didn't emit pollutants like the gasoline-based cars of the time. Electric cars quickly became popular with urban residents, especially among women. They were perfect for short trips around the city. As more people gained access to electricity in the 1910s, it became easier to charge electric cars, adding to their popularity.

Many innovators at that time took note of the electric vehicle's high demand and started exploring to improve the technology. For example, Ferdinand Porsche, founder of the sports car company by the same name, developed an electric car called the P1 in 1898 (Figure 5.1). Around the same time, he created the world's first hybrid electric car which is powered by electricity and a gas engine. Thomas Edison thought electric vehicles were the superior technology and worked to build a better electric vehicle battery. Even Henry Ford partnered with Edison to explore options for a low-cost electric car in 1914.

Figure 5.1 Ferdinand Porsche's first car produces in 1989.

However, Henry Ford's mass-produced Model T (Figure 5.2) had caused a blow to the electric car in 1908, which was gasoline-powered and widely available and affordable. By 1912, the price of gasoline car was $650, while an electric roadster was sold for $1,750. The same year, Charles Kettering introduced the electric starter, eliminating the need for the hand crank and giving rise to more gasoline-powered vehicle sales.

Figure 5.2 Henry Ford's Model T Car.

Other developments contributed to the decline of the electric vehicle include

- In 1920s, the U.S. had a better system of roads connecting cities, and Americans wanted to go out and explore.
- The discovery of Texas crude oil, and gas became cheap and readily available for rural Americans, and filling stations began popping up across the country.
- A very few Americans outside of cities had electricity at that time.
- Cheap, abundant gasoline and continued improvement in the internal combustion engine hampered demand for alternative fuel vehicles.

In the end, electric vehicles all but disappeared by 1935. Over the next 30 years or so, electric vehicles entered a sort of dark ages with little advancement in the technology.

Late 1960s, soaring oil prices, gasoline shortages and peaking with the 1973 Arab Oil Embargo, created a growing interest in lowering the U.S.'s dependence on foreign oil and finding homegrown sources of fuel. Congress took note and passed the Electric and Hybrid Vehicle Research, Development, and Demonstration Act of 1976, authorizing the Energy Department to support research and development in electric and hybrid vehicles. However, the vehicles developed in the 1970s still suffered from drawbacks of limited performance, low speeds (45 miles per hour) and low range (40 miles) before needing to be recharged. In the 20 years since the long gas lines of the 1970s, interest in electric vehicles had mostly died down.

The first turning point (suggested by many) was the introduction of the Toyota Prius, released in Japan in 1997 (Figure 5.3). Prius became the world's first mass-produced hybrid electric vehicle. In 2000, the Prius was released worldwide, and it became an instant success with celebrities, helping to raise the profile of the car. To make the Prius a reality, Toyota used a nickel metal hydride battery, supported by the research of Energy Department. Since then, rising gasoline prices and growing concern about carbon pollution have helped make the Prius the best-selling hybrid worldwide.

Another event that helped reshape electric vehicles was the announcement in 2006 by Tesla Motors, a mass production luxury electric sports car with a range more than 200 miles on a single charge. In the short time, Tesla has won wide acclaim for its cars and has become the largest auto industry employer in California. Tesla's announcement and its success spurred many big automakers to accelerate work on their own electric vehicles. In late 2010, the Chevy Volt and the Nissan LEAF (Figure 5.4) were released in the market. The first commercially available plug-in hybrid, the Volt has a gasoline engine that supplements its electric drive once the battery is depleted, allowing consumers to drive on electric for most trips and gasoline to extend the vehicle's range. In contrast, LEAF is an all-electric vehicle powered by only an electric motor.

Figure 5.3 Toyota Prius released in Japan in 1997.

	Chevy Volt	Nissan Leaf
Price	$41,000	$32,780
Availability	Dec., 2010	Dec., 2010
Power	Electric/gas	Electric
Range	25-30 mi. (40-48 km) on electricity; 350 mi. (563 km) including gas engine	100 mi. (161 km) per charge, depending on weather and driving conditions

Figure 5.4 Chevy Volt and the Nissan LEAF.

Over the next few years, other automakers began rolling out their electric vehicles with one of the early limitations of charging the vehicles on the go. Automakers and other private businesses also installed their own chargers at key locations.

At the same time, new battery technology began hitting the market, helping to improve a plug-in electric vehicle's range. In addition, the investment in battery research and development has helped cut electric vehicle battery costs simultaneously improving the vehicle batteries' performance (power, energy and durability). This in turn has helped lower the costs of electric vehicles, making them more affordable for consumers.

Consumers now have more choices than ever when it comes to buying an electric vehicle. As gasoline prices continue to rise and the prices on electric vehicles continue to drop, electric vehicles are gaining in popularity.

5.2 Role of EVs in Energy Sector

Industry, transportation, residence, and commerce are the four important sectors of energy consumption in the world. The concerns about fossil fuel depletion and its impact on greenhouse warming have motivated the governments to find the alternative resources. As the transportation sector has a considerable amount of energy consumption (about 30%), it has gained more attention in terms of replacing the gasoline vehicles with other types of vehicles such as electric, hydrogen, etc. Approximately 90% of transportation energy requirement is met by petroleum. Therefore, the transportation electrification can significantly reduce the dependency of fossil fuels and can reduce the carbon footprint effectively.

Several types of electric vehicles (EVs) available in the market namely Hybrid electric vehicles (HEVs), plug-in hybrid electric vehicles (PHEVs), plug-in electric vehicles (PEVs), and battery electric vehicles (BEVs) can play a critical role in reaching emission goals. In HEVs, the batteries cannot be charged by power grid. There are two energy resources for battery charging that include gasoline and regenerative braking. The energy of braking system is converted to heat in combustion engine vehicles to charge the battery of HEVs. Also, the gasoline can be converted directly to electric energy to charge the battery using an electricity generator. Under the light load and low-speed conditions, the electric motor is used to drive the wheels, while both electric and gasoline motors are used under the heavy load and high-speed conditions. The PHEVs can be powered by both gasoline and electric energies. The battery of PHEVs can be charged during regenerative braking, like HEVs, and the external power grid. The PHEVs can be plugged into the power grid and charged completely. The use of both energy resources extends the driving range of PHEVs. The PEVs are powered fully by the electricity. They are plugged into an external power grid to charge the batteries.

The financial perspective of EVs is important for people who buy EV. A comprehensive study where the fuel and maintenance costs of electric and gas vehicles are compared yields that the operation cost of EVs will be less than that of gas vehicles. However, the purchasing cost of EVs is usually higher. The comparison of an EV and a gas vehicle with similar characteristics, for instance, Nissan Leaf as an EV and Honda Civic as a gas vehicle, shows that the price of EV (is much higher than that of the gas vehicle. Therefore, the governments need to provide attractive incentives to EV buyers to increase the penetration of EVs in their countries. Hence, many countries have designed several incentives and subsidies to encourage the people to buy the EVs. In the U.S., for example, the federal government grants a tax credit up to $7500 depending on EV battery capacity. Moreover, the citizens of the U.S. were given subsidies from their state as well. The residents in Canada, for instance, in Ontario province, can receive up to C $14000 to buy an EV based on vehicle type and battery capacity. Other countries such as France, UK, Sweden, Japan, etc., have also designed incentives to encourage their citizens. Due to the government incentives and technological improvement of EVs (specially in batteries), it is forecasted that the penetration of EVs may increase in the future.

EVs are high-power consumers that should be supplied by power grid. For instance, the battery capacity of Nissan Leaf, is 30 kWh which is equal to 3–5 times the daily electricity demand of a household. The power grid will be facing challenges and stresses due to large penetration of EVs into the grid. However, these plugin electric vehicles can also utilize as distributed storage systems (of course with consent of EV owner) which can help integrating renewable energy integration. These may well support renewable energy fluctuation and reduce the externalities cost of renewable energy in the electricity system. On the demand side, the integration of electric vehicles into the grid will flat the load curve and thereby reducing the power supply cost. These electric vehicles offer an opportunity to improve and cut down the operation cost if accommodated with bi-directional power flow controllers and with the participation of PEV owners.

5.3 Grid Integration of Plugin Electric Vehicles - V2G Technology

It can be forecasted that the large adoptions of EVs will bring a huge challenge to the existing grid system. The controlled loads resulting from EV charging would cause a significant negative impact on grids such as the power quality degradation and the transformer overloading etc. It also involves a large investment in upgrading grid utilities and deploying distributed generators. Eventually, the vehicle grid integration can provide an effective and economic solution to integrate the additional charging loads into grids.

The Vehicle to Grid (V2G) technology can achieve the bidirectional energy exchange between EVs and grids. Thus, it can store the surplus power during lower-demand periods and feeding them back to grids during higher-demand periods to achieve the balance of the supply and demand of electricity. In this situation, EVs are not only the pure electric loads, but also served as the distributed energy storage system. In consequence, the V2G mode can utilize the EV fleet to provide spinning reserve, frequency regulation and so forth for grids. Moreover, EV users could benefit from rewards based on the provision of grid services. However, the degradation on EV batteries would be incurred due to the discharging requirements. Despite this, the cost-benefit analysis for EVs participating in ancillary services to grids suggests that it is worth trying the V2G operation from the economic perspective, especially with the advancement of battery technology and the rapid increase of the renewable energy integration.

The scenario of incorporating EVs in grids is as the efficient integration of information and energy. The bottom line of this integration is how to build an intelligent communication link to achieve the coordinated energy exchange between EVs and grids to achieve the win-win ecosystem.

5.3.1 Charging Modes of PEVs

According to the power rating of the chargers, the battery charging mode can be classified into slow and fast charging mode. Their basic characteristics for PEV charging are summarized in Table-5.1 These charging modes can be applied in different fields in accordance with their features.

Table 5.1 Basic characteristics of slow and fast charging.

Feature	Type	
	Slow charging mode	Fast charging mode
Advantages	✓ Easy for charging posts installation ✓ Home charging	✓ Relatively short replenishing time
Drawbacks	λ Rather long replenishing time	λ Hard for charging posts installation λ Remarkable negative impacts on grids

Slow Charging Mode: The slow charging mode employs the AC, and it is designed in different standards. In China, the home power system offers 220 V of AC, with 16 or 32 A for PEV slow charging. Therefore, the charging power is either 3.52 kW 7.04 kW. However, the charging power in America varies from 1.44 kW to 19.2 kW. It is convenient to recharge PEVs by directly using the home supply via the slow charger. The slow charging points are the popular method and can be installed at residential garage or roadside parking. The issue with slow charging is for a PEV with a 24-kWh battery, it will take about 17 h for charging it from empty to full with a rates of 1.44 kWh. This requires the PEV users have enough free time for the charging tasks. Apart from overnight charging at home, the slow charging mode is also applicable to the workplaces and public places.

Fast Charging Mode: The fast-charging mode uses the DC instead of AC. The rating of fast chargers is up to 50-kW or more to achieve a relatively short charging time. Tesla built 120-kW supercharger stations for fast charging. The supercharger can charge Model S with a 90-kWh battery from empty to 80% in about 40 minutes.

These fast-charging options are more high-tech than a simple household outlet due to extremely high currents and voltages. So, the fast chargers should be equipped with a tethered cable rather than the ordinary household cable. Although the less consuming time is attained for PEV charging in this situation, it may trigger extreme surges in demand and threaten the stability and security of the electric grid.

5.3.2 Power Interaction Modes between PEVs and Grid

The power interaction between PEVs and grids could be uncoordinated or coordinated depending on the adopted charging strategies. The uncoordinated charging mode of PEVs operates regardless of the supply and demand status of distribution networks and is more likely to jeopardize the grid stability and reliability when many PEVs are being charged. The coordinated mode is based on V2G technique developed for integrating more PEVs into the existing power system with the aim of improving the stability and reliability of the distribution network. The power flow direction of three operating modes and their alternative terms are provided in Table-5.2

Table 5.2 Power flow direction of three operating modes and their alternative terms.

Operating mode	Power flow direction	Alternative term
Uncoordinated charging	From grid to PEV	● Dumb charging ● Uncontrolled charging ● Conventional charging
Unidirectional V2G	From grid to PEV	● Coordinated charging ● Regulated charging ● Smart charging
Bidirectional V2G	From grid to PEV From PEV to grid	● Coordinated charging/ discharging ● Regulated charging/discharging ● Smart charging/discharging

Uncoordinated Charging Mode: The uncoordinated charging mode refers to the conventional charging mode where PEVs start charging immediately once they are plugged in. In this case, the plugged-in time is entirely determined by PEV users, and there are no PEV aggregators to regulate the PEV charging profile. Hence, the charging processes of PEVs are completely dominated by the owner's wish. It has been investigated that most of the uncoordinated charging events occur in the home-charging situation, in which PEV users charge their vehicles as soon as they arrive home during early evening hours. However, the load level of power grid is high at this moment. PEV charging activities during this period would significantly stress the distribution system.

There are serval aspects of PEVs, such as vehicle pattern, vehicle penetration, vehicle load demand, plugged-in/plugged-out time, vehicle charging location, vehicle charging pattern and so forth have impact on distribution network. Nonetheless, it is likely to result in the collapses of power grid if a high penetration level of PEVs is intense especially during load-peak periods. So, a coordinated power interaction relation between PEVs and grids should be established to mitigate the charging problem associated with of large-scale PEV penetrations.

Coordinated Charging Mode: This mode can be broadly divided into unidirectional and bidirectional V2G. The unidirectional V2G mode is the so-called coordinated charging mode, and the bidirectional V2G mode is the so-called coordinated charging/discharging mode. In the uncoordinated charging mode and the unidirectional V2G mode, the energy only flows from the grid to PEVs. Whereas in the bidirectional V2G mode the energy not only flows from the grid to PEVs but also flows from PEVs to the grid.

The concept of Vehicle-to-Home (V2H) is a small version of V2G technology which allows a PEV to supply homes with power generated from its battery. With the help of V2H technology, PEVs can be severed as a battery storage system to feed any electrical appliance where the PEV is parked. Despite this, they don't have the capability to operate as an off-line uninterruptible power supply (UPS). The V2G technology can utilize PEVs to achieve the reactive power compensation which makes the grid to operate in economical mode. The Vehicle-to-Vehicle (V2V) concept is developed for a local PEV community that can charge or discharge PEV battery among them. These concepts enable PEVs act as distributed sources for the power grid within the V2G technology.

5.3.3 V2G Technologies

Unidirectional V2G: This technology has the capability to modulate the charging power of PEVs to better correspond to the grid requirements with the assistance of communication technology. A smart infrastructure, capable of supporting two-way information flow between PEVs and grids is indispensable. A majority of PEVs have been equipped with the on-board unidirectional chargers. It is easy to implement unidirectional V2G in comparison to bidirectional V2G.

At peak hours, unidirectional V2G allows the third party to propel PEV users to charge their vehicles during the entire off-peak period, to avoid congestion. Therefore, the unidirectional V2G services can absorb the surplus power in the valley periods to charge the batteries full. Nevertheless, it is incapable of sending energy from PEV batteries back into the grid due to absence of the bidirectional power converters. The extra degradation on PEV batteries would rarely be incurred in this technology.

A properly designed unidirectional chargers can charge PEVs with the surplus power of grids to mitigate the impact of large-scale PEVs charging. This in turn provide an effective support to the distribution and transmission system with the active load management of PEVs. This also can help consume the abundant renewable energy sources such as solar and wind energy by the coordinated charging strategy. Additionally, unidirectional V2G would provide a foundation for the implementation of bidirectional V2G in the future.

Bidirectional V2G: This refers to a technology that allows for the reversal of electric energy stored in PEV batteries. It facilitates the balance of generation and demand along with satisfying the charging demands of PEVs. A two-way communication system which is an essential component of this technology helps to feed the stored energy from PEV batteries to the electric grid during the higher-demand periods. The feedback component of bidirectional V2G helps to provide spinning reserve, frequency regulation and other ancillary services for power grid. In bidirectional V2G, many PEVs serve as a controllable fleet under a PEV aggregator. This fleet has the similar function to the distributed energy storage systems capable of providing ancillary services to the grid. Hence, the PEV discharging processes are indispensable for energy feedback.

This bidirectional technology will lead to the extra adverse impact on the health of PEV batteries. However, these incurred costs might be compensated by the reasonable incentives offered by the V2G service providers. PEV aggregators should build an interest compensation mechanism to promote PEV users to participate in bidirectional V2G. Under such a mechanism, PEV users may benefit from rewards based on the provision of grid services and energy trading revenue.

It is worth noting that the bidirectional V2G has a greater advantage than unidirectional V2G on the integration of all kinds of resources including renewable energy and DGs. In the recent past bidirectional V2G has been tested in several projects. But this technology is still in the experimental stage. Several technical and regulatory issues need to be resolved before it can be widely and effectively used.

5.3.4 Methods of V2G Implementation

As an effort to integrate EVs into power system, the existing research efforts are divided into centralized and decentralized methods. In centralized techniques, the charging/discharging processes of EVs are directly controlled to improve the operation efficiency of grids. In decentralized methods, the charging/discharging behaviors of EV users are coordinated with an incentive strategy to fulfil the objectives of grids. Various objectives of V2G strategy include minimizing power losses, minimizing load variances, maximizing load factors, minimizing generation costs etc.

Centralized method: In the centralized scheduling method, PEV aggregators control and optimize charging based on the grid status and PEV charging demands for a special target. Hence, the controllers should be equipped on the PEV chargers to adjust the charging rates of PEVs in this scenario. Figure-5.5 shows the general process to implement centralized scheduling method for PEV coordinated charging.

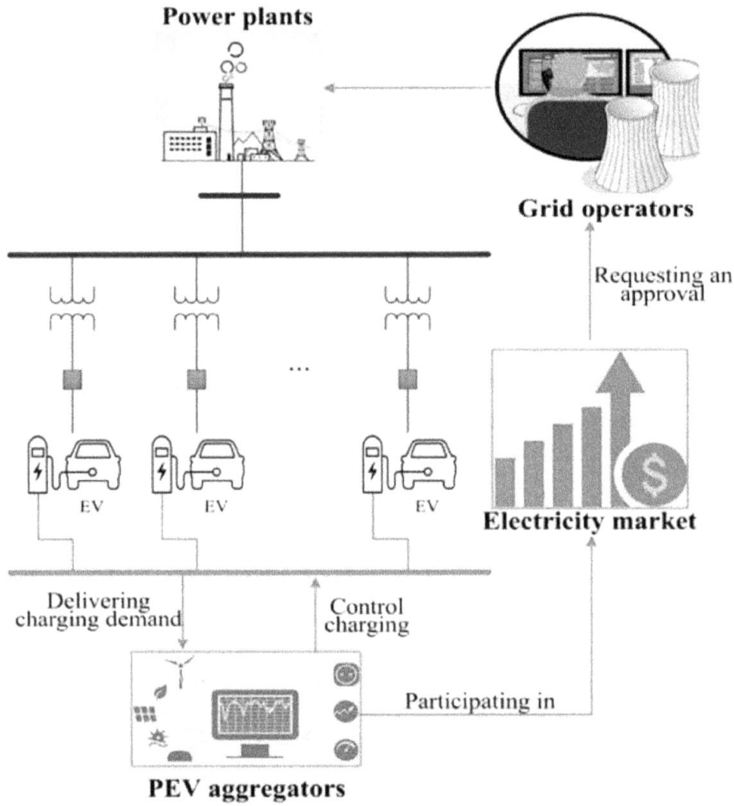

Figure 5.5 General framework for centralized scheduling method.

PEV aggregators are the crucial bonds in integrating PEVs in grid and participation in electricity markets. The various activities/activities involved as follows:

- PEV aggregator obtains the information regarding the identity number of the charging post, the nominal capacity of PEV batteries, the plugged-out/plugged-in time, and the plugged-out/plugged-in *SoC* of PEV batteries.

- Based on the information the aggregator computes optimal charging/discharging strategy of each PEV to determine their buy/sell bids in accordance with the received charging demands and the forecasted market behaviors.
- The bidding instruction are reported to grid operators for approval ahead of schedule.
- The grid operators will check whether the biding strategy is a safe and economic operation for power system.
- If yes, PEV aggregators can proceed to the electricity market; if not, grid operators will ask PEV aggregators to change their bidding strategies until a protocol is reached.
- Finally, PEV aggregators will charge each PEV in accordance with the optimal load profile of the fleet.
- On the other hand, PEV aggregators will check out with each PEV user by the online settlement system.

As illustrated in Figure 5.5, there are three major stakeholders in centralized approache, namely grid operators, PEV aggregators and PEV users. They have extremely different considerations in their respective positions. From the view of grid operators, they focus primarily on the grid performance improvement and secondarily the total operational costs reduction. PEV owners require a more convenient and affordable recharging system. It may result in severe conflicts with grid operators on the charging periods. In this context, PEV aggregators are employed to coordinate both the sides. PEV aggregator's perspective, is to control the PEV chargers to maximize the total operational profits along with satisfactory charging demands of PEV owners under the grid restrictions. Hence, they should design a reasonable compensation strategy to make more PEV users to participate in the centralized scheduling to provide V2G ancillary services. The critical issue for PEV aggregators is to develop the so-called load shifting or valley filling strategy to alleviate the demand pressure of power grid during the load peak hours.

The centralized approaches require the PEV users to give up their charging authorities to control the PEV charging process. PEV aggregators can make a generous profit by scheduling PEVs to charge at valley hours and discharge at peak hours. The critical issue is that this action will lead to the extra degradation on PEV batteries. Nonetheless, it may still not be enough from the benefits of PEV users. On this account, the battery ageing costs for PEVs taking part in V2G should be evaluated to compensate PEV users for the extra battery degradation.

Decentralized scheduling method: The decentralized scheduling method refers to an incentive strategy, wherein, the charging behaviors of PEVs will be indirectly coordinated by the electricity price mechanism. In a simple case grid operators or PEV aggregators broadcasts the price information to each PEV user to shift the PEV loads from peak demand hours to low demand hours. Hence, it can be easily implemented in the short time. Figure-5.6 illustrates the general process to implement decentralized scheduling methods of the V2G technology.

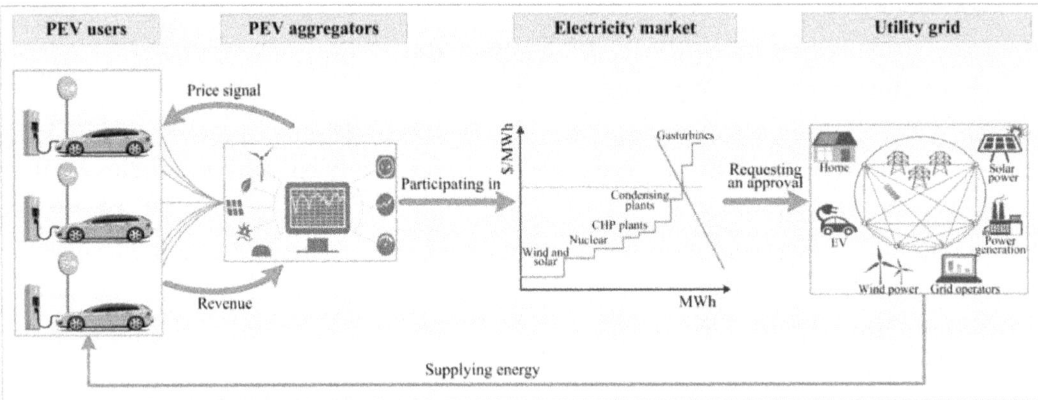

Figure 5.6 Implementation decentralized scheduling method of V2G.

The steps involved in the decentralized method are as follows:

- PEV aggregators collect or forecast the PEV charging demands during each period to determine the bidding strategy.
- The bidding information will be transmitted to grid operators for approval in advance just like the centralized scheduling method.
- When the agreement is reached, PEV aggregators will broadcast the price to each PEV user.
- Consequently, each PEV user will schedule their vehicles charging as per their own willingness such as reduced cost etc.

In the decentralized scheduling approach, PEV aggregators manage the charging of PEVs. Each PEV user will respond to the price signals as per their wish. And the bidding strategy will be delivered to the PEV aggregators for regulated charging.

Figure 5.6 illustrates that there are also three main participants in decentralized approaches, namely, grid operators, PEV aggregators and PEV users like centralized method of implementation. As per the operating principle, the decentralized scheduling approach is a price-signal based method. This should assume that each PEV user responds to the price signal and make an enough intelligent charging/discharging decision. The PEV aggregator forecast the charging demands from PEVs to make the bidding decision ahead of schedule; on the other hand, they ought to establish the communication network to inform PEV users of the price information. Finally, with respect to grid operators, their primary responsibility is to evaluate whether the bidding strategy of each aggregator will endanger the safe operation of power grid. In addition, they also desire the operating costs are reduced and power efficiency is improved.

The PEV users in the decentralized method preserve individual authority on the PEV charging processes according to the price scheme broadcast by the energy coordinator. However, it is possible that a part of PEV users may not be sensitive to the electricity price mechanism, or they do not receive this kind of price information at all. Consequently, in this situation the unexpectedly plugged-in PEVs will also pose an additional stress on power system.

5.4 PEV Charging Technologies

EVCS are primarily refueling stations that provide electricity to charge electric vehicles. The charger point is integrated into the cable, charging port, and interface panel. Depending on a grid configuration, various parameters such as voltage rating, frequency rating, and transmission standards determine the power outlet configuration. The Electric Power Research Institute (EPRI), Society of Automotive Engineers (SAE), and the International Electrotechnical Commission (IEC) have a substantial role to play in categorizing charging modes and levels in each country as well as managing the differences in safety standards. Various charging types are outlined in established standards, including alternating current (AC) Level-1, alternating current (AC) Level-2, and direct current fast charging (DCFC).

5.4.1 Modes of Charging

Modes/levels of charging PEVs are mainly classified as Level-1, Level-2 and Level-3 based on the nature and capacity of charging stations.

Level-1 charging: Level 1 (L1) charge cables are included with every EV. The device is universally compatible, does not require installation fees, and is pluggable into any standard grounding 120V /240V outlet. The maximum current passed through these chargers is 16 amps. This charging point provides a maximum power outlet of 1.9 kW and takes between 8 and 16 h to charge the battery according to its capacity fully. SAE J1772 connector is used to connect the electric vehicle. L1 charging can reduce charging costs even further when combined with a tariff-based charging.

Level-2 charging: Level 2 (L2) charging stations are most used in public and residential locations. The charging stations use a single-phase power supply of 240 V with a maximum current flow capacity of 40 A for residential and commercial installations and a three-phase AC power supply of 400 V with a maximum current flow capacity 80 A for public stations. EVs with the industry-standard J-plug are universally compatible with SAE J1772 charging stations. L2 charging can provide a quick charge along the way. Overcurrent voltage and overcurrent protection are built into L2 charging systems

Level 3 charging: Level 3 (L3) is the fastest charging method for electric vehicles. Most DC fast charging stations are located in malls, government buildings, movie theaters, airports, and refueling stations. By using DC charging technology, these stations are designed to provide a similar user experience to that of traditional filling stations. The charge time for a battery from 0 to 80% is usually 15 to 20 min with DC fast charging, the remaining 20% of the battery will always be charged in slow mode, regardless of the charging level. DC power is delivered to the electric vehicle by converting AC power from an off-board charger. Charging voltages for L3 typically range from 200 to 600 V, and power outputs range from 36 to 240 kW.

DCFC can be divided into two categories based on SAE standards: DC Level 1 and DC Level 2. In addition to their 36-kW power output, DC L1 charging stations also have a current flow capacity of 80 A. Chargers with DC L2 power output have a current flow capacity of 200 A and a power output of 90 kW. DC charger connectors having the SAE/IEC J1772/IEC 62,196–3 standard is recommended by SAE and IEC. DC fast-charging stations have limitations due to high installation costs. The primary standards and protocols for EV charging are described in Table-5.3.

Table 5.3 Electric vehicle charging standards and protocols.

EV charger topology	Charging connector	Charging communication	Charging power quality	Charging safety
IEC 61851-1	IEC 62196-1	ISO 15118/IEC 61850	IEEE 1547	IEC 60529
IEC 61851- 21	IEC 62196-2	SAE J2847/SAE J2836	SAE J2894	IEC 60364- 7-722
IEC 61851- 22	IEC 62196-3	SAE J2293-2/OCPP	IEC 1000-3-2	ISO 6469-3
IEC 61851- 23	SAE J1772	OCPI/OSCP/	NEC 690	SAE J1766
IEC 61851- 24	IEEE 1901	OpenADR	SAE J2380	SAE J2464

5.4.2 Methods of PEV Charging

There are various ways to charge a battery and control its current. Electric vehicles use rectifiers to convert AC into DC for charging their batteries. The methods to transfer charge, include inductive charging, conductive charging, and battery swapping. The classification of various charging methods is presented in Figure 5.7.

Conductive charging (CC): The benefits of conductive charging are its economic viability, ability to charge quickly, simplicity of operation, and high efficiency. In addition to onboard and offboard charging systems, conductive charging has also been categorized. Chargers installed onboard an electric vehicle, such as AC-DC converters, are usually slow chargers that charge the vehicle entirely inside. Offboard chargers, on the other hand, provide rapid charging. EV range can also be increased by reducing the vehicle's weight using offboard chargers. **Overnight Depot Charging is a** facility that provides overnight depot charging offers fast and slow charging options. The charge point is typically located at the end of the lines. It is used for charging at night. **Pantograph Charging is a** charging infrastructure is used for applications requiring higher battery capacity and more power, such as buses and trucks. Pantograph charging can be further categorized as (a) Top-down Pantograph where the charging setup is located on top of the bus stop and (b) Bottom-up Pantograph where the buses equipped with charging equipment.

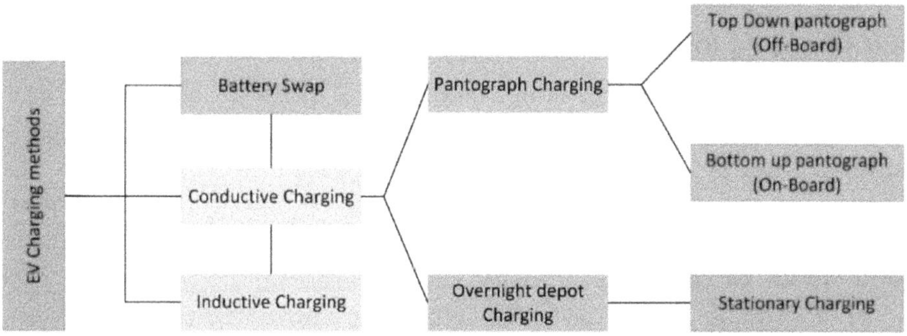

Figure 5.7 Classification of PEV charging methods.

Inductive/wireless power transfer (WPT): A two-coil system is used in this technology based on electromagnetic induction. The receiving coil is installed in the vehicle, while the charging coil is placed on the road surface. A standard connector is unnecessary for the charger (but standard coupling technology is required), and the charger can be operated while driving. This is not very efficient due to large distance between the transmitter and receiver coils and eddy current loss if the transmitter coil is not switched off. Figure-5.8 shows a typical inductive charging setup.

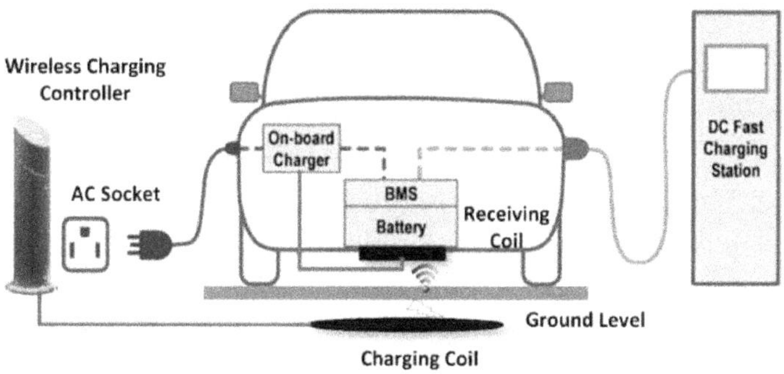

Figure 5.8 A typical inductive charging setup.

Battery Swap Station (BSS): Batteries are swapped through a method called Battery Exchange", which involves renting the battery from the BSS owner monthly". The BSS allows slow charging and gives batteries a longer life cycle. Solar and wind power, locally generated renewable energy sources, can easily be integrated into the BSS system. In this technique the battery can be replaced, or the vehicle can be refueled quickly. Other advantages of this system are contribution to the V2G initiative. Since the BSS owner owns the EV batteries, charging the EVs at this station is more expensive than burning fuel for the ICE engine due to leasing fees. This techniques multiple batteries need to be purchased and a large space for storage, which could be expensive in an area with high traffic.

5.4.3 PEV Charging Infrastructure

An adequate infrastructure allowing customers to recharge their electric vehicles would make it easier to adopt electric vehicles Fast, reliable, and convenient technology will make it easier for consumers to adopt electric vehicles. Adopting electric vehicles promotes environmentally-friendly transportation and reduces our dependence on fossil fuels. The increasing energy demand can be met by a large-scale deployment of renewable energy sources, coordinated with the increasing number of PEVs. A complete PEV charging infrastructure involves power infrastructure, control and communication infrastructure, and charging ports and connectors meeting various standards, as depicted in Figure 5.9 and Figure 5.10.

Connectors for AC Charging: A universal AC connector has not been agreed upon by EV manufacturers. The connector's pin-out, size, and shape vary depending on the charger's electric vehicle manufacturer, country, and power level. Different regions have different mains voltages and frequencies because of their specific AC mains voltages and frequencies. As a standard AC connector, most voltage-controlled devices have a few large pins for voltage sensing and a few smaller pins for communication. Here are the four types of AC connectors that are currently in use around the world:

Connector type 1: This connector is used to charge the vehicle with single-phase AC power. The connector is a round configuration composed of five pins, including two AC lines, two signal lines, and one earth pin for protection. The maximum voltage and current rating are 120 V or 240 V, and 80 A respectively.

Connector type 2: This connector can charge *the* device with *both* AC and DC power. Moreover, it is 3-φ AC compatible. A single-phase can have a maximum voltage rating of 230 V and a maximum current rating of 80 A; the maximum voltage rating for a three-phase *supply* can be 400 V at 63 A.

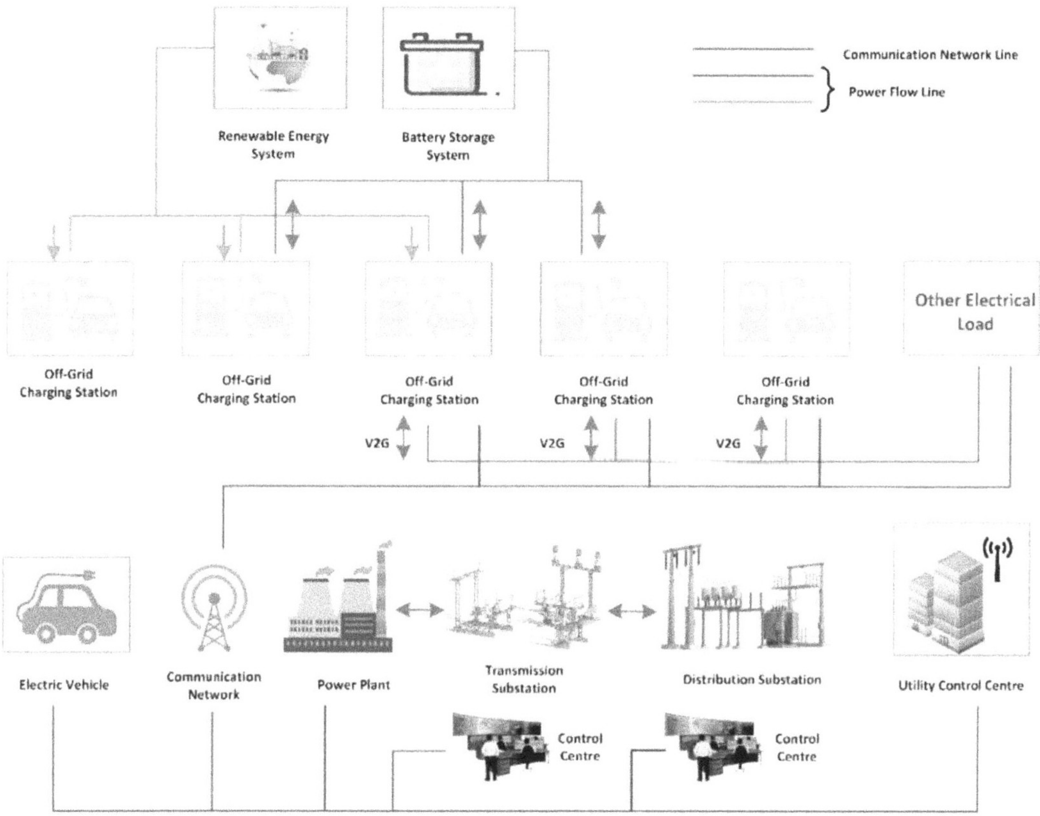

Figure 5.9 Schematic diagram of charging infrastructure for electric vehicles.

	GB/T 20234 DC	CHAdeMO	CCS Combo 1	CCS Combo 2
	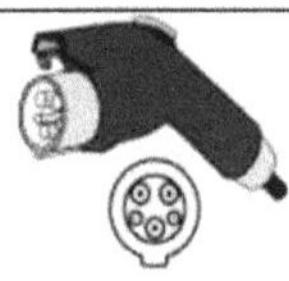	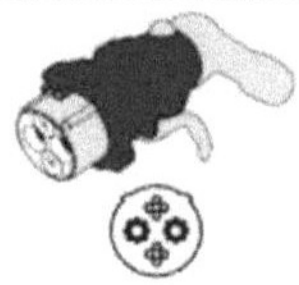		
Charging Type	AC Charging	AC Charging	DC Fast Charging	DC Fast Charging
No. Of Pins	5	4	7	9
Capacity	Up to 187.5 kW	200 kW to 400 kW	50 kW - 100 kW	200 kW
Voltage	750 V to 1000 V	500 V	600 V	1000 V
Current Rating	250 A	125 A to 400 A	Up to 200 A	Up to 350 A
Compatibility	Yes	Yes	No	Yes
	SAE J1772 Type 1	**IEC 62196-2 Type 2**	**GB/T 20234 C**	**Tesla Supercharger**

Charging Type	AC Charging	AC Charging	AC Charging	DC Fast Charging
No. Of Pins	5	7	7	6
Capacity	7.68kW	12.8kW	12.16kW	140kW
Voltage	120V to 240V	400V	380V	480V
Current Rating	Up to 80 A	Up to 63 A	10A	250A
Compatibility	Yes	Yes	Yes	Yes

Figure 5.10 Schematic of charging ports and connectors according to various standards.

US Tesla connector: This connector was designed by Tesla specifically for use in the United States. This connector can use both single-phase AC and DC power. This charger can charge at a maximum of 17.2 kW at 240 V AC.

Connectors for DC Charging: According to the manufacturer, their power output varies from 50 kW to 500 kW. In addition to more excellent power capability, power conversion and control become more expensive and bulkier. DC fast chargers are typically deployed offboard to avoid interfering with the power lines. Another reason is safety considerations. As power converters and power handling components grow, passenger safety becomes a key concern. Currently, DC connectors have five major variants and are discussed as follows.

CCS combos 1 and 2: An electric vehicle charging interface initiative is a registered organization formed by Audi, BMW, Daimler, Mennekes, Opel, Phoenix Contact, Porsche, TUV SUD, and Volkswagen called CharIN e.V. ChargeIN is the organization that oversees Combo-type connectors. The Combined Charging System connector has the main advantage of performing AC and DC charging simultaneously. The connectors are defined by several standards, including IEC 62196–1, IEC 62196–2, and IEC 62196–3. Their maximum power handling capacity is 350 kW, and the connectors can handle voltages from 200 V to 1 kV up to 350 A.

CHAdeMO: In 2010, Toyota Motor Corporation established the CHAdeMO Association to collaborate with Nissan Motor Co. Ltd., Toyota Motor Corporation, Nissan Motor Co. Ltd., Fuji Heavy Industries Ltd., and Tokyo Electric Power Company. CHAdeMO is a part of IEEE standards (IEEE Standard 2030.1.1TM-2015) and IEC standards (61851-23-6), as well as industry standards (61851–23, -24, and 62196–3). CHAdeMO, the first DC standard with a power handling capacity between 200 kW and 400 kW, enables V2X (vehicle to X, where 'X' may refer to the vehicle, the grid, building etc.) with its version 1.1 protocol.

Tesla DC connector: Tesla's superchargers are equipped with their own charging connector in the US. The Tesla connector has the unique feature which enables AC and DC charging with same connector and pins. It can also connect CHAdeMO charging stations to the connector. A 120-kW power rating is available for each of these connectors.

GB/T China connector: The standard 20234.3–2015 defines a DC charging connector in China. The Controller Access Network (CAN) protocol communicates with connectors like this. This connector has the unique property of charging two batteries simultaneously i.e., the low- voltage auxiliary battery and the high-voltage main battery. This product can deliver up to 250 A of current at a nominal voltage ranging 750 V to 1 kV.

5.4.4 Control and Communication Infrastructure

The control and communication system controls and monitors PEV charging system. Charging an electric vehicle increases the power demand for the power system which may lead to voltage and frequency instability. Hence there a need to collect the information from various stakeholders of the grid and process the data to maintain reliability, stability and economic operation of the grid.

Electric vehicle charging control: Figure-5.11 Shows how electric vehicle charging is controlled based on mobility, coordination, and control structures. The controls for EV charging involve the electric grid, EV charging stations, and EVs. Section 5.3.4 can be referred for details.

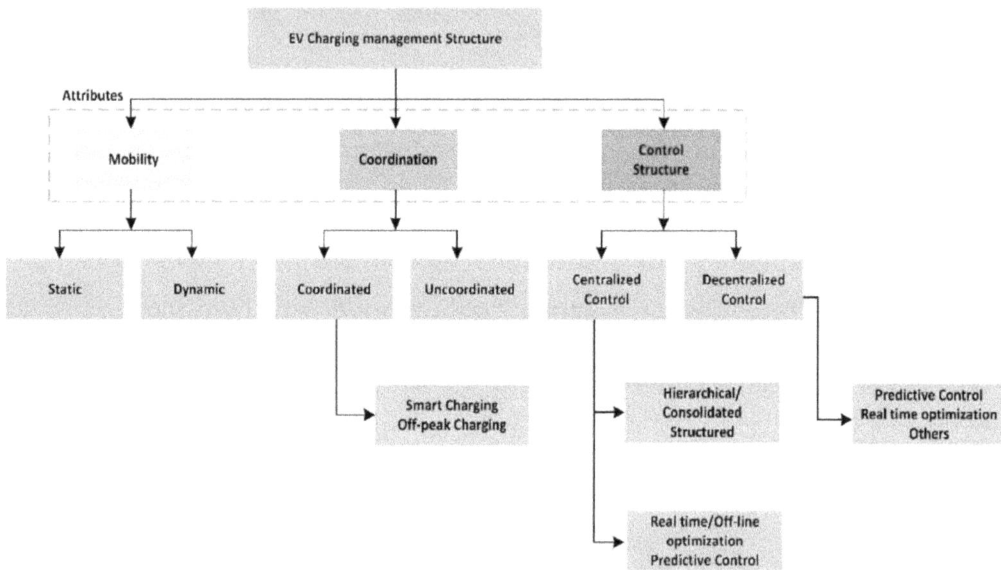

Figure 5.11 EV charging systems are classified into different control strategy.

Communication Network for Electric Vehicle Charging: A successful charging management system for electric vehicles depends on effective communication between EVs, EVSEs, and the grid as shown in Figure 5.12. Various communications protocols are classified into wired and wireless technologies. The use of electric vehicles can be integrated into different private areas, including home area networks (HAN), industrial areas networks (IAN), building areas networks (BAN), and neighborhood areas networks (NAN), or field areas networks (FAN). Control and monitoring the charging/discharging of electric vehicles and other domestic electricity usage are handled through these networks.

Wireline Communication: Wireline technologies are effective for long-distance data transfers i.e. in large cities. One of the most popular protocols in data communication in wireline technology is Power Line Communication (PLC). This technology uses the same power line for the power transmission and data transmission. Due to its robustness and reliability, it can be used in the event of interference. The PLC concept is used in several protocols, including Home Plug 1.0, Home Plug turbo, Home Plug AV, HD-PLC, and UPA. Optical and DSL protocols are also a part of wireline communication. The significant advantage of optical communications is their significantly higher data rates (several Gbps) and transmission ranges (many kilometers) compared to PLC. Optical communication is also resilient against electromagnetic interference. The DSL protocol uses digital communication over a telephone line, therefore, the existing set up of infrastructure can be used for this purpose as in case of PLC protocol.

Wireless Communication: Providing data exchange between vehicles and charging stations requires wireless communication which completes the communication path. PEVs rely on this to provide information concerning their charging status. In wireless communication networks, electrical devices are connected using hierarchical mesh structures. The hierarchical mesh structure is made up of wireless LAN devices. This uses four popular wireless technologies, namely Zigbee, Cellular, WIFI, and satellite networks.

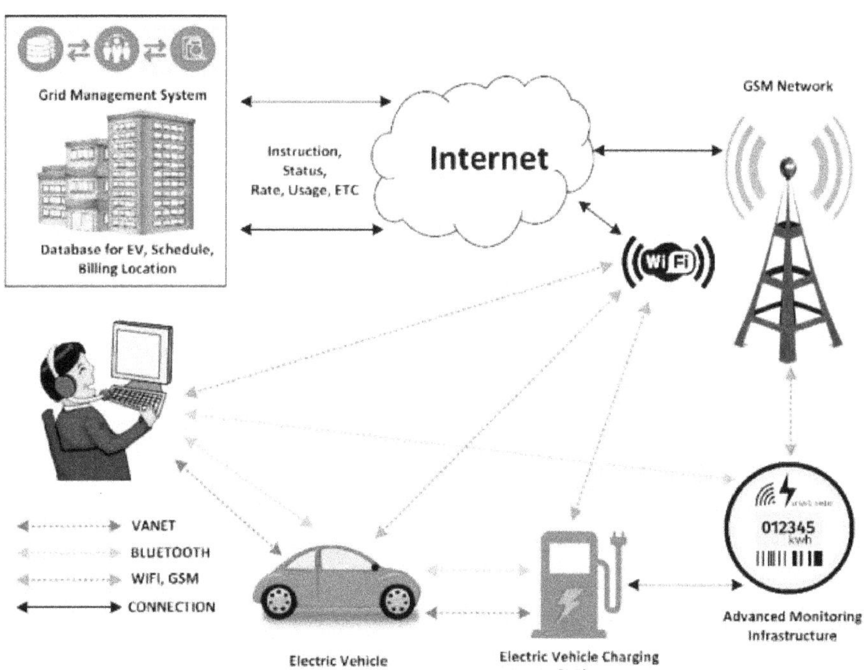

Figure 5.12 Communication network for electric vehicle charging systems.

5.5 Challenges Associated with Grid Integration

One of the major negative impacts of V2G technology is the limited predictability of the requirements of users and the availability of EVs for V2G operations. The diverse and non-periodic charging periods of EVs to the grid pose a significant/serious technical challenge in implementing V2G services. The other major challenges associated with providing grid integration are grid stability, power quality, communication issues, interoperability conditions, cyber-security, and cost shown in Figure 5.13.

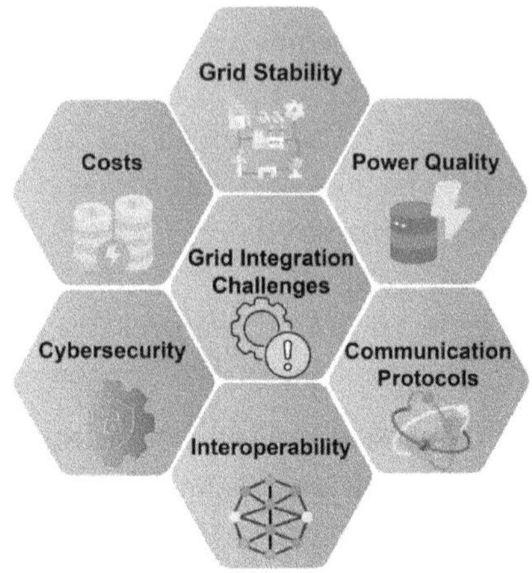

Figure 5.13 Major challenges associated with grid integration.

5.5.1 Grid Stability

The bidirectional power flow enabled by V2G integration could disrupt normal grid frequency regulation and stability if it is not managed carefully. Scaling up EV adoption may create overloading issues on cables, distribution feeder lines, and transformers in areas that mandate infrastructure upgrades like installing additional generation capacity. Moreover, the repeated battery cycling involving EV storage for grid services risks accelerated battery capacity reduction and replacement costs that would affect utilities and vehicle owners. Evaluating and mitigating the impacts of reliability, equipment loading, and asset lifespan will be critical considerations for ensuring the viability of large-scale V2G integration.

5.5.2 Power Quality

Integrating EVs into the grid through V2G technology introduces potential power quality challenges. The fluctuating charging and discharging patterns of large numbers of EVs can lead to harmonics on the system, voltage fluctuations for customers, and even instability in grid frequency if not controlled. The variable output of renewable resources also degrades power quality upon V2G integration. This also poses a regulatory challenge to maintain the IEEE-519 power quality standards while charging and discharging the EV. Implementing robust power quality control systems helps safeguard high-quality electricity supplied to the grid.

5.5.3 Communication Protocols

The limited communication ranges, irregular connections, high mobility, and large volumes of data in V2G networks pose critical challenges to achieving fast authentication while guaranteeing transaction reliability and efficiency. V2G integration necessitates communication protocols that allow EVs to interface with the grid and provide data on battery charge status and availability. Such protocols require standardization to guarantee interoperability between the disparate vendor systems involved. Communication protocols especially the OCPP, have significant challenges in the integration of V2G technology. OCPP does not natively support advanced bidirectional energy flows and V2G specific features in its current forma; yet it does provide an overarching framework for communicating between charge stations and central management systems. OCPP extension is required to enable real-time data exchange and compatibility with various original equipment manufacturers (OEMs) support and energy management systems (EMS). Moreover, cybersecurity errors in OCPP can leave systems vulnerable to data breaches and unauthorized access, posing a threat to user privacy and the stability of the grid

5.5.4 Interoperability

V2G integration requires interoperability between systems from various vendors, which can be challenging to achieve at a large scale in a critical infrastructure. Standards and agreements between the various parties are usually required to tackle this issue. The Open Charge Point Interface/Protocol (OCPI/OCPP) has emerged as a common protocol to facilitate communication between different charging providers. It establishes a hierarchy for EV charging stations, with location at the top level, then charging ports, then connectors. This allows all players in the charging ecosystem to have an interoperable system to manage access and billing across different charge point installations. A key advantage of OCPI/OCPP is that it is open-source and independent; any service provider or charge point operator may implement the protocol free of charge to enable roaming access across charge stations. Although it has benefits, it still needs to be implemented for widespread EV adoption. OCPI/OCPP also allows bilateral agreements between parties and integration through larger roaming hubs.

5.5.5 Cyber Security

As current centralized electricity generation and management is gradually replaced with distributed generation and smart grids, system operators encounter various challenges in maintaining network security while sustaining operational efficiency. Smart appliances and smart meters in the smart grid generate a substantial amount of real-time data. It is necessary to tackle not just the obstacles associated with V2G infrastructure and technology, but also the complexities in communication, including concerns regarding the privacy and security of EV users. This introduces novel implications for both the user and the pace at which EVs are adopted. Although EVs possess these capabilities, they are also prone to face security and privacy risks. Recently EVs have experienced data breaches, stealing energy, financial fraud, and even critical health and safety problems. EV sensors can be misused in many ways.

5.5.6 Cost

V2G integration incurs expenses related to infrastructure and control mechanisms. These costs must be weighed against potential economic gains from grid services and reduced Green House Gas emissions. Optimal scheduling requires balancing technology capabilities and financial incentives with efficiency limits. Scheduling bidirectional power flow appears crucial to profitability. Careful cost-benefit and sensitivity analysis should be done for future grid investment and policy decisions. Still, challenges remain in managing vehicle owner participation, distributed software coordination, and equipment degradation over V2G system lifetimes spanning decades. Performing vigorous cost-benefit analyses is key to determining the ultimate value proposition of V2G technology. One key consideration is the cost of additional hardware for the V2G configuration.

5.6 Solutions for the Challenges of Grid Integration

The critical requirement for smooth operations is the development of standardized communication protocols between vehicles, chargers, and the grid to enable interoperability and security. There is also a need to build out bidirectional charging infrastructure to facilitate electricity flow in both directions. Economic and sustainability factors also play a major role in consumer acceptance. Emerging technologies such as blockchain, quantum computing, AI and machine learning (ML), advanced bi-directional chargers, wireless chargers, 6G, internet of things (IoT), solid-state batteries can overcome these technical challenges associate with V2G market.

5.6.1 Power Electronics

Power electronics offer pathways to enhance the efficiency of conversion systems within EV charging infrastructure, facilitating improved V2G capabilities. They allow bidirectional power flow for EVs to function as mobile storage and grid service providers. Power electronics can furnish reactive power support to stabilize the grid and bolster efficiency. They also enable load balancing to evenly distribute demand and mitigate peaks. Additionally, they permit active power control to regulate EV output for balanced power quality. For further improved quality output of power transfer, auxiliary converters can enable the support to integrate vehicle and the grid structure. The bidirectional AC-to-DC converter (BADC) has versatile role to play in V2G, G2V, and V2H modes to regulate, rectify, and deliver power through them. Bidirectional DC-to-DC converter (BDC) can operate on V2G, and G2V only. Advanced power electronics and control methods have the potential to smooth out power fluctuations and enhance quality. Intelligently controlled bidirectional power flow enables EVs to both charge from and discharge to the grid as needed, to aid stabilization and optimize renewables.

Additionally, unified designs that tie together vehicles, battery storage, renewables, and power quality compensation capabilities can provide collaborative grid services for issues like voltage deviations, harmonics mitigation, or load balancing through a shared smart infrastructure. Power electronics systems integrated into VPPs, to manage Energy Storage Systems for charge/discharge of EVs, provides economic advantages and offers ancillary services to the grid. By optimizing conversion and power flows, these technologies help maximize the value and reliability of V2G integration.

5.6.2 Advanced Control Systems

Advanced control systems have the capability to regulate EV charging and discharging to maintain grid stability and power quality. By modulating EV power output, advanced controls ensure that V2G services are safe, dependable, and cost-effective. These systems also facilitate coordinating V2G with other grid services like frequency/voltage regulation and spinning reserves for effective grid integration. Advanced controls employing AI and ML can also optimize V2G systems by analyzing data and enhancing performance over time. Robust control mechanisms are crucial for extracting maximum value from V2G platforms while avoiding disruptions.

5.6.3 Flexibility through Energy Storage Systems (ESS)

Many ESS are designed to respond rapidly with flexible responses to changing power needs. This adaptability is crucial for optimal performance and efficient operation. V2G does this through improving energy efficiency, lowering carbon intensity, and supporting variable renewable energy options. V2G involves EVs providing power from their batteries to the grid during peak demand hours. This also enhances system reliability. ESS can accumulate surplus renewable energy from sources like solar PV and wind, enabling EVs to serve as mobile storage reservoirs and grid service providers. Storage can also furnish backup power to EV charging stations, ensuring dependable, cost-effective V2G capabilities. The storage facilitates V2G services with renewables, allowes EVs to harness clean energy and stabilize the grid as demand response assets. Strategic energy storage deployment unlocks the potential of EVs to both utilize and support sustainable resource. Studies have shown that deploying EVs and photovoltaics can eliminate 80% of CO_2 emissions from the environment at the source level. The fast-charging tendency of infrastructure with 350 kW battery storage systems has the potential to enhance savings in terms of money.

5.6.4 Securing Communication

Communication systems enable the optimization of V2G services by using AI and ML algorithms to analyze data and improve the performance of V2G systems. V2G infrastructure and EVSE environment is subjected to multiple attack vectors such as false data injection attacks (FDIA), man-in-the-middle (MITM) attacks, denial-of-service (DoS), denial-of-distributed-service (DDos), and side-channel attacks. To secure such enterprise of EVSE infrastructure, one has to review the CIAA properties of these V2G services that could be minimally impacted by such attacks. Advanced ML algorithms can provide short-term solutions for surmounting these challenges. Precisely forecasting necessary load, pricing, renewable generation, and customer behaviors is essential for devising optimal grid operation and control strategies. By leveraging AI and ML to enhance prediction capabilities, utilities can improve the coordination of residential EV charging in the near term. However, longer-term solutions will still be required for sustainable management as EV adoption expands.

Emerging cryptographic standards and decentralized identity frameworks utilizing concepts such as PKI, Zero-Trust (ZT), and distributed ledger technology can help overcome authentication challenges. Solutions based on these standards will likely enable fast, highly secure, decentralized interoperable authentication frameworks. They can be optimized for the unique demands of grids, with millions of mobile EVs. Blockchain's secure, transparent distributed ledger capabilities can help surmount technical obstacles to grid-integrated V2G systems. By facilitating tamper-proof, decentralized communication between EVs and the grid, blockchain enables reliable transactions and interactions.

Blockchain's cryptography and consensus mechanisms offer pathways to actualize robust vehicle-grid coordination amid the complexities of large-scale distributed resources. It was proposed to employ blockchain to construct a secure energy trading platform for V2G infrastructure. It examines interoperability and security challenges inherent to V2G systems. To address these issues, the authors suggest leveraging blockchain's immutable and transparent distributed ledger capabilities. This enables robust and efficient communication between EVs and the electrical grid. The proposed decentralized system unlocked V2G's potential through enhanced security and coordination. Although there is a tradeoff of increased memory usage and latency, implementing blockchain technology in the system provides enhanced security and transparency.

References

1. Ahmadian, B. Mohammadi-Ivatloo and A. Elkamel, "A Review on Plug-in Electric Vehicles: Introduction, Current Status, and Load Modeling Techniques," in *Journal of Modern Power Systems and Clean Energy*, vol. 8, no. 3, pp. 412-425, May 2020, doi: 10.35833/MPCE.2018.000802.
2. Wei Wu, Boqiang Lin, "Benefits of electric vehicles integrating into power grid, Energy, Volume 224, 2021, 120108, ISSN 0360-5442, https://doi.org/10.1016/j.energy.2021.120108.
3. https://www.energy.gov/articles/history-electric-car
4. J. García-Villalobos, I. Zamora, J.I. San Martín, F.J. Asensio, V. Aperribay, Plug-in electric vehicles in electric distribution networks: A review of smart charging approaches, Renewable and Sustainable Energy Reviews, Volume 38, 2014, Pages 717-731, ISSN 1364-0321
5. RC Green, L Wang, M. Alam, The impact of plug-in hybrid electric vehicles on distribution networks: a review and outlook, Renewable Sustainable Energy Rev, 15 (2011), pp. 544-553
6. MD Galus, MG Vayá, T Krause, G. Andersson, The role of electric vehicles in smart grids, Wiley Interdiscip Rev Energy Environ, 2 (2013), pp. 384-400
7. J Mullan, D Harries, T Bräunl, S. Whitely, The technical, economic and commercial viability of the vehicle-to-grid concept, Energy Policy, 48 (2012), pp. 394-406
8. M Yilmaz, PT. Krein, Review of the impact of vehicle-to-grid technologies on distribution systems and utility interfaces, IEEE Trans Power Electron, 28 (2013), pp. 5673-5689
9. JDK Bishop, CJ Axon, D Bonilla, M Tran, D Banister, MD. McCulloc, Evaluating the impact of V2G services on the degradation of batteries in PHEV and EV, Appl Energy, 111 (2013), pp. 206-218
10. W Kempton, J. Tomić, Vehicle-to-grid power implementation: from stabilizing the grid to supporting large-scale renewable energy, J Power Sources, 144 (2005), pp. 280-294
11. A Nebel, C Kruger, F. Merten, Vehicle to grid and demand side management—an assessment of different strategies for the integration of electric vehicles, In: IET conference renewable power generation (RPG 2011). IET: Edinburgh (2011):143
12. SB Peterson, JF Whitacre, J. Apt, The economics of using plug-in hybrid electric vehicle battery packs for grid storage, J Power Sources, 195 (2010), pp. 2377-2384
13. JAP Lopes, FJ Soares, PMR. Almeida, Integration of electric vehicles in the electric power system, Proc IEEE, 99 (2011), pp. 168-183
14. C Guille, G. Gross, Design of a conceptual framework for the V2G implementation, In: 2008 IEEE energy 2030 conference. IEEE: Atlanta, GA (2008), pp. 1-3

15. S Han, S. Member, S Han, K. Sezaki, Development of an optimal vehicle-to-grid aggregator for frequency regulation, IEEE Trans Smart Grid, 1 (2010), pp. 65-72
16. S. S. G. Acharige, M. E. Haque, M. T. Arif, N. Hosseinzadeh, K. N. Hasan and A. M. T. Oo, "Review of Electric Vehicle Charging Technologies, Standards, Architectures, and Converter Configurations," in *IEEE Access*, vol. 11, pp. 41218-41255, 2023, doi: 10.1109/ACCESS.2023.3267164.
17. Ishtiak Mahmud, Mohtarima Begum Medha, M. Hasanuzzaman, Global challenges of electric vehicle charging systems and its future prospects: A review, Research in Transportation Business & Management, Volume 49, 2023, 101011, ISSN 2210-5395
18. Kang Miao Tan, Vigna K. Ramachandaramurthy, Jia Ying Yong, Integration of electric vehicles in smart grid: A review on vehicle to grid technologies and optimization techniques, Renewable and Sustainable Energy Reviews, Volume 53, 2016, Pages 720-732, ISSN 1364-0321
19. K. M. Bhargavi, N. S. Jayalaksmi, S. Malagi and V. K. Jadoun, "Integration of Plug-in Electric Vehicles in Smart Grid: A Review," *2020 International Conference on Power Electronics & IoT Applications in Renewable Energy and its Control (PARC)*, Mathura, India, 2020, pp. 214-219, doi: 10.1109/PARC49193.2020.236595.
20. Yanchong Zheng, Songyan Niu, Yitong Shang, Ziyun Shao, Linni Jian, Integrating plug-in electric vehicles into power grids: A comprehensive review on power interaction mode, scheduling methodology and mathematical foundation, Renewable and Sustainable Energy Reviews, Volume 112, 2019, Pages 424-439, ISSN 1364-0321, https://doi.org/10.1016/j.rser.2019.05.059.
21. K. Uddin, T. Jackson, W.D. Widanage, *et al.,* On the possibility of extending the lifetime of lithium-ion batteries through optimal V2G facilitated by an integrated vehicle and smart-grid system, Energy, 133 (2017), pp. 710-722
22. D. Wang, J. Coignard, T. Zeng, *et al.* Quantifying electric vehicle battery degradation from driving vs. vehicle-to-grid services, J Power Sources, 332 (2016), pp. 193-203
23. S.M. Shariff, D. Iqbal, M.S. Alam, F. Ahmad, A state of the art review of electric vehicle to grid (V2G) technology IOP Conf. Ser.: Mater. Sci. Eng., 561 (1) (2019), Article 012103, 10.1088/1757-899X/561/1/012103
24. Z. Ali, A. Mahmood, S. Khatoon, W. Alhakami, S.S. Ullah, J. Iqbal, S. Hussain, A generic internet of things (IoT) middleware for smart city applications Sustainability, 15 (1) (2023), p. 743, 10.3390/su15010743
25. S. Mateen, M. Amir, A. Haque, F.I. Bakhsh, Ultra-fast charging of electric vehicles: A review of power electronics converter, grid stability and optimal battery consideration in multi-energy systems Sustain. Energy Grids Netw., 35 (2023), Article 101112, 10.1016/j.segan.2023.101112

CHAPTER

6

Demand Side Management

6.1 Introduction

Demand-Side Management (DSM) refers to a set of techniques and strategies used to optimize electricity consumption by shifting or reducing demand in response to supply constraints, economic factors, or environmental concerns. Due to integration of a large number of small, intermittent, and weather-dependent renewable distributed generators (DGs), it has become a challenge for power system operators to maintain demand and supply balance for the stable operation of the power system. In this situation, flexible demand-side resources which respond to price, incentive, or emergency signals sent by the utilities can play significant role in maintaining this balance. Designing any demand response (DR) program requires deep knowledge of customers' behaviour toward energy consumption. A major challenge in designing load-specific DR programs is to know the accurate usage behaviour of those loads by the vast majority of customers. It has been observed that many energy management benefits are achievable if customers are informed about the usage of individual loads. Obtaining this knowledge through individual submeters/sensors for individual loads in customers' vicinity is not a feasible solution as it requires huge investment for the purchase and installation of submeters/sensors and the related infrastructure. With the advent artificial intelligence (AI), and the Internet of Things (IoT), DSM has become a critical aspect of modern energy management. This chapter provides a detailed technical exploration of DSM principles, implementation strategies, modelling and benefits and suitable for graduate and postgraduate students in power systems engineering.

Figure 6.1 shows the framework of demand side management in a smart grid. The objectives of demand-side management generally revolve around improving overall energy efficiency, reducing peak demand, enhancing grid reliability, and achieving environmental sustainability. Here are some common objectives of demand-side management:

- Reduction of peak load;
- Improvement of the economic feasibility and efficient operation of the grid;
- Keeping supply and demand balance;
- Improve the overall energy efficiency;
- Increasing the economic viability of electricity distribution network;
- Enhancing the stability and reliability of the electricity grid.

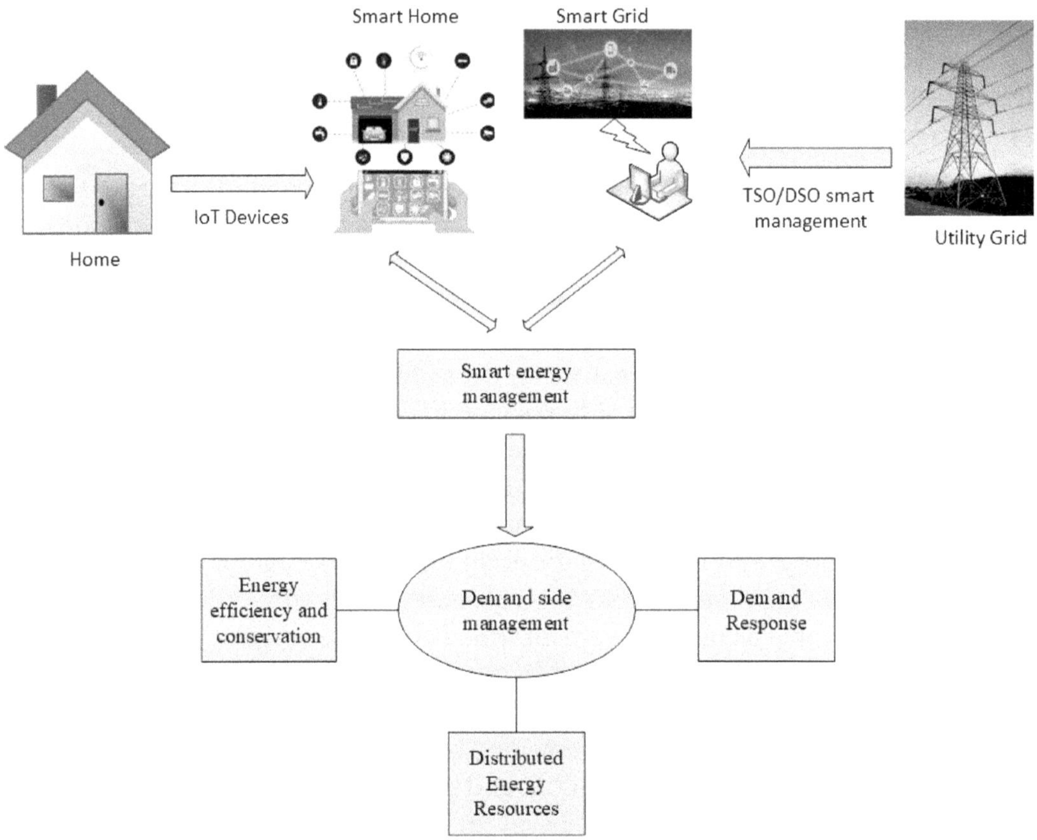

Figure 6.1 Smart grid framework and DSM.

6.2 Evolution of DSM

DSM has evolved to its present structure over several decades while adapting to technological, economic, and environmental needs. The history of DSM can be categorized into the following phases:

Before the 1970s, electricity grids operated in a traditional centralized manner, focusing solely on supply-side management. The primary approach to address the increased energy demand was just to install more power generation plants to generate additional electricity without considering impact on the overall efficiency. The system operated in a one-directional flow, where utilities supplied power, and consumers passively consumed it without awareness or control over their usage. There was no emphasis on optimizing energy consumption, nor were there mechanisms to dynamically regulate demand. The power grid functioned as a rigid, centralized system, relying solely on increasing supply to meet growing energy needs. Consumer participation in energy management was virtually non-existent, as the dominant approach prioritized large-scale generation over efficiency and adaptability.

During the 1970s and 1980s, a series of oil crises underscored the importance of energy conservation and efficiency. In response, utilities implemented the first Demand-Side Management (DSM) programs to reduce energy consumption and mitigate the high costs of peak-hour generation. Strategies such as peak shaving and load shifting were introduced, encouraging consumers to adjust their electricity usage to off-peak hours, thereby reducing burden on the grid.

In the 1990s and early 2000s, Demand Response (DR) programs gained momentum as utilities began offering financial incentives to consumers who reduced their energy usage during peak hours. The introduction of smart meters represented a significant technological advancement, enabling real-time monitoring and greater control over energy consumption. This period paved the way for the development of more sophisticated energy management strategies.

Since the 2010s, the emergence of smart grids and Artificial Intelligence (AI) has transformed Demand-Side Management (DSM). The increasing integration of renewable energy sources, such as solar and wind, has created a need for more dynamic and responsive energy demand management solutions. AI and Internet of Things (IoT)-based demand response systems now facilitate automated, real-time energy optimization. Furthermore, advancements in Electric Vehicle (EV) charging management and Vehicle-to-Grid (V2G) systems have reshaped energy distribution and consumption, contributing to a more sustainable and efficient power grid.

DSM continues to evolve as part of next-generation smart grids, making energy consumption more efficient, flexible, and sustainable. Table 6.1 presents various evolution stages of DSM.

Table 6.1 Evolution Stages of DSM.

Stage	Key Features	Technology Used
Stage 1: Load Management (1970s–1980s)	Peak shaving, load shifting, industrial load control	Manual switching, time-based tariffs
Stage 2: Demand Response (1990s–2000s)	Incentive-based DSM, voluntary load reduction	Smart meters, real-time pricing
Stage 3: Smart Grid DSM (2010s–Present)	AI-driven automation, renewable integration, IoT-based DSM	AI, IoT, blockchain, V2G, energy storage

6.3 Need of Demand Side Management (DSM)

In the modern smart grid era, Demand-Side Management (DSM) plays a crucial role in ensuring the stability, efficiency, and sustainability of power systems. As electricity consumption continues to rise, advanced DSM strategies are essential to balance supply and demand while optimizing energy use. In the modern smart grid era, DSM is essential due to following factors as listed below:

Growing Electricity Demand: One of the primary drivers of DSM is the growing demand for electricity. Rapid urbanization, the increasing adoption of electric vehicles (EVs), and the expansion of data centres and industrial operations place significant stress on power grids. Without effective demand management, these rising loads could lead to grid instability and higher operational costs.

Renewable Energy Integration: Another key factor is the integration of renewable energy sources such as solar and wind. Unlike traditional power generation, renewables are inherently intermittent, making it challenging to match supply with demand. DSM helps mitigate these fluctuations by dynamically adjusting consumption patterns, thereby reducing reliance on fossil-fuel-based backup power plants.

Grid Reliability & Stability: Grid reliability and stability are also major concerns addressed by DSM. By efficiently distributing electrical loads, DSM reduces the risk of blackouts, voltage fluctuations, and other disruptions. This proactive approach ensures a more resilient and responsive power infrastructure.

Energy Cost Reduction : Cost reduction is another advantage of DSM, particularly through Time-of-Use (TOU) pricing and incentive programs. These strategies encourage consumers to shift their electricity usage to off-peak hours, lowering overall energy costs and alleviating strain on the grid during peak demand periods.

Environmental Sustainability: From an environmental perspective, DSM supports sustainability efforts by minimizing dependence on fossil-fuel-based power plants.

By optimizing energy consumption and integrating cleaner energy sources, DSM significantly contributes to reducing carbon emissions and promoting a greener future.

Technological Advancements: Advancements in technology have further enhanced DSM capabilities. Smart meters, the Internet of Things (IoT), and AI-driven automation enable real-time monitoring and control of energy usage. These innovations allow utilities and consumers to implement demand response programs with greater precision and efficiency.

Electrification of Transportation: Finally, the electrification of transportation presents new challenges and opportunities for utility and consumers. The widespread adoption of EVs introduces substantial charging loads that could overwhelm power grids if unmanaged. DSM strategies ensure that EV charging is scheduled during off-peak hours, while Vehicle-to-Grid (V2G) systems enable bidirectional energy flow, allowing EVs to supply power back to the grid when needed.

As the energy landscape continues to evolve, DSM remains a fundamental component of modern power systems. By leveraging advanced technologies and strategic energy management practices, DSM supports a more reliable, cost-effective, and sustainable electricity network for the future.

6.4 DSM Strategies and Policies

Reducing the generation side peak demand is very costly; a study conducted in found that at least 10% of supply costs only result in 1% of annual hours. Demand side management provides an affordable solution for handling such a problem. By altering the consumer's energy consumption pattern, Demand side management lowers the overall peak load demand and improves grid stability. This lower energy consumption costs and carbon emission.

Various demand side management strategies are presented in Figure 6.2. These management strategies can be applied to mitigate the energy demand of buildings and improve their energy efficiency. The key strategy is to move to on-demand consumption instead of on-demand generation. The three main pillars of DSM are strategic energy conservation and energy efficiency, Demand Response (DR), and Distributed Energy Resources (DER).

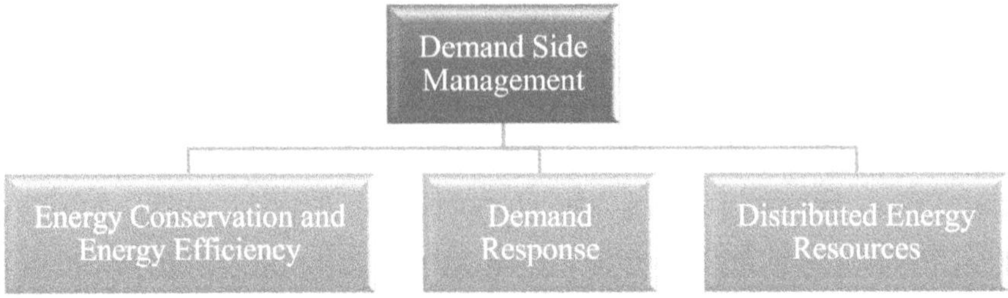

Figure 6.2 Demand side management strategies.

6.4.1 Energy Conversion and Energy Efficiency

Energy conservation and energy efficiency are distinct yet interconnected strategies aimed at optimizing energy use. Energy conservation involves reducing overall energy consumption through behavioural changes and prudent practices, such as turning off lights and appliances when not in use. It emphasizes minimizing waste and curbing unnecessary energy demand. Proposed the energy audit method to conserve energy efficiently. There was no more intelligent approach to energy management for significantly reducing demand in a reliable manner. On the other hand, energy efficiency focuses on getting more output from the same amount of input, achieved through technological advancements and improved processes these studies address energy management without considering the configuration of smart energy systems. They use a transitional approach for energy conservation. It aims to enhance the performance of energy-consuming systems, appliances, and industrial operations. While conservation addresses the need to lower overall energy demand, efficiency aims to achieve the same outcomes with less energy input. Together, these approaches form a holistic approach to sustainable energy management, contributing to environmental preservation, cost reduction, and the promotion of responsible resource use. The technological advancements include:

Energy-Efficient Appliances: Advances in the design and engineering of appliances, such as refrigerators, air conditioners, washing machines, and lighting fixtures, have led to significant reductions in energy consumption while maintaining or even improving performance. Using energy-efficient equipment and buildings, as well as promoting consumers' energy-conscious behaviour, to reduce energy usage. Typical instances are switching to energy-saving lights from incandescent bulbs and switching to variable-speed air conditioning from standard air conditioning. Enhancing and performing routine maintenance on electrical equipment by recovering heat from waste, improving maintenance techniques, using contemporary equipment with optimum designs, and implementing cogeneration.

Energy Storage Solutions: Innovations in energy storage technologies, including batteries and other storage systems, enable the capture and utilization of excess energy during periods of low demand for later use during peak demand times, thereby improving overall system efficiency and reliability.

Building Automation Systems: The integration of advanced sensors, controls, and automation technologies into building systems allows for more precise management of energy usage, including heating, cooling, lighting, and ventilation, based on real-time occupancy and environmental conditions.

6.4.2 Demand Response (DR)

Demand Response (DR) is a key component of Demand Side Management (DSM) that enables consumers to adjust their electricity usage in response to grid conditions, price signals, or incentives from utilities. DR programs aim to enhance grid reliability, improve market efficiency, and integrate more renewable energy resources. By modifying demand patterns, DR helps balance supply and demand without requiring additional power generation, making it a cost-effective solution for energy management.

Demand response works through a structured mechanism where consumers either voluntarily or automatically adjust their electricity consumption. A simple schematic of a household demand response strategy is shown in Figure 6.3. The following steps describe how DR functions in a power system:

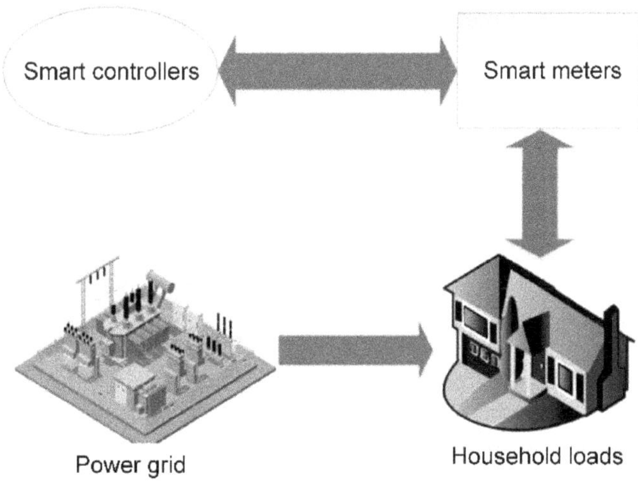

Figure 6.3 Demand response strategy.

Signal Initiation: Utilities or grid operators monitor the power demand and identify peak load conditions, excess supply, or price fluctuations.

Communication with Consumers: Utilities send DR signals via smart meters, IoT devices, mobile apps, or automated energy management systems.

Consumer Response: Consumers react by either reducing their load, shifting usage to off-peak hours, or increasing consumption during periods of excess renewable generation.

Compensation Mechanism: Consumers receive incentives, rebates, or reduced tariffs for participating in DR programs.

Demand Response is classified into three major categories i.e.

(A) Price Based Demand Response (PBDR),

(B) Incentive Based Demand Response (IBDR) and

(C) Ancillary Services Demand Response (ASDR).

This is presented in Figure 6.4 with further subclassifications.

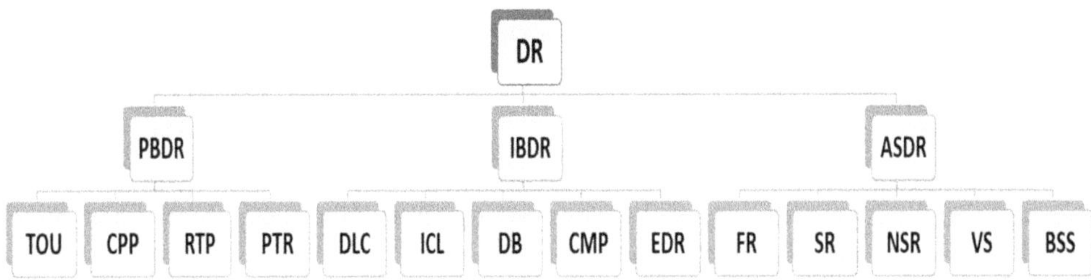

Figure 6.4 Classification of Demand Response programs.

A. Price-Based Demand Response (PBDR)

Price-Based Demand Response (PBDR) is a demand-side management (DSM) strategy where electricity prices vary based on supply and demand conditions. Consumers respond to these price signals by adjusting their electricity usage, reducing peak demand, and optimizing energy consumption patterns. This strategy enhances grid stability, promotes energy efficiency, and reduces the need for additional power generation during peak hours.

PBDR relies on economic incentives rather than direct control by utilities. Consumers receive real-time or scheduled price information, encouraging them to shift or reduce their electricity usage during high-price periods and take advantage of lower rates during off-peak hours. Price-Based Demand Response can be classified into the following categories:

- Time-of-Use (TOU) Pricing
- Critical Peak Pricing (CPP)

- Real-Time Pricing (RTP)
- Peak Time Rebate (PTR)

Time-of-Use (TOU) Pricing

TOU pricing divides the day into different time blocks, each with a fixed electricity price. The price varies based on peak, off-peak, and sometimes mid-peak periods.

- **Peak hours:** Higher electricity prices to discourage consumption.
- **Off-peak hours:** Lower electricity prices to encourage energy use when demand is low.
- **Mid-peak hours (optional):** Intermediate pricing between peak and off-peak periods.

Consumers can plan their electricity usage to avoid peak-hour pricing by shifting flexible loads such as dishwashing, water heating, and EV charging to off-peak hours.

Advantages

- Simple and predictable pricing structure.
- Encourages behavioural changes without requiring advanced technology.
- Reduces peak demand and strain on the grid.

Challenges

- Requires consumer awareness and willingness to shift loads.
- Fixed time slots may not reflect real-time grid conditions.

Critical Peak Pricing (CPP): CPP introduces significantly higher electricity prices during extreme peak demand events, typically occurring a few times per year. The utility notifies customers in advance about these critical peak events. Critical Peak Pricing (CPP) operates by identifying periods of extreme electricity demand, such as during heatwaves or severe cold spells, when the power grid is under significant strain. During these critical peak events, utilities implement substantially higher electricity rates to encourage consumers to reduce their usage, thereby alleviating stress on the grid. Outside of these critical peak periods, electricity prices remain comparable to Time-of-Use (TOU) rates or may even be lower, ensuring that consumers benefit from cost savings when demand is not at its peak.

Advantages

- Highly effective at reducing peak loads during extreme demand conditions.
- Allows utilities to manage supply shortages efficiently.
- Encourages proactive energy conservation.

Challenges

- Requires consumers to react to event-based pricing on short notice.
- May cause discomfort if essential appliances must be turned off during peak hours.

Real-Time Pricing (RTP): RTP dynamically adjusts electricity prices based on real-time wholesale market rates or grid conditions. Prices fluctuate hourly or even at smaller intervals, directly reflecting supply-demand changes. Unlike fixed or pre-scheduled pricing structures, RTP allows electricity rates to change hourly or even at sub-hourly intervals, reflecting the actual supply-demand balance. During periods of high electricity demand, prices rise, encouraging consumers to reduce their usage, while during periods of low demand, prices drop, incentivizing increased consumption.

Utilities provide real-time price updates through mobile apps, emails, or smart meter notifications. This transparency allows users to make informed decisions about their electricity usage, helping them optimize costs. Many households and businesses integrate automated energy management systems, such as smart appliances and AI-driven load control mechanisms, which respond dynamically to price changes. These systems automatically adjust energy consumption patterns to take advantage of lower-priced periods, improving efficiency and cost savings.

Advantages

- Provides the most accurate price signal based on actual grid conditions.
- Encourages consumers to optimize energy usage using automation and smart appliances.
- Enhances renewable energy integration by shifting demand to times of high solar/wind generation.

Challenges

- Requires advanced smart grid infrastructure.
- Consumers may find it difficult to adjust manually without automation.
- Price volatility may cause uncertainty for consumers.

Peak Time Rebate (PTR): Unlike TOU, CPP, and RTP, which penalize consumption during peak hours, PTR rewards consumers who reduce their electricity usage during designated peak periods. Consumers receive financial incentives or rebates for lowering their consumption compared to their typical usage baseline. Utilities first identify these high-demand periods, which typically occur during extreme weather conditions or when the power grid is under significant strain. Customers are then encouraged to lower their consumption below a predefined baseline during these peak times. Participants who successfully reduce their electricity usage receive financial incentives in the form of rebates or bill credits, effectively rewarding them for contributing to grid stability.

Advantages

- Encourages voluntary participation without penalizing consumers.
- Easy to implement alongside TOU or CPP.
- Increases consumer engagement in demand response programs.

Challenges

- Requires accurate baseline consumption data.
- Utilities must fund rebates, which may impact revenue models.

The performance comparison of various demand side managements under this catogery is presented in Table 6.2.

Table 6.2 Comparison of PBDR Categories.

Pricing Type	How Prices Change	Consumer Response	Effectiveness in Peak Reduction	Technology Requirement
TOU Pricing	Fixed schedule	Behavioural shift	Moderate	Low
CPP	Event-based	Peak avoidance	High	Moderate
RTP	Hourly/dynamic	Automated shift	High	High
PTR	Rebate incentive	Voluntary shift	Moderate	Low

B. Incentive-Based Demand Response (IBDR)

Incentive-Based Demand Response refers to programs where consumers voluntarily reduce or shift their electricity consumption in exchange for financial incentives. Unlike price-based demand response, which relies on dynamic pricing to influence behaviour. IBDR directly compensates participants for reducing demand during peak periods or grid emergencies. These programs help utilities maintain grid reliability, prevent blackouts, and reduce operational costs.

IBDR programs are typically managed by utilities, system operators, or third-party aggregators, who notify participants when demand reduction is needed. Participants may be industrial, commercial, or residential customers, and their responses can be manual (e.g., turning off non-essential equipment) or automated through smart grid technology.

IBDR is broadly categorized into Direct Load Control (DLC), Interruptible/Curtailable Load (ICL), Demand Bidding/Buyback (DB), Capacity Market Programs (CMP) and Emergency Demand Response (EDR). Each category operates differently, offering unique benefits and challenges.

Direct Load Control (DLC): Direct Load Control programs allow utilities to remotely control specific customer appliances, such as air conditioners, water heaters, and industrial loads, to reduce electricity demand during peak hours or emergencies. Customers enrol voluntarily and receive incentives, such as bill discounts or rebates, in exchange for allowing temporary control over their appliances.

Interruptible Curtailable Load (ICL): ICL programs target large industrial and commercial consumers, who agree to reduce their electricity consumption upon request from the utility. In return, they receive lower electricity rates or financial incentives. Unlike DLC, customers manually reduce their load based on prior agreements.

Demand Bidding/Buyback Programs (DB): Under this program, large consumers or aggregators submit bids to reduce electricity consumption in response to price signals or grid requirements. If the bid is accepted, they must reduce their load accordingly and receive payments for the curtailed electricity.

Capacity Market Programs (CMP): Consumers commit to reducing load at predetermined times to help utilities meet capacity requirements. They receive payments based on their availability rather than actual reductions. If they fail to meet commitments, penalties may apply.

Emergency Demand Response (EDR): During grid emergencies, utilities request voluntary load reductions from consumers. Participants receive financial incentives if they successfully reduce demand, but there are no penalties for non-participation.

A comaprative analysis of various methods under IBDR is presented in Table 6.3

Table 6.3 Comparison of IBDR Categories.

Category	Target Consumers	Control Method	Incentive Structure	Response Time	Implementation Challenges
Direct Load Control (DLC)	Residential, Small Commercial	Remote Utility Control	Bill Discounts, Rebates	Seconds to Minutes	Consumer reluctance, infrastructure costs
Interruptible/Curtailable Load (ICL)	Large Industrial, Commercial	Manual Consumer Action	Lower Tariffs, Incentive Payments	Minutes to Hours	Potential productivity loss, scheduling complexity
Demand Bidding/Buyback	Large Commercial, Industrial	Market-Based Bidding	Compensation Based on Market Prices	Minutes to Hours	Requires forecasting, complex participation
Capacity Market Programs	Large Industrial, Commercial	Pre-committed Load Reduction	Fixed Capacity Payments	Hours to Days	Penalties for non-compliance, demand prediction required
Emergency Demand Response (EDR)	Industrial, Commercial, Residential	Voluntary Load Reduction	Incentives for Participation	Minutes to Hours	Uncertain participation rates, consumer awareness needed

C. Ancillary Service Demand Response (ASDR)

Ancillary services in the power system refer to the support services necessary to maintain grid stability, reliability, and efficiency. Demand response (DR) plays a crucial role in providing ancillary services by allowing consumers to adjust their electricity usage in response to grid conditions. Unlike traditional demand response programs that focus on peak demand reduction, ancillary services DR supports real-time grid operations by responding to frequency fluctuations, voltage stability issues, and reserve capacity requirements.

Ancillary services demand response works through automated control systems, where participating consumers allow utilities or grid operators to manage their loads dynamically. These programs are typically utilized by large industrial and commercial consumers who can provide fast and reliable load adjustments. The integration of smart grid technologies, such as Advanced Metering Infrastructure (AMI) and automated demand response (ADR) systems, enables real-time communication and rapid response to grid signals. ASDR is broadly classified into the following categories:

Frequency Regulation (FR): Frequency regulation demand response helps maintain the stability of the power grid by adjusting load consumption in response to frequency deviations. Power grids operate at a standard frequency (e.g., 50 Hz or 60 Hz), and any deviation can lead to instability or even blackouts. DR participants provide rapid load adjustments, either increasing or decreasing consumption, to help balance supply and demand in real-time.

Spinning Reserves (SR): Spinning reserves refer to the backup generation or demand reduction capacity that can be activated within seconds to minutes in response to unexpected supply shortages or grid contingencies. For example, demand response participants can offer a more cost-effective and environmentally friendly alternative by reducing their load instead of increasing generation.

Non-Spinning Reserves (NSR): Non-spinning reserves are those spinning reserves which involve resources that are not online and require a slightly longer response time (typically 10–30 minutes). For example, Demand response participants commit to reducing or shifting load within the required timeframe to support grid stability.

Voltage Support (VS): Voltage support services ensure that the power system maintains proper voltage levels across the grid. Demand response contributes to voltage support by adjusting reactive power loads or reducing energy consumption in those areas experiencing voltage instability. For examples, large commercial buildings with smart inverters and capacitor banks adjust their power factor to support voltage stability, reducing stress on the grid.

Black Start Support (BSS): Black start services refer to the capability of restoring power to the grid after a blackout without relying on external power sources. Demand response plays a role by reducing initial demand during the grid restoration process, allowing critical power infrastructure to be restarted efficiently. For example, industrial consumers with on-site backup generators disconnect from the grid and operate independently during black start events, reducing strain on the power restoration process. Table-6.4 presents a comparison of various ASDR schemes.

Table 6.4 Comparison of Ancillary Services Demand Response Categories.

Category	Purpose	Response Time	Example Participants	Key Benefit
Frequency Regulation	Maintain grid frequency stability	Seconds	Industrial refrigeration, pumps	Ensures real-time supply-demand balance
Spinning Reserves	Provide immediate backup capacity	Seconds-minutes	HVAC systems, commercial buildings	Prevents sudden grid failures
Non-Spinning Reserves	Backup capacity with delayed response	10–30 minutes	Industrial processes, manufacturing	Cost-effective compared to spinning reserves
Voltage Support	Maintain voltage stability	Seconds-minutes	Smart inverters, capacitor banks	Reduces grid stress and power losses
Black Start Support	Assist in grid restoration	Minutes-hours	Backup generators, large facilities	Aids in faster recovery from blackouts

6.4.3 Distributed Energy Resources

A Distributed Energy Resource (DER) is a small-scale power generation system connected to the electrical grid. As DER adoption grows, it offers several benefits, such as lower costs, better reliability, higher efficiency, improved power quality, and greater energy independence.

DERs are classified into Distributed Generation (DG) and Electric Energy Storage, as shown in Figure 6.5. DERs use both renewable and conventional energy sources. Renewable sources include solar, wind, hydro etc. while conventional sources include diesel generators, gas turbines, microturbines, and combustion turbines. Despite being unpredictable and intermittent, Renewable Energy Sources (RES) like solar, wind, biomass, geothermal, and small hydro turbines are becoming more popular.

Energy storage systems (ESS) convert electricity into mechanical, electrochemical, electromagnetic, thermodynamic, or chemical energy. Different storage methods, their costs, principles, and benefits. Even with the rise of renewables, conventional sources still dominate due to their availability and established infrastructure.

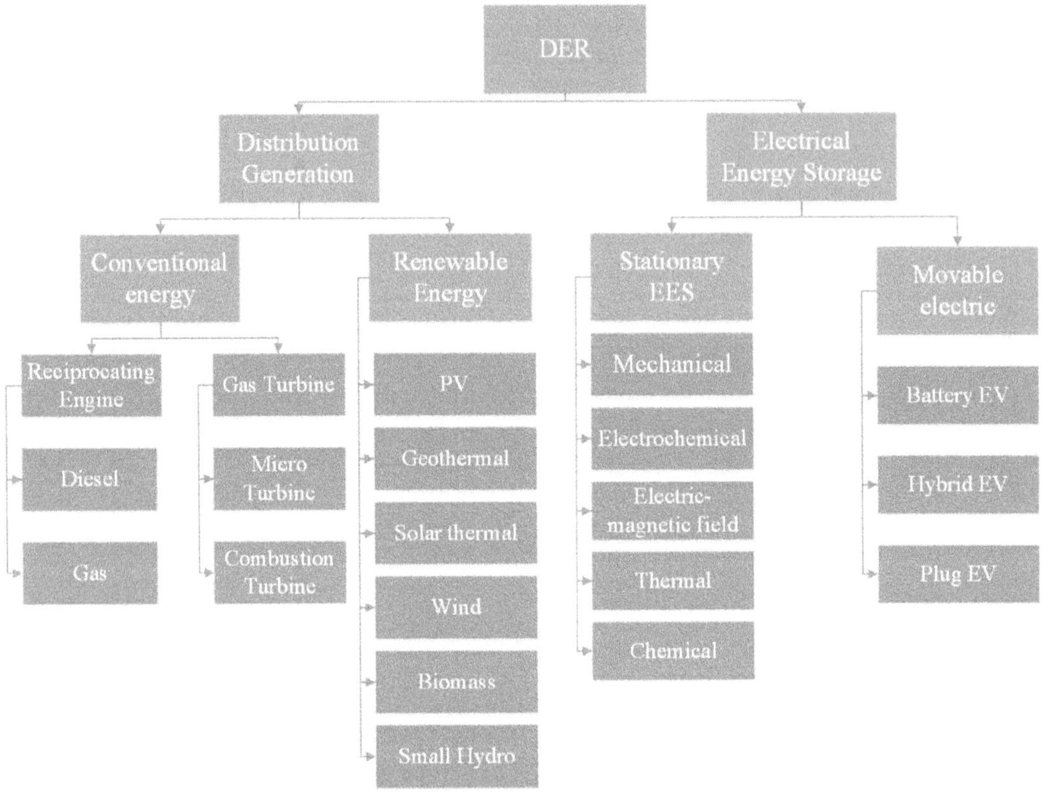

Figure 6.5 Classification of DER.

6.5 Implementation of DSM

Demand side management (DSM) involves several load shaping functions to manage electricity consumption efficiently which include (Figure-6.6):

- Peak Shaving (clipping)
- Valley Filling
- Load Shifting
- Strategic Conservation
- Load Growth
- Flexible Load Curve

Figure 6.6 presents various load shaping functions used in DSM. These techniques are also known as demand management or load management strategies. Technical aspects of each strategy is discussed in following subsections.

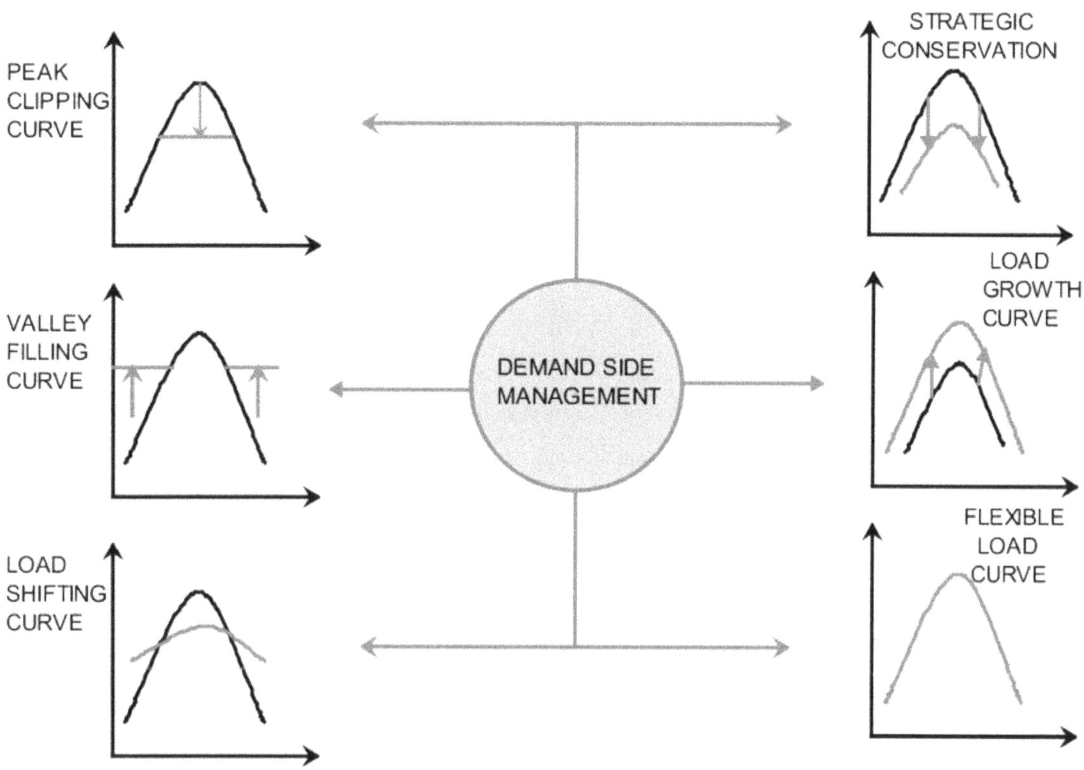

Figure 6.6 Demand side management implementation techniques.

6.5.1 Peak Shaving

Peak shaving is a critical demand-side management (DSM) technique aimed at reducing electricity demand during peak hours. It helps utilities and consumers optimize energy usage, lower electricity costs, and enhance grid stability. This technique involves reducing energy consumption to flatten peak loads, thereby minimizing the need for additional power generation capacity. This technique is helpful when capital costs for the installation of new power generation units are quite high. Conceptual representation of peak shaving is shown in Figure 6.7.

This plot visually represents the concept of peak shaving. The red dashed curve shows the original power demand with a sharp peak. The blue curve represents the demand after peak shaving. The dotted grey line indicates the new, lower peak level. This can be effectively implemented by integrating smart technologies such as IoT, AI, and battery storage and involving consumers. Peak shaving can achieved by the following techniques. And Table 6.5 presents the performance comparison various Peak Shaving methods.

Reducing load: Non-essential loads are turned off or shifted to off-peak periods.

Utilizing Energy Storage Systems: Battery storage discharges stored energy to supply the increased demand.

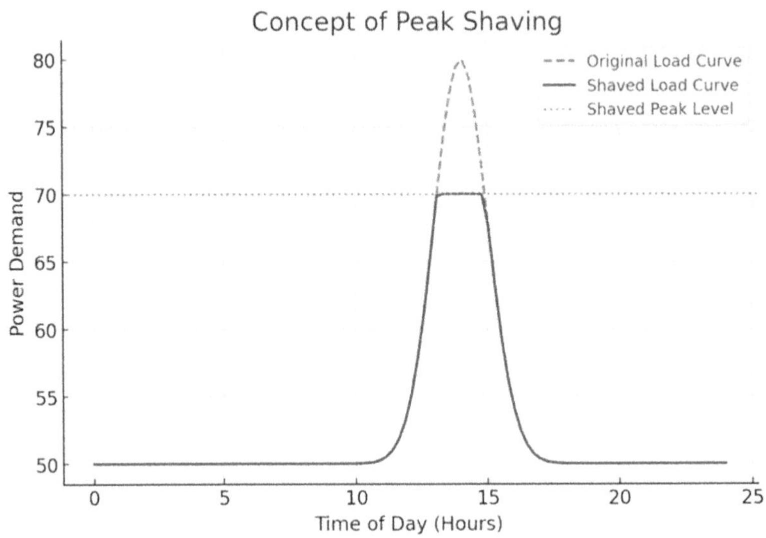

Figure 6.7 Principle of peak shaving.

Using On-site Generation: Distributed generation (e.g., solar, diesel generators) supplements grid power.

Incentive Programs: Consumers adjust their electricity consumption in response to price signals or incentives.

6.5.2 Valley Filling

Valleys are the periods of low demands experienced regarding base loads. Higher levels of power generation during such periods may be at a loss if demands are not raised due to unused energies. Valley filling is a demand-side management (DSM) strategy aimed at increasing electricity consumption during off-peak periods by encouraging consumers to use their appliances during these periods to avail of reduced tariffs. It refers to increasing load consumption during off-peak periods and thereby flattening the load curve. This technique helps optimize grid utilization, improve efficiency, and reduce dependency on costly peak-hour generation. Figure-6.8 presents a typical valley filling concept.

Table 6.5 Comparison of Peak Shaving Strategies.

Method	Description	Advantages	Challenges
Load Reduction	Turning off non-essential loads	Cost-effective, immediate response	May impact operations
Energy Storage	Batteries discharge stored energy	Reliable, smooth transition	High upfront cost
On-Site Generation	Using local power sources (e.g., solar)	Reduces grid dependency	Fuel/maintenance costs (if generators)
Demand Response	Consumers adjust usage based on price patterns	Incentive-based, no infrastructure cost	Requires consumer's active participation

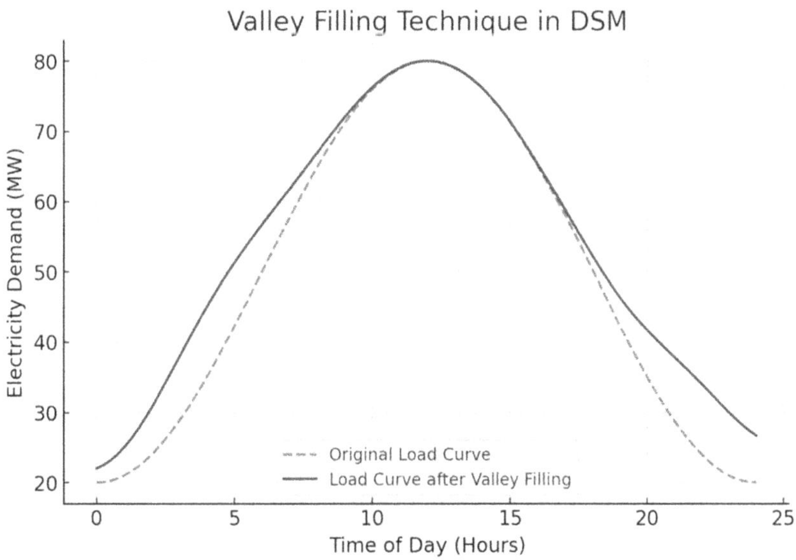

Figure 6.8 Valley Filling Concept.

Valley filling helps integrating Renewable Energy Sources (RES) like solar and wind. Given the nature of these energy sources and the costs involved, curtailing generation may not be the best option. Instead, energy storage and flexible (transferable) loads can help absorb excess power. The following techniques are used for Valley filling

Demand Curve Adjustment: During low-demand hours (valley periods), additional loads are encouraged to consume power, raising the baseline consumption level.

Load Management Techniques: Automated systems, such as smart grids and IoT-enabled devices, control and distribute loads dynamically between renewable and conventional generation to feed the connected load.

Energy Storage Systems (ESS): Excess electricity generated during low-demand periods is stored in batteries or pumped hydro storage and used later when needed.

Time-of-Use (TOU) Pricing: Consumers are incentivized with lower electricity rates during off-peak hours to shift their energy-intensive activities to these times.

Industrial and Commercial Load Participation: Industries and large commercial users schedule high-energy processes (e.g., water heating, refrigeration, or manufacturing) during valley periods.

6.5.3 Load Shifting

Load shifting deals with the shifting of loads from peak to off-peak periods. This facilitates the decrease in peak demand, but the total electricity consumption remains the same. Its applications include customer load shifting using stored electricity and or shifting flexible loads from high demand hours to low demand hours (e.g., shifting the washing machine loads from 9:00 AM (peak period) to 2:00 PM (off-peak period)). The load shifting eases utilities' work by increasing system stability and reliability. This technique helps balance the load on the electrical grid, reducing the need for additional generation capacity and enhancing grid stability.Various studies have explored load shifting from different perspectives, including optimization techniques, consumer behaviour, and its impact on energy markets.

Load shifting offers several significant benefits in demand-side management. One of the primary advantages is the reduction of peak demand, which helps lower stress on power grids and minimizes the risk of blackouts. By redistributing energy consumption to off-peak hours, utilities can maintain a more stable and efficient power supply. Additionally, load shifting results in cost savings for both consumers and utility providers. By taking advantage of lower electricity rates during off-peak periods, consumers can reduce their energy bills, while utilities benefit from decreased operational costs and reduced strain on infrastructure.

Another key benefit of load shifting is its role in increasing the utilization of renewable energy sources. By aligning energy demand with periods of high solar and wind power generation, this strategy helps integrate renewable energy more effectively into the grid. As a result, reliance on fossil fuel-based power plants during peak hours is reduced, leading to a lower carbon footprint and a more sustainable energy system.

Despite these benefits, several challenges must be addressed for successful implementation. One major challenge is consumer engagement, as load shifting requires active participation and trust in demand-side management programs. Without consumer willingness to adjust their energy usage patterns, the effectiveness of load shifting can be limited. Additionally, infrastructure requirements pose another obstacle, as the implementation of smart meters, demand response systems, and automation technologies is necessary to facilitate efficient load shifting.

Regulatory and policy barriers also present challenges, as some regions lack the necessary policies and incentives to promote load shifting effectively. Without proper incentives, consumers and businesses may be reluctant to adopt this energy management strategy. Furthermore, utilities may experience economic impacts, as shifting demand away from peak hours can lead to potential revenue losses. If not managed properly, this could create financial challenges for utility providers, requiring adjustments to pricing structures and business models.

6.5.4 Strategic Conversion

Strategic conservation refers to the systematic reduction of energy consumption while maintaining operational efficiency and service quality. It employs utility-designed programs to induce a change in the shape of the load curve. These programs target the customers by offering them certain benefits to improve equipment efficiency, reduce waste, and change their consumption patterns. It is a proactive approach in Demand Side Management (DSM) that involves implementing energy-saving measures, upgrading technologies, and optimizing usage patterns. The main goal is to reduce energy demand without compromising productivity or comfort.

Several studies highlight the effectiveness of strategic conservation in energy management. A chapter by Grigoras et al in 2020 discussed the role of smart meters in tracking real-time energy usage, allowing for the implementation of effective conservation strategies. By providing consumers with detailed insights into their energy consumption patterns, smart meters enable informed decision-making that leads to reduced energy waste and optimized usage. For example, a commercial building installs a smart energy management system that adjusts lighting, heating, and cooling based on occupancy and weather conditions. By doing so, the building reduces its overall energy consumption by 20% annually without affecting comfort levels. Strategic conservation works through various energy efficiency measures and behavioural changes which include:

Energy-Efficient Appliances: Upgrading to devices with higher energy ratings (e.g., LED lighting, smart HVAC systems).

Building Automation: Using smart sensors and automation systems to optimize energy usage.

Process Optimization: Adjusting industrial and commercial processes to use energy more efficiently.

Demand Response Programs: Encouraging consumers to shift energy-intensive tasks to non-peak hours.

Renewable Integration: Utilizing solar, wind, and battery storage to reduce dependency on the grid.

6.5.5 Load Growth

It refers to increasing the loads through strategic additions beyond valley filling through intelligent systems and processes and more efficient equipment. Unlike demand reduction strategies, load growth focuses on managing and optimizing the expansion of energy consumption while ensuring grid stability, efficiency, and sustainability. The goal is to accommodate the rising electricity demand without overloading the grid or increasing reliance on inefficient energy sources. An example of load growth management is seen in the widespread adoption of electric vehicles (EVs). As EV penetration increases, utilities must expand their charging infrastructure, introduce demand response programs, and encourage off-peak charging to prevent grid overload.

Load growth strategies in DSM involve a combination of energy efficiency measures, demand response programs, and infrastructure expansion to meet increasing energy needs. It works through the following approaches.

Optimized Infrastructure Expansion: Utilities plan and invest in grid upgrades, such as transmission and distribution network enhancements, to accommodate growing demand.

Integration of Renewable Energy: The increase in electricity consumption is managed by incorporating more renewable energy sources, such as solar and wind, reducing dependency on fossil fuels.

Electrification of Sectors: Industries, transportation (electric vehicles), and heating systems transition from conventional energy sources to electricity, contributing to controlled load growth.

Smart Grid Implementation: Advanced metering, automation, and AI-driven forecasting help utilities anticipate demand patterns and ensure efficient energy distribution.

Incentivized Load Management: Governments and utilities encourage off-peak electricity usage through pricing schemes, reducing strain on the grid during high-demand periods.

Load growth in Demand Side Management (DSM) brings several significant benefits that contribute to economic, technological, and environmental advancements. One of the primary advantages is economic development, as a stable and sufficient electricity supply supports industrial and commercial expansion. With a reliable energy infrastructure, businesses can operate efficiently, attract investments, and contribute to job creation, fostering overall economic growth.

Another crucial benefit is enhanced grid utilization. By optimizing the use of existing energy generation and distribution systems, load growth ensures that infrastructure operates at its maximum efficiency. This reduces energy wastage and enhances the overall performance of the power grid. Additionally, increased load growth enables better integration of renewable energy sources. A larger, well-managed energy demand can accommodate more sustainable power generation, reducing reliance on fossil fuels and promoting a cleaner environment.

Furthermore, load growth encourages technological advancements by driving investment in smart grid technologies and energy-efficient appliances. As electricity demand increases, there is a greater incentive to develop innovative solutions that optimize energy consumption and improve grid reliability.

Lastly, better utility revenue management is another key advantage. With a clearer understanding of future electricity consumption trends, energy providers can plan more effectively, balance costs, and ensure long-term financial stability. By leveraging these benefits, DSM strategies can create a more resilient and sustainable energy ecosystem.

6.5.6 Flexible Load Curve

A Flexible Load Strategy in Demand-Side Management (DSM) refers to the ability of consumers to adjust their electricity consumption patterns in response to grid conditions, energy prices, or external incentives. This approach allows demand to be dynamically modified to enhance grid efficiency, minimize costs, and integrate renewable energy sources more effectively. Flexible loads are primarily found in industries, commercial buildings, and residential sectors where energy-intensive processes, heating/cooling systems, and smart appliances can be controlled dynamically. The consumers who are willingly offering flexibility in their load usage pattern are identified for being offered various incentives for their cooperation.

For example, consider an industrial facility that operates large-scale machinery. By integrating demand response programs and automation systems, the facility can reduce its power consumption during peak demand hours and shift energy-intensive operations (e.g., cooling, heating, or charging electric vehicles) to off-peak periods. Similarly, in residential settings, smart thermostats and home energy management systems can adjust heating and cooling schedules based on real-time electricity pricing, optimizing energy usage without compromising comfort.

The flexible load strategy operates by leveraging demand response programs, energy storage, and smart automation to manage electricity consumption dynamically based on real-time grid requirements. One of the primary mechanisms is load shifting, where consumers adjust their electricity usage from peak hours to off-peak periods. By doing so, grid congestion is reduced, leading to improved stability and efficiency. Another essential approach is load shedding, which involves temporarily reducing or turning off non-essential loads during peak demand periods to prevent potential grid instability.

Additionally, the strategy includes load boosting, where certain loads, such as energy storage charging or water heating, are increased when there is an excess supply of renewable energy. This helps in maximizing the utilization of surplus power and reducing waste. Furthermore, real-time demand response plays a crucial role, as smart grids and automation systems dynamically adjust loads based on signals from utilities, energy prices, or frequency variations in the power grid. This adaptive control mechanism ensures a balanced and efficient energy distribution, benefiting both consumers and power providers.

Implementing a flexible load strategy in Demand Side Management (DSM) offers numerous advantages. One of the key benefits is grid stability, as it helps balance supply and demand, preventing blackouts and frequency fluctuations. By dynamically adjusting energy consumption, the grid can operate more reliably, ensuring a steady and secure power supply.

Another significant advantage is cost savings for both consumers and utilities. Consumers can reduce their electricity expenses by shifting their usage to periods when electricity prices are lower. At the same time, utilities benefit by avoiding costly infrastructure expansions, as flexible load management reduces the strain on the power grid during peak hours.

A flexible load strategy also plays a crucial role in enhancing renewable energy utilization. As solar and wind power generation fluctuates based on weather conditions, aligning energy consumption with periods of high renewable output allows for better integration of these clean energy sources. This helps to maximize their contribution to the power grid while minimizing curtailment.

Furthermore, the strategy contributes to reducing carbon emissions by lowering dependence on fossil fuel-based peak power plants. By shifting loads to times when cleaner energy sources are available, the overall energy mix becomes more sustainable, supporting global efforts to combat climate change.

6.6 Mathematical Formulation of DSM Optimization

Demand Side Management (DSM) optimization involves formulating an objective function subject to constraints that govern electricity consumption, cost, grid stability, and user comfort. The formulation of a DSM optimization problem depends on the specific objectives, such as cost minimization, peak demand reduction, or energy efficiency improvement.

6.6.1 General DSM Optimization Model

A DSM optimization problem can be expressed as:

$$\min F(x) \quad \text{subject to } G(x) \leq 0, H(x) = 0 \qquad .....(6.1)$$

where

F(x) is the objective function, representing total cost, peak demand, or energy efficiency to be optimized.

G(x) represents the inequality constraints, such as power limits, voltage stability, and user comfort constraints.

H(x) represents the equality constraints, such as power balance and demand-response participation rules.

6.6.2 Common Objective Functions in DSM Optimization

Cost Minimization: The most common DSM objective is electricity bill minimization for consumers or operational cost minimization for utilities. The total electricity cost over a time horizon T can be represented as:

$$\min \sum_{t=1}^{T} P_t \cdot L_t \qquad .....(6.2)$$

where

P_t = Electricity price at time t (price-based demand response).

L_t = Total load consumption at time t.

If demand response incentives are included, the objective function becomes:

$$\min \sum_{t=1}^{T} P_t \cdot L_t - R_t \qquad .....(6.3)$$

where R_t represents the reward given to consumers for reducing peak demand.

Peak Demand Reduction (Load Shifting): DSM can aim to flatten the demand curve by reducing peak consumption and shifting loads to off-peak hours. This can be formulated as:

$$\min \max_{t \in T} L_t \qquad .....(6.4)$$

which minimizes the highest peak demand over the time horizon T. Alternatively, Peak-to-Average Ratio (PAR) can be minimized:

$$min \frac{max\ L_t}{\frac{1}{T}\sum_{t=1}^{T} L_t} \qquad .....(6.5)$$

where

$max\ L_t$ = maximum load over time.

$\frac{1}{T}\sum_{t=1}^{T} L_t$ = average load over time.

Energy Efficiency Optimization: DSM can also be formulated to maximize energy efficiency by minimizing total energy consumption:

$$\min \sum_{t=1}^{T} L_t \qquad .....(6.6)$$

This function ensures that the overall energy demand is reduced while meeting necessary load requirements.

6.6.3 Constraints of DSM Optimization

DSM is governed by multiple constraints, including technical (load balance, power consumption limits), economic (cost minimization), environmental (renewable energy integration), behavioural (consumer comfort), and regulatory (market policies). These constraints impact in optimizing energy consumption, improving grid reliability, and integrating renewable energy. Each of these constraints plays a crucial role in determining the effectiveness of DSM strategies. Mathematical models incorporating these constraints help to optimize DSM planning and ensures benefits to both utilities and consumers while maintaining grid stability. Each of these constraints can be expressed mathematically to illustrate their effect on DSM strategies as follows.

Load balance constraint: One of the fundamental constraints in DSM is maintaining the balance between electricity supply and demand at any given time. This constraint ensures that total electricity consumption does not exceed the available supply.

$$\sum_{i=1}^{N} P_i^t \leq P_{availabale}(t), \ \forall t \qquad .....(6.7)$$

where,

$P_i(t)$ = Power consumption of consumer i at time t.

$P_{available}(t)$ = Total available power supply at time t.

N = Total number of consumers.

This constraint prevents overloading the power system by ensuring that total demand does not surpass the generation capacity.

Power Consumption Constraint (Load Shifting and Curtailment Limits): Load shifting and demand curtailment strategies in DSM must adhere to power consumption limits to ensure that the shifted demand does not exceed the new constraint.

$$P_i^{min} \leq P_i \leq P_i^{max}(t) \quad .....(6.8)$$

where

P_i^{min} = Minimum allowable power consumption for consumer.

P_i^{max} = Maximum allowable power consumption for consumer.

This ensures that the consumer's energy consumption remains within acceptable limits, preventing excessive reductions that could disrupt operations.

Cost Constraint (Consumer's Energy Bill Limit): Economic constraints limit the extent to which consumers can participate in DSM programs, as they seek to minimize their electricity bills.

$$C_{total} = \sum_{t=1}^{T} P(t).C(t) \leq C_{max} \quad .....(6.9)$$

where,

C_{total} = Total energy cost for the consumer.

C(t) = Price per unit energy at time t.

P(t) = Power consumption at time t.

C_{max} = Maximum allowable electricity cost for the consumer.

This constraint ensures that particular DSM strategy do not lead to excessive energy costs for consumers.

Renewable Energy Integration Constraint: DSM is often used to improve the integration of renewable energy sources, but this must consider the variability of renewable generation.

$$P_{demand}(t) \leq P_{renewable}(t) + P_{grid}(t) \quad .....(6.10)$$

where,

$P_{demand}(t)$ = Total power demand at time t.

$P_{renewable}(t)$ = Available renewable energy at time t.

$P_{grid}(t)$ = Additional power supplied from the grid at time t.

This constraint ensures that demand-side adjustments do not exceed the available renewable energy supply, promoting sustainable energy use.

Consumer comfort constraint: For DSM programs involving load reduction in residential or commercial buildings, constraints related to consumer comfort must be maintained (e.g., temperature in HVAC systems).

$$T_{min} \leq T(t) \leq T_{max}, \quad \forall t \qquad .....(6.11)$$

where

T(t) = Temperature at time t.

T_{min}, T_{max}= Acceptable temperature range for consumer comfort.

This constraint ensures that energy-saving measures, such as reducing HVAC loads, do not compromise user comfort beyond acceptable limits.

Demand Response Participation Constraint: Not all consumers can always participate in DSM programs due to operational requirements.

$$X_i \in \{0, 1\}, \forall I \qquad .....(6.12)$$

where $X_i = 1$, if consumer i is participating in DSM, otherwise $X_i = 0$.

This binary constraint ensures that only eligible consumers are included in DSM strategies.

Energy storage constraint: For DSM programs utilizing battery storage systems, the state of charge (SoC) must remain within operational limits.

$$E_{min} \leq E(t) \leq E_{max}, \forall t \qquad .....(6.13)$$

where

E(t) = Energy stored in the battery at time t.

E_{min} = Minimum allowable stored energy.

E_{max} = Maximum allowable stored energy.

This ensures that battery storage is neither overcharged nor completely depleted, allowing for efficient energy shifting.

Regulatory and Market Constraints: DSM programmes must comply with energy market regulations and policies.

$$\sum_{i=1}^{N} P_i^{reduced} \geq P_i^{target}, \forall t \qquad .....(6.14)$$

where

$P_i^{reduced}$ = Demand reduction by consumer i.

P_{target} = Minimum required demand reduction set by regulatory authorities.

This ensures that DSM programs meet policy objectives for demand reduction and grid stability.

Battery constraints: The manufacturer's recommended battery level range must be maintained, leading to the implementation of the following constraints.

$$SOC_{min}(t) \leq SOC(t) \leq SOC_{max}(t) \quad .....(6.15)$$

$$SOC(t) = \frac{E_{bat}^{t}}{E_{bat}^{cap}} \quad .....(6.16)$$

Battery maximum charging and discharging power limit can be represented as:

$$0 \leq \frac{P_{bat}^{ch}(t)}{\eta_{ch}} \leq P_{max}^{ch} \quad .....(6.17)$$

$$0 \leq P_{bat}^{dch}(t).\eta_{ch} \leq P_{max}^{dch} \quad .....(6.18)$$

where

$SOC_{min}(t), SOC_{max}(t)$ = Minimum and maximum state of charge of battery at time t.

$E_{bat}^{cap}, E_{bat}^{t}$ = Capacity of battery and the energy of battery at any time t in KWh.

$P_{bat}^{ch}(t), P_{bat}^{dch}$ = Battery's charging and discharging power respectively at time t.

$P_{max}^{ch}, P_{max}^{dch}$ = Maximum battery's charging and discharging power respectively.

Power safety constraint: This constraint sets an upper limit on the total energy allocated during any given period, ensuring that it never exceeds the maximum available energy from the grid.

$$\sum_{i=1}^{N} \sum_{j=1}^{m} P_{ij}^{t} \leq P_{grid}^{max}(t) \ , \forall i, j \quad .....(6.19)$$

where

P_{ij}^{t} = Total energy required by all running appliances at time t.

$P_{grid}^{max}(t)$ = Maximum energy from grid at that time t.

Prioritization of appliance constraints: In this case, DSM optimization prioritizes appliances based on their importance. A priority index (PI) is used to classify loads, where the PI is inversely proportional to the appliance's load factor and directly proportional to its peak demand.

$$PI = \frac{P_{max}}{loadfactor} \quad .....(6.20)$$

where

PI = Priority Index.

P_{max}= Peak demand of appliance.

6.6.4 Algorithms used for DSM Optimization

Optimization algorithms for solving DSM optimization problem can be broadly classified into five main categories. Each of these categories has distinct approaches to solve DSM-related optimization problems.

- Classical Optimization Algorithms
- Uncertainty-Based Optimization Algorithms
- Evolutionary or Meta-Heuristic Computation Algorithms
- Game Theory-Based Optimization Algorithms
- Soft Computing-Based Optimization Algorithms

Classical Optimization Algorithms: Classical optimization techniques use mathematical formulations to solve DSM problems efficiently. These algorithms typically require deterministic inputs and work well for problems with clearly defined constraints and objectives.

- **Linear Programming (LP):** Used for optimizing load scheduling and demand response by solving linear cost functions.
- **Non-Linear Programming (NLP):** Suitable for complex DSM problems where objective functions and constraints are non-linear.
- **Mixed-Integer Programming (MIP):** Combines both continuous and discrete variables, widely used in DSM for load management.

Uncertainty-Based Optimization Algorithms: These algorithms handle DSM problems where uncertainties exist, such as fluctuating electricity prices and unpredictable renewable energy generation.

- **Stochastic Programming:** Incorporates probabilistic models to optimize DSM solutions under uncertainty.
- **Robust Optimization:** Ensures optimal DSM solutions that remain feasible under worst-case scenarios.
- **Fuzzy Logic-Based Optimization:** Handles imprecise and vague information to improve DSM decision-making.

Evolutionary or Meta-Heuristic Computation Algorithms: Meta-heuristic algorithms are inspired by natural processes and provide near-optimal solutions for complex DSM problems that are difficult to solve using classical methods. Few of them are illustrated as follows.

- **Genetic Algorithms (GA):** Mimics natural selection to optimize load scheduling and energy efficiency.
- **Particle Swarm Optimization (PSO):** Inspired by the movement of birds, used for optimizing demand-side energy resources.

- **Artificial Bee Colony (ABC) Optimization:** Mimics the behaviour of bees to distribute and manage energy loads efficiently.

Game Theory-Based Optimization Algorithms: Game theory-based methods are used in DSM where multiple consumers or stakeholders interact and compete for energy resources.

- **Cooperative Game Theory:** Encourages consumers to participate in DSM programs by sharing cost benefits.
- **Non-Cooperative Game Theory:** Models situations where consumers independently optimize their energy usage based on electricity pricing.
- **Stackelberg Game Theory:** Used in energy markets where utilities set pricing strategies and consumers adjust their demand accordingly.

Soft Computing-Based Optimization Algorithms: Soft computing techniques use intelligent and adaptive approaches to optimize DSM solutions.

- **Artificial Neural Networks (ANNs):** Used for load forecasting and predicting energy consumption patterns.
- **Fuzzy Logic Systems:** Helps in decision-making under uncertain conditions in DSM programs.
- **Machine Learning-Based Optimization:** Uses data-driven models to optimize energy scheduling and demand response programs.

6.7 Challenges of DSM

Demand Side Management (DSM) plays a crucial role in optimizing energy consumption and ensuring grid stability. However, several challenges hinder its effective implementation. These challenges arise due to technical, economic, regulatory, and consumer-related factors. Overcoming these challenges requires a combination of consumer education, advanced technology, supportive policies, and innovative business models. Below is a detailed discussion of the key challenges DSM faces, along with suitable examples and technical issues. By addressing these issues, utilities and consumers can maximize the benefits of DSM, leading to a more efficient and sustainable energy system. Demand Side Management (DSM) plays a critical role in optimizing electricity consumption, integrating renewable energy, and reducing peak demand. However, several challenges arise during its implementation. Below, DSM challenges are explained under different categories with suitable examples :

Load Profile of Appliances: Different appliances have distinct energy consumption patterns that impact DSM implementation. Some appliances, such as refrigerators, require a continuous power supply, while others, such as washing machines, can be scheduled flexibly. Modelling appliance-specific load profiles for accurate demand forecasting and scheduling is complex.

Renewable Energy Integration: Renewable energy sources (e.g., solar, wind) are intermittent and unpredictable, creating difficulties in maintaining grid stability and demand-supply balance. Solar energy generation is highest at noon but residential electricity demand peaks in the evening. Wind power fluctuates based on weather conditions, making it hard to ensure continuous power supply. Energy storage systems (batteries) are required to balance the supply, but they are expensive and have efficiency losses.

Load Categorization and Accurate Forecasting: Classifying loads into shiftable, non-shiftable, and interruptible categories is crucial for DSM effectiveness. However, forecasting demand accurately requires large datasets and sophisticated machine learning models. Industrial loads can be scheduled in advance, while residential loads are unpredictable. Accurate forecasting errors can lead to power outages or wastage. Time-series analysis and AI-driven forecasting models must process large data volumes to predict future demand patterns.

Constraints in DSM Implementation: DSM programs are subject to constraints such as consumer comfort, grid stability, and technical feasibility. Consumers may not be willing to delay appliance usage beyond a specific time (e.g., delaying a washing machine for only 2 hours). The total energy available at any time must not exceed the maximum grid capacity. Mathematical optimization models must incorporate multiple constraints to ensure realistic DSM implementation.

Dynamic Pricing Complexity: Dynamic pricing mechanisms (e.g., Time-of-Use, Real-Time Pricing, Critical Peak Pricing) encourage consumers to shift their usage, but not all consumers respond effectively. Some customers may not have the flexibility to shift loads despite price incentives. Industries may benefit from lower off-peak prices, but residential users may not have enough automation to take advantage. Pricing algorithms must dynamically adjust based on demand fluctuations and consumer participation rates.

Customer Categorization: Different customers (residential, commercial, industrial) have varying energy consumption patterns, requiring tailored DSM strategies. Industrial users can participate in demand response programs by shifting production schedules. Residential users have unpredictable energy usage patterns, making control more difficult. DSM programs need different strategies for each category, requiring detailed customer profiling and behaviour analysis.

Consumer Behaviour and Participation: Many DSM programs rely on consumer participation, but lack of awareness, resistance to change, and inconvenience limit adoption. Consumers may not respond to demand response programs due to a lack of trust in utilities. Households with manual appliances may find it difficult to shift loads compared to those with smart automation. Behavioural models and incentive structures must be designed to encourage participation and make DSM programs user-friendly.

Optimization Techniques for DSM: Finding an optimal solution for DSM involves balancing cost, efficiency, and grid stability, that requires advanced optimization algorithms. Metaheuristic optimization (e.g., genetic algorithms, particle swarm optimization) may take longer time to converge. Game theory-based DSM approaches require extensive computational resources. DSM optimization must be computationally efficient while considering real-time constraints.

Regulatory and Policy Challenges: DSM implementation depends on government policies, incentives, and regulations, which vary across regions. In some countries, there are no incentives for consumers to participate in demand response programs. Regulatory barriers prevent utilities from implementing real-time pricing models. DSM policies must align with market structures, environmental goals, and economic feasibility.

Cyber Security Concerns in DSM: DSM relies on smart grids, IoT devices, and cloud-based platforms, making it vulnerable to cyberattacks. Hacking of smart meters can manipulate energy consumption data. A cyberattack on the demand response system could lead to large-scale grid failures. Secure communication protocols, encryption, and AI-based threat detection are required to safeguard DSM systems.

6.8 Solutions to the Challenges of DSM

Demand Side Management (DSM) implementation faces various technical, economic, and regulatory challenges. The following solutions address these barriers under different key categories:

6.8.1 AI and Machine Learning (ML) based Demand Forecasting

Accurate demand forecasting is essential for DSM optimization, allowing utilities to predict energy consumption patterns and manage loads effectively. Few of the latest solution methods are given as below.

Deep Learning Techniques: Recurrent Neural Networks (RNNs), Long Short-Term Memory (LSTM), and Transformer models analyse historical consumption patterns and external factors (e.g., weather, price fluctuations).

Hybrid Models: AI-based forecasting combined with statistical techniques (e.g., ARIMA, SARIMA) enhances accuracy and reliability.

Reinforcement Learning (RL): Adaptive learning mechanisms help utilities dynamically optimize energy distribution based on real-time demand response feedback.

Smart Meter Integration: AI-driven smart meters provide real-time consumption of data, improving short-term and long-term demand prediction.

6.8.2 Battery Storage Solutions to Balance Renewable Energy Fluctuations

Renewable energy sources like solar and wind are intermittent, requiring efficient storage solutions to ensure a stable energy supply. Different electricity storage management strategies are listed as below.

Advanced Lithium-Ion and Solid-State Batteries: High-efficiency energy storage solutions with better charge retention and longevity.

Vehicle-to-Grid (V2G) Technology: Electric vehicles (EVs) contribute to grid stability by supplying stored energy during peak demand periods.

Hybrid Energy Storage Systems (HESS): Integration of supercapacitors and battery systems optimizes storage performance and enhances response time.

AI-Based Energy Storage Optimization: Machine learning algorithms predict optimal charging/discharging times based on energy prices and grid demand.

6.8.3 Increased Consumer Participation in DSM by Strong Strategies

Consumer participation is crucial for DSM success, requiring effective engagement strategies such as:

Dynamic Pricing Awareness: Educating consumers about Time-of-Use (TOU) and real-time pricing to encourage off-peak energy use.

User-Friendly DSM Apps: Mobile applications providing real-time consumption insights and incentives for load shifting.

Incentive-Based Demand Response (IBDR): Monetary rewards or bill discounts for customers participating in DSM programs.

Behavioural Analytics: AI-driven insights personalize DSM recommendations based on user consumption habits.

6.8.4 Secured Real-Time Communication Networks to Counter Cyber Attacks

Cyber Security is a significant concern in smart grid and DSM implementations, requiring robust security solutions such as:

Blockchain-Based Energy Trading: Decentralized ledgers enhance security, transparency, and reliability in energy transactions.

AI-Powered Intrusion Detection Systems (IDS): Machine learning detects anomalies and prevents cyberattacks on smart grids.

Secure Communication Protocols: Adoption of end-to-end encryption (TLS/SSL) and multi-factor authentication (MFA) for data exchanges.

Decentralized Identity Management: Distributed identity frameworks prevent unauthorized access to DSM control systems.

References

1. Eto, Joseph H, The Past, Present, and Future of U.S. Utility Demand-Side Management Programs, report, energy markets and policy Berkeley lab, 12/1996.
2. Lampropoulos, I.C., Kling, W.L., Ribeiro, P.F., & van den Berg, J. (2013), History of demand side management and classification of demand response control schemes. *2013 IEEE Power & Energy Society General Meeting*, 1-5.
3. Gellings, C.W. Evolving practice of demand-side management. J. Mod. Power Syst. Clean Energy 5, 1–9, 2017.
4. Bakare, M.S., Abdulkarim, A., Zeeshan, M. *et al.* A comprehensive overview on demand side energy management towards smart grids: challenges, solutions, and future direction. *Energy Inform* 6, 4 (2023).
5. Saleem, M.U., Usman, M.R., Usman, M.A., Politis, C., 2022. Design, deployment and performance evaluation of an IoT based smart energy management system for demand side management in smart grid. IEEE Access vol. 10, 15261–15278.
6. Iqbal, S., Sarfraz, M., Ayyub, M., Tariq, M., Chakrabortty, R.K., Ryan, M.J., et al., 2021. A comprehensive review on residential demand side management strategies in smart grid environment. Sustainability vol. 13, 7170.
7. Naimah K., Arirohman, I.D., Zen, M.R., Wicaksono, R.M., Soelami, F.N., Rahmatullah, A., et al., 2023. Analysis of Electrical Energy Consumption in Office Buildings of the Institute Technology of Sumatra in Energy Conservation and Efficiency Efforts. J. Edukasi Elektro vol. 7, 181–194.
8. Nathaphan, S., Therdyothin, A., 2023. Effectiveness evaluation of the energy efficiency and conservation measures for stipulation of Thailand energy management system in factory. J. Clean. Prod. vol. 383, 135442.
9. Zhao, W., Li, H., Wang, S., 2023. Energy performance and energy conservation technologies for high-tech cleanrooms: State of the art and future perspectives. Renew. Sustain. Energy Rev. vol. 183, 113532.
10. Kuang, Y., Tan, R., Zhang, Z., 2023. Saving energy by cleaning the air?: Endogenous energy efficiency and energy conservation potential. Energy Econ. vol. 126, 106946.
11. Subhasis Panda, Sarthak Mohanty, Pravat Kumar Rout, Binod Kumar Sahu, Mohit Bajaj, Hossam M. Zawbaa, Salah Kamel, Residential Demand Side Management model, optimization and future perspective: A review, Energy Reports, Volume 8, 2022, Pages 3727-3766, ISSN 2352-4847, https://doi.org/10.1016/j.egyr.2022.02.300.

12. Joskow, Paul L., and Catherine D. Wolfram. 2012. "Dynamic Pricing of Electricity." *American Economic Review* 102 (3): 381–85.DOI: 10.1257/aer.102.3.381

13. Faruqui, Ahmad & Sergici, Sanem & Akaba, Lamine. (2012). Dynamic Pricing of Electricity for Residential Customers: The Evidence from Michigan. Energy Efficiency. 6. 10.2139/ssrn.2072658.

14. Borenstein, S., & Holland, S. P., On the efficiency of competitive electricity markets with time-invariant retail prices, 2003, The RAND Journal of Economics 36 (3), 469-493.

15. Robert Walton, “ACEEE: Demand response can save 10% off peak demand — if utilities use it”, Report by American Council for an Energy-Efficient Economy, feb. 13 2017.

16. Ali Fattahi, Sriram Dasu, Reza Ahmadi (2022) Peak-Load Energy Management by Direct Load Control Contracts. Management Science 69(5):2788-2813,

17. Aalami, H. & Moghaddam, M. & Yousefi, G. Reza. (2009). Demand response modeling considering Interruptible/Curtailable loads and capacity market programs. Applied Energy. 87. 243-250. 10.1016/j.apenergy.2009.05.041.

18. Subhasis Panda, Sarthak Mohanty, Pravat Kumar Rout, Binod Kumar Sahu, Shubhranshu Mohan Parida, Indu Sekhar Samanta, Mohit Bajaj, Marian Piecha, Vojtech Blazek, Lukas Prokop, A comprehensive review on demand side management and market design for renewable energy support and integration, Energy Reports, Volume 10, 2023, Pages 2228-2250, ISSN 2352-4847

19. Platt G, Paevere P, Higgins A, Grozev G (2014) Electric vehicles: new problem or distributed energy asset? In: Distributed generation and its implications for the utility industry. Elsevier, Amsterdam, pp 335–355.

20. Oskouei MZ, Şeker AA, Tuncel S, Demirbaş E, Gozel T, Hocaoğlu MH, Abapour M, Mohammadi-Ivatloo B (2022) A critical review on the impacts of energy storage systems and demand-side management strategies in the economic operation of renewable-based distribution network. Sustainability 14(4):2110.

21. Kothari, D.P., Nagrath, I.J., 2003. Modern Power System Analysis. Tata McGraw-Hill Education.

22. Dahiru AT, Daud D, Tan CW, Jagun ZT, Samsudin S, Dobi AM. A comprehensive review of demand side management in distributed grids based on real estate perspectives. Environ Sci Pollut Res Int. 2023 Jul;30(34):81984-82013. doi: 10.1007/s11356-023-25146-x. Epub 2023 Jan 18. PMID: 36652076; PMCID: PMC10349782.

23. Zhao, L., Zhang, Y., & Zhang, H. (2022). Deep reinforcement learning-based joint load scheduling for household multi-energy system. *Applied Energy*, 314, 118933.

24. Aslam, S., Javaid, N., Asif, M., Iqbal, U., Iqbal, Z., & Sarwar, M. A. (2018). A mixed integer linear programming based optimal home energy management scheme considering grid-connected microgrids.

25. Zhang D, Shah N, Papageorgiou LG (2013) Efficient energy consumption and operation management in a smart building with microgrid. Energy Convers Manag 74:209–222.

26. Iqbal, S., Sarfraz, M., Ayyub, M., Tariq, M., Chakrabortty, R. K., Ryan, M. J., & Alamri, B. (2021). A Comprehensive Review on Residential Demand Side Management Strategies in Smart Grid Environment. *Sustainability*, 13(13), 7170.
27. Grigoras, G., Ivanov, O., Neagu, B.C., & Kar, P. (2020). Smart Metering Based Strategies for Improving Energy Efficiency in Microgrid Architectures, Control and Protection Methods (pp. 487-508). Springer, Cham.
28. Chojecki, A., Rodak, M., Ambroziak, A., & Borkowski, P. (2020). Energy management system for residential buildings based on fuzzy logic: design and implementation in smart-meter. IET Smart Grid, 3(2), 254–266.
29. K. Karunanithi, S. Saravanan, B.R. Prabakar, S. Kannan, C. Thangaraj, Integration of Demand and Supply Side Management strategies in Generation Expansion Planning, Renewable and Sustainable Energy Reviews, Volume 73, 2017, Pages 966-982, ISSN 1364-0321,
30. Chen Y, Cheng X, Huang K, Liu J, Yi T (2020) Research on residents' electricity behaviour analysis and control strategy optimization. Int J Glob Warm 22(1):91–110.
31. Zhang, Linghao & Tang, Yong & Zhou, Te & Tang, Chao & Liang, Huihui & Zhang, Jie. (2022). Research on flexible smart home appliance load participating in demand side response based on power direct control technology. Energy Reports. 8. 424-434. 10.1016/j.egyr.2022.01.219.
32. Gyamfi, S., & Krumdieck, S. (2011). Price, environment and security: Exploring multi-modal motivation in voluntary residential peak demand response. *Energy Policy*, 39(5), 2993–3004.
33. Subhasis Panda, Sarthak Mohanty, Pravat Kumar Rout, Binod Kumar Sahu, Mohit Bajaj, Hossam M. Zawbaa, Salah Kamel, Residential Demand Side Management model, optimization and future perspective: A review, Energy Reports, Volume 8, 2022, Pages 3727-3766, ISSN 2352-4847,
34. Zhu Z, Tang J, Lambotharan S, Chin WH, Fan Z (2012) An integer linear programming based optimization for home demand-side management in smart grid. Paper presented at the 2012 IEEE PES innovative smart grid technologies (ISGT).
35. Althaher S, Mancarella P, Mutale J (2015) Automated demand response from home energy management system under dynamic pricing and power and comfort constraints. IEEE Trans Smart Grid 6(4):1874–1883.
36. Behrangrad M, Sugihara H, Funaki T (2010) Analyzing the system effects of optimal demand response utilization for reserve procurement and peak clipping. Paper presented at the IEEE PES general meeting.
37. Kwon S, Ntaimo L, Gautam N (2018) Demand response in data centers: integration of server provisioning and power procurement. IEEE Trans Smart Grid 10(5):4928–4938.
38. Liu R-S, Hsu Y-F (2018) A scalable and robust approach to demand side management for smart grids with uncertain renewable power generation and bi-directional energy trading. Int J Electr Power Energy Syst 97:396–407.
39. Hasaranga W, Hemarathne R, Mahawithana M, Sandanuwan M, Hettiarachchi H, Hemapala K (2017) A fuzzy logic based battery SOC level control strategy for smart

micro grid. Paper presented at the 2017 third international conference on advances in electrical, electronics, information, communication and bio-informatics (AEEICB).

40. Javaid N, Javaid S, Abdul W, Ahmed I, Almogren A, Alamri A, Niaz I (2017b) A hybrid genetic wind driven heuristic optimization algorithm for demand side management in smart grid. Energies 10(3):319.
41. Logenthiran T, Srinivasan D, Phyu E (2015) Particle swarm optimization for demand side management in smart grid. Paper presented at the 2015 IEEE Innovative Smart Grid Technologies-Asia (ISGT ASIA).
42. Zhao J, Lv L, Sun H (2015) Artificial bee colony using opposition-based learning. In: Genetic and evolutionary computing. Springer, pp 3–10.
43. Lokeshgupta, B. and Sivasubramani, S. (2019), Cooperative game theory approach for multi-objective home energy management with renewable energy integration. IET Smart Grid, 2: 34-41
44. Marzband, M., Javadi, M., Domínguez-García, J.L. and Mirhosseini Moghaddam, M. (2016), Non-cooperative game theory based energy management systems for energy district in the retail market considering DER uncertainties. IET Gener. Transm. Distrib., 10: 2999-3009.
45. Li, J., Ma, G., Li, T. *et al.* A Stackelberg game approach for demand response management of multi-microgrids with overlapping sales areas. *Sci. China Inf. Sci.* **62**, 212203 (2019).
46. Macedo M, Galo J, De Almeida L, de C. Lima A (2015) Demand side management using artificial neural networks in a smart grid environment. Renew Sustain Energy Rev 41:128–133.
47. Ibrahim O, Bakare MS, Amosa TI, Otuoze AO, Owonikoko WO, Ali EM, Adesina LM, Ogunbiyi O (2023) Development of fuzzy logic-based demand-side energy management system for hybrid energy sources. Energy Convers Manag 10:100354.
48. P. Balakumar, T. Vinopraba, K. Chandrasekaran, Deep learning based real time Demand Side Management controller for smart building integrated with renewable energy and Energy Storage System, Journal of Energy Storage, Volume 58, 2023, 106412, ISSN 2352-152X,
49. Choi E, Cho S, Kim DK. Power Demand Forecasting using Long Short-Term Memory (LSTM) Deep-Learning Model for Monitoring Energy Sustainability. 2020; 12(3):1109.
50. Ngo, NT., Pham, AD., Truong, T.T.H. *et al.* Developing a hybrid time-series artificial intelligence model to forecast energy use in buildings. *Sci Rep* 12, 15775 (2022).
51. Pannakkong W, Vinh VT, Tuyen NNM, Buddhakulsomsiri J. A Reinforcement Learning Approach for Ensemble Machine Learning Models in Peak Electricity Forecasting. *Energies*. 2023; 16(13):5099.
52. Yadav, K. (2023). A Review on Smart Metering Using Artificial Intelligence and Machine Learning Techniques: Challenges and Solutions. In: Kumar Sharma, D., Sharma, R., Jeon, G., Kumar, R. (eds) Data Analytics for Smart Grids Applications—A Key to Smart City Development. Intelligent Systems Reference Library, vol 247. Springer, Cham.

CHAPTER

7

Smart Grid Monitoring Systems

7.1 Introduction

A smart grid monitoring system is a sophisticated network of sensors, data acquisition devices, and communication technologies deployed throughout an electricity grid, allowing for real-time monitoring of voltage, current, power flow, and other critical parameters, enabling efficient management and optimization of power generation, transmission, and distribution, often utilizing advanced analytics and AI to predict potential issues and take proactive measures to maintain grid stability and reliability. A Smart Grid Monitoring System (SGMS) comprised of sensors installation at consumer premises, distribution transformers, substations and weather stations to collect the real-time data. Smart meters installed at consumer premises provides the details of energy consumption, the sensors installed at distribution transformer provides temperature and loading levels, and the substation sensors provide voltage, current and power factor values at critical points of the grid. And the data from weather station to know the impact on load and generation. A communication network, comprising of optical fiber or cellular network provides two-way data exchange between smart meters and the utility. **A Real-time Monitoring Systems for** Displaying live grid conditions and alerts for potential issues, **A predictive Analytics which uses** historical data and algorithms to anticipate grid problems and to optimize operations and **Data Dashboards to** present key metrics and insights to grid operators are the other key component of SGMS. The various technologies deployed for the monitoring of an electrical grid include, Supervisory Control And Data Acquisition (SADA) systems, Advanced Metering Infrastructure (AMI) and Wide Area Monitoring Systems (WAMS) which will be discussed in the following sections of this chapter.

7.2 Supervisory Control and Data Acquisition (SCADA)System

A SCADA (supervisory control and data acquisition) is an automation control system that is used in industries such as energy, oil and gas, water, power, and many more. The system has a centralized system that monitors and controls entire sites, ranging from an industrial plant to a complex of plants across the country. A SCADA system works by operating with signals that communicate via channels to provide the user with remote controls of any equipment in a given system. It also implements a distributed database, or tag database, that contains tags throughout the plant. These tags represent a single input or output value that is monitored or controlled by the SCADA system in the centralized control room. The tags are stored in the distributed database as value-timestamp pairs. It's very common to set up the SCADA systems to also acquire metadata, such as programmable logic controller (PLC) register paths and alarm statistics. While these systems simplify a given infrastructure, their components are quite complex. Figure-7.1 depicts how a basic SCADA system looks like. And the essential composing parts of a SCADA system include:

- Remote Terminal Units (RTUs)
- Master Terminal Unit (MTU)
- Human Machine Interface (HMI) systems
- Historian
- Communication infrastructures

Remote Terminal Unit: is responsible for collecting real-time data and information from sensors that are connected to the physical environment using link LAN/WAN. RTUs forward information to MTU. These are additionally in charge of conveying the present status data of physical devices associated with the system. Apart from RTUs, Programmable Logic Controllers (PLCs) and Intelligent Electronic Devices (IEDs) are also used to interface the sensors/actuators through input and output modules. They act as a more modern alternative or a complement of a setup with RTUs. IEDs can control several different aspects of a piece of a sensor/actuators compared to PLC, which are developed for a specific task. IEDs are easier to configure and require less wiring as compared to traditional RTUs. Generally, IEDs have a communication port, and they can communicate to a substation PLC directly or act as a gateway towards the SCADA server.

Master Terminal Unit is the central monitoring station. It is incharge of controlling and commanding the RTU machine over communication links. It also responds to messages from RTU and processes and stores them for succeeding communication.

Human Machine Interface provides a communication interface between SCADA hardware and software components. It is responsible for controlling SCADA operational information, for example, controlling, monitoring, and communication between several RTU and MTU in the form of text, statistics, or other comprehensible content.

Historian is used for accumulating two-way communication data, events, and alarms between the SCADA control center. It can be described as a centralized database or a server located at a distant location. Historian is queried to populate graphical trends on the HMI.

Communication network provides communication services between various components in the SCADA network framework. The medium utilized can be either wireless or wired. Presently, wireless media is generally used as it interfaces geologically circulated areas and less available zones to communicate effortlessly.

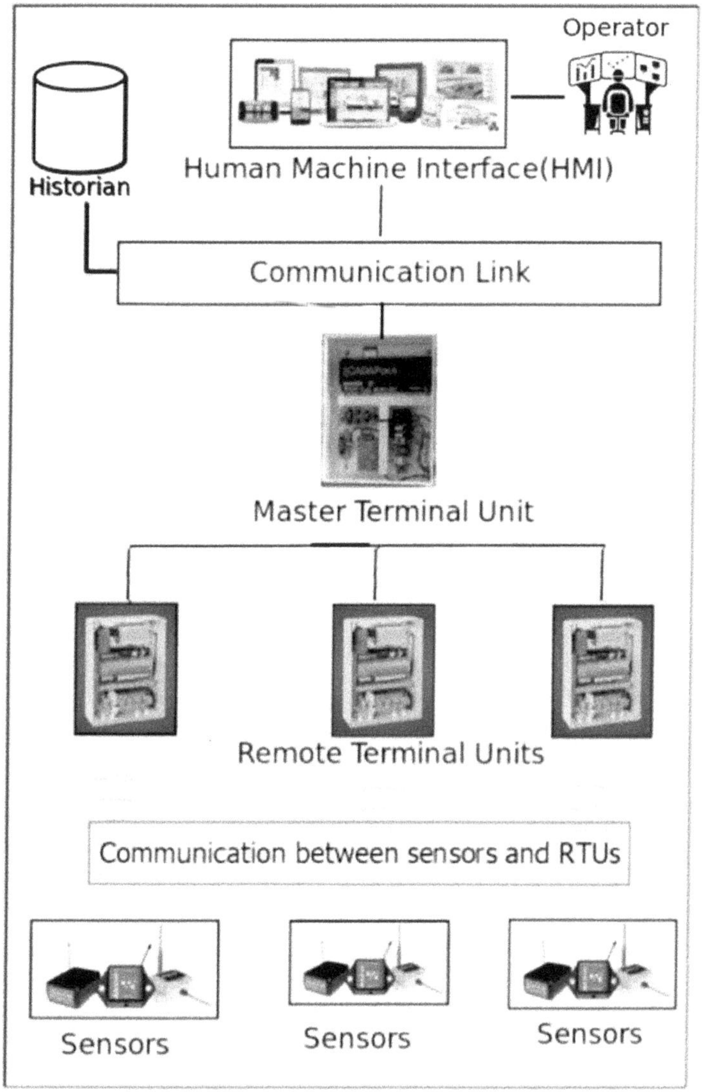

Figure 7.1 A basic SCADA system.

7.2.1 SCADA Generations

The advancement of communication paradigm is divided into four primary ages, such as the first era: Monolithic, second era: Distributed, third era: Networked, fourth era: IoT-based SCADA.

Monolithic SCADA systems: It refers to those systems which work in an isolated environment and do not have any connectivity to the other systems. The motive of these systems is to work in a solitary way. Large minicomputers were used for SCADA system computing. PDP-11series, which was developed by Digital Equipment Corporation, is an example of a first-generation SCADA system. In this architecture, RTUs communicate to MTU using Wide Area Networks (WAN), as shown in Figure-7.2. However, the WAN protocols used at that time were completely different than the current WAN protocols. The communication protocols were proprietary, which can be used only to connect RTUs with proprietary MTU from the same vendor.

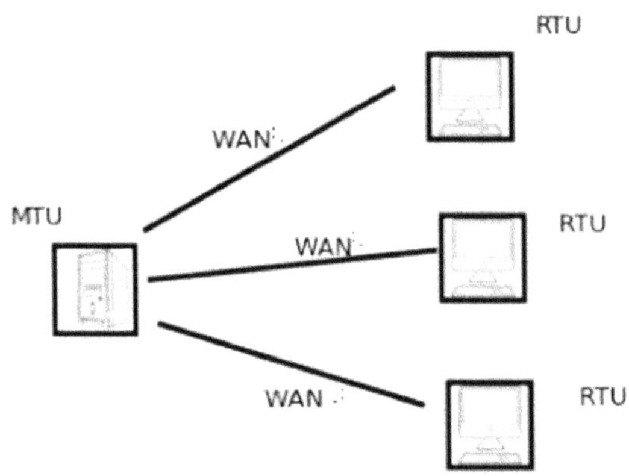

Figure 7.2 A Monolithic SCADA systems.

These protocols were limited to permit scanning, control, and data exchange between MTU and RTUs. In the absence of network connectivity, the connection between MTU and RTUs was done at the bus level (e.g. using RS-232 communication standards) or using proprietary adapter plugging into the CPU backplane. Connecting different vendor RTUs to MTU was an impossible task resulting in an urgent requirement for the open standards. In some cases, to provide redundancy to the SCADA system, an equally equipped system, working as a backup system was connected to the master system.

Distributed SCADA systems: These systems were inter-connected and confined inside small range networks like Local Area Networks (LAN), as shown in Figure-7.3. However, the WAN/LAN protocols used in this generation were completely different than the current WAN/LAN protocols. Distributed architecture is used in the case of multiple clients and stations. The information was shared using the LAN. However, some of the LAN protocols used were proprietary nature, which again kept a restriction on the systems connected to a LAN to work as a distributed MTU. WAN was used to the intercommunication between RTUs and MTU. The LAN protocols have a range limited to the local environment. The total cable length limit between systems on the network was limited to 600 feet, limiting multiple system connections in a room itself. Similar to the monolithic SCADA, distributed SCADA systems were also confined to proprietary hardware, software, network protocols, and peripheral devices supplied by the vendor. All the devices that are connected to SCADA LAN were unable to communicate with external devices using other communication protocols. Their communication was restricted to proprietary protocols supplied by vendors. In shorts, distributed SCADA systems were more open at MTU but still lack the capabilities at the RTU. The security of SCADA systems was not of concern in this generation also.

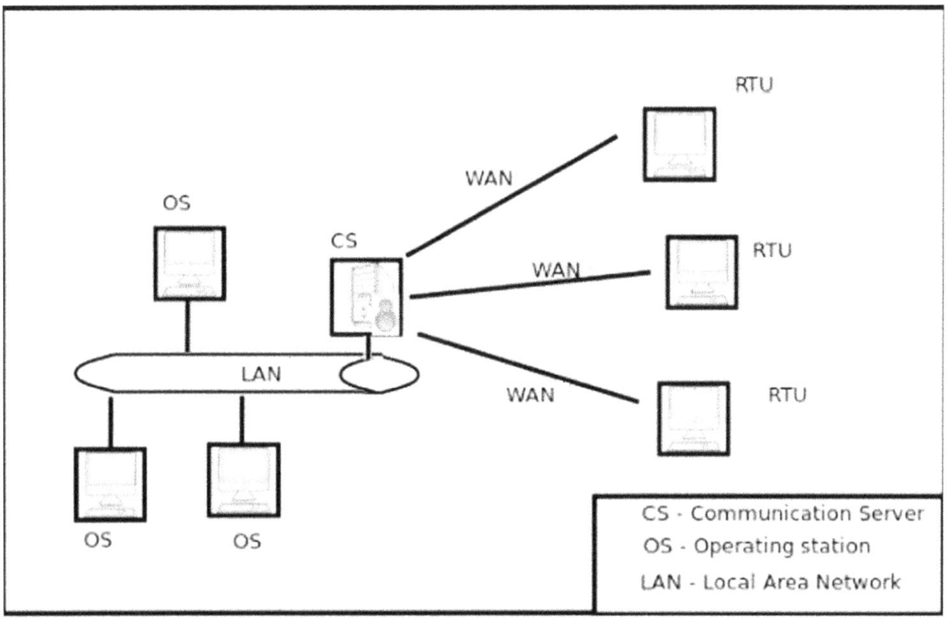

Figure 7.3 Architecture of a typical Distributed SCADA system.

Networked SCADA systems: It utilizes networks and the web broadly because of the standardization and cost-effective solutions for large-scale systems. This is also referred to as a modern SCADA system. In this design, SCADA systems may be geographically distributed. However, Networked SCADA is closely related to Distributed SCADA, with a significant difference in the usage of open protocols and standards for communication rather than proprietary protocols resulting in distributing MTU functionality across a WAN shown in Figure-7.4.

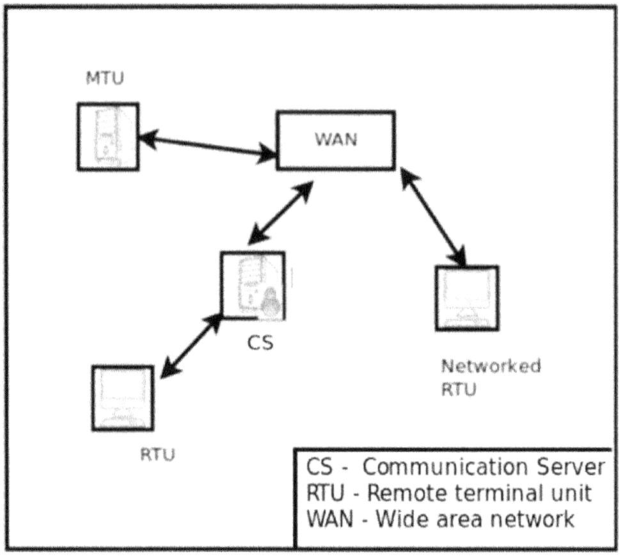

Figure 7.4 An architecture of a typical Networked SCADA system.

Due to the use of open standards, third-party peripheral devices can be connected to the network. The significant game-changing improvement in networked SCADA was the Internet Protocol's use for the communication between MTU and RTUs, resulting in disaster survivability.

IoT-based SCADA: The industries have been utilizing the power of technology to build, monitor, and control the systems. IoT innovation and economically accessible cloud computing with SCADA systems have considerably lessened its infrastructure and deployment costs. Moreover, the integration and maintenance are also easy as compared to the previous generations. Industries 4.0 is an example of a fourth-generation SCADA system, as shown in Figure-7.5. It includes distributed cognitive computing, Cyber Physical System (CPS) IoT, and cloud computing. SCADA systems already share a few characteristics of IoT, e.g., data access, manipulation, and visualization. IoT differs in terms of interoperability, scalability, and capability of big data analytics. The collection and control of all data are done using an open communication standard. The collected data is stored on clouds and extraction to get valuable insights from data. Industrial IoT (IIoT)or Industry 4.0 refers to the developments in fourth-generation SCADA systems. IIoT is described as IoT in industries. It is a network of devices with a significant focus on transfer, control of critical information, and large data insights. Therefore, to inculcate IIoT in SCADA, several devices, protocols need to be integrated into the existing system.

IIoT has also improved its resilience by identifying anomalous behavior using data-driven techniques.

Over the evolution of SCADA from a monolithic architecture to an IoT-based SCADA, these systems rely on effective information communication to collect, analyze and display data from heterogeneous devices using different protocols and networks over wired or wireless network. However, a wide heterogeneity in the technologies, protocols, and proprietary architectures increases complexity, cost, and efforts to achieve seamless interconnection and sometimes results in faulty communication.

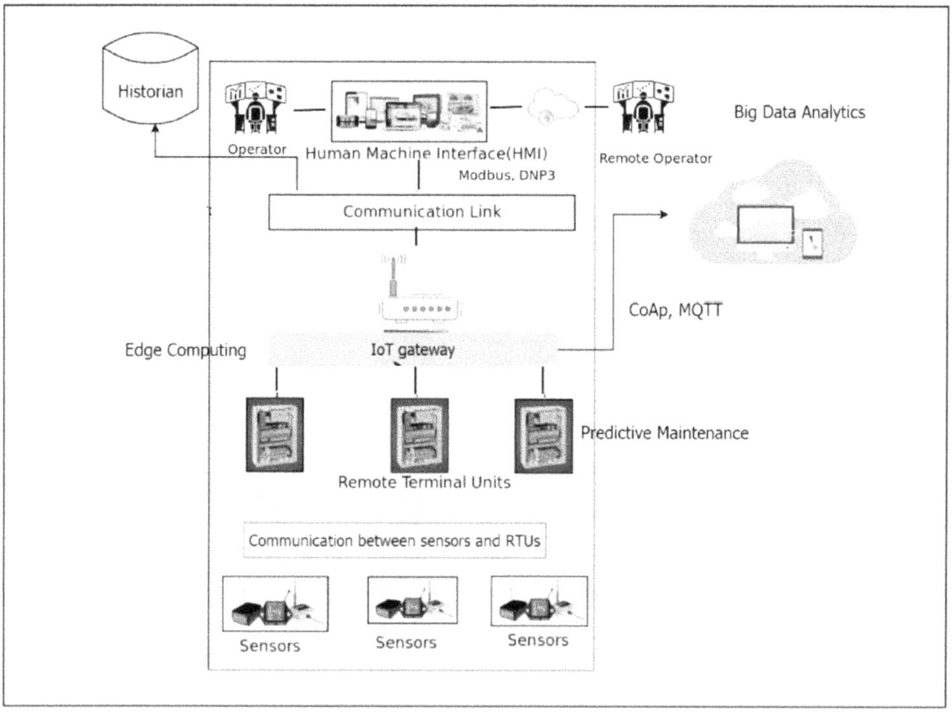

Figure 7.5 IoT based SCADA system.

A series of standards were introduced to provide a homogeneous system utilizing open communication protocols. Open Platform for Communication (OPC) foundations established Object Linking and Embedding (OLE) for Process Control to develop common standards for open connectivity of industrial automation devices and systems in 1996. Later, OLE for Process Control was released as an Open Platform for Communication (OPC) to adapt to other applications' standards. OPC is built on a client-server architecture. In 2008, OPC Unified Architecture (OPC UA) was released, which is a platform-independent service-oriented architecture. It supported interoperability for Industrie 4.0 and IoT. It integrates all the functionality of the individual OPC Classic specifications into a single extensible framework that supports functional equivalence to all classical OPC and a secure, extensible, and platform-independent, unlike classical OPC. The standard for OPC-UA is IEC 62541.

7.2.2 Integration of SCADA into Smart Grid

Integrating SCADA into the Smart Grid is interesting topic, and can be connected by electrical, communications and data networks, which allows for distributed and central aggregation of information and control over the entire electrical utility network as depicted in Figure-7.6. SCADA empowers the electricity consumer by interconnecting energy management systems to authorize the customer to manage their own demand of energy and control costs. It allows the grid to be self-healing by automatically responding to power quality issues, power outages, and power system faults. SCADA optimizes the grid assets by monitoring and optimizing those assets while minimizing operations and maintenance costs. The Smart Grid, intelligence and control need to exist along the entire power supply chain. This includes the electricity generation and transmission from beginning to delivery end-points at the customer's side and includes both fixed and mobile devices in the SG architecture.

Digital communications take place over a diversity of technologies, devices, and protocols that comprise wireless and wired telephone, voice and data dispatch radio, fiber optics networks, power line carriers, distribution line carriers, and satellite. SCADA software allows for dynamic grid management that involves monitoring the required number of control points. To be fully efficient and operational, monitoring occurs for every segment of the power line and a piece of equipment in the distribution system, in addition to allowing the customers to observe and control their own devices and energy consumption. This results in considerable amount of data to be analyzed, organized and used for both automated and manual decision software that can be classified into two basic categories: decentralized and back office.

The decentralized software is important due to the significance of the SCADA devices and data collection and computation ability, which prevent a centralized data acquisition solution. The Intelligent Electronic Devices (IEDs) will evolve to mitigate the organization, collection, and data analysis necessary for performing decision making, data routing, and other control actions that may be necessary based on the data received. This functionality subsists either as part of the device's firmware or via RTU configurable functions and settings within each device.

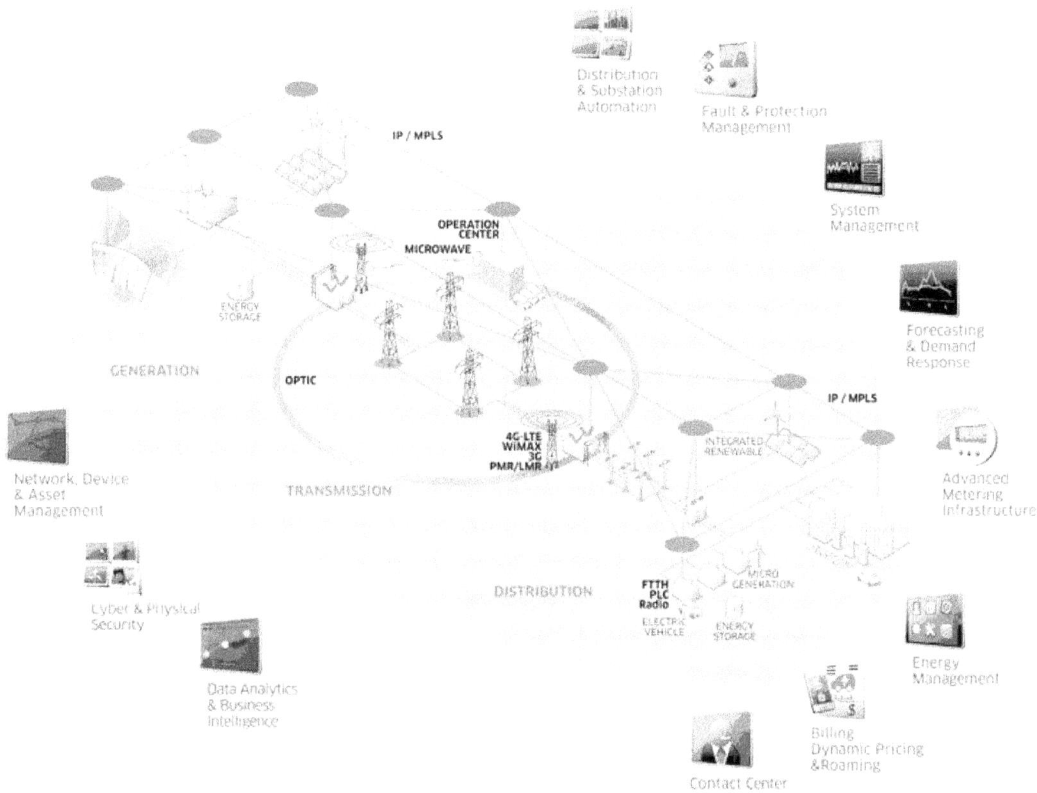

Figure 7.6 SCADA integration into Smart Grid.

7.2.3 Applications of SCADA

SCADA system is widely used in a power grid to collect, analyze and observe its data effectively. A grid deals with power generation, transmission, distribution, and renewable energy sectors, and monitoring and control are the main aspects in all these areas. Thus, the SCADA implementation of power system improves the overall efficiency of the system, by saving cost and time. This can be achieved by an optimizing operation, loss minimization, supervising and controlling the generation and transmission systems. SCADA function in the power system network provides greater system reliability and stability for integrated grid operation. The power grid automation system offers contingency based fast Load Shedding, Power Control and SCADA functionality for the electrical system. These applications such Generation, Distribution, Solar PV generation and Wind generation etc., may be supplied by different vendors. Some of them are discussed in this section.

SCADA in Power Generation: With the use of Programmable Logic Controllers (PLC) hardware and advanced communication links along with SCADA software and hardware in power generating stations, delivering an optimum solution for each and every operation with flexible and advanced control structures. Figure-7.7 shows the SCADA in power generation where it supervises several operations.

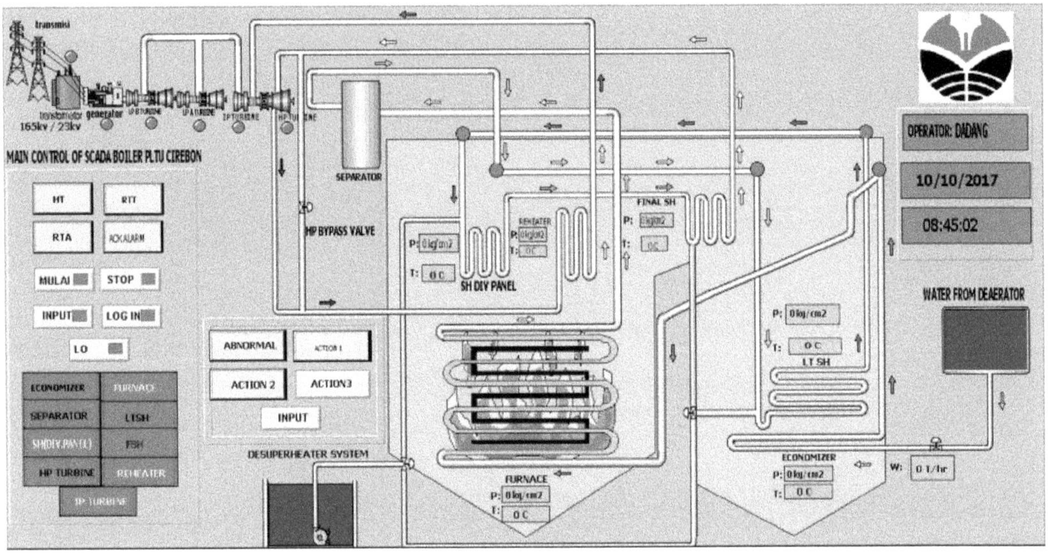

Figure 7.7 SCADA system of Cirebon steam power plant.

The functions of SCADA in power generation include

- Continuous monitoring of Speed and Frequency.
- Geographical monitoring of coal delivery and water treatment processes.
- Supervising the status of circuit breakers, protective relays.
- Generation planning.
- Active and reactive power control.
- Turbine protection.
- Load scheduling.
- Historical data processing related to generation.
- Transformer and tap-changer control based on the status of the electrical network.

SCADA for Power Distribution System: Distribution management systems deal with electric power from distribution substation to the different loads with the use of medium and low voltage cables and transmission lines. Most of the power distribution or utility companies rely on manual labor to carry out the distribution tasks like load shedding, hourly parameter check, fault diagnosis, etc. The implementing SCADA to the power distribution not only reduces the manual labor operation but also facilitates smooth automatic operations with minimizing disturbance. Figure-7.8 shows the structure of SCADA in power distribution system where it collects the whole data from various electrical substations and does the processing. Programmable logic controllers in electrical substations continuously monitor the substation components and corresponding transfer that to centralized PC-based SCADA system. In the event of any power outages or failure, this SCADA allows detecting the fault type and location, therefore without waiting for the calls from customers. SCADA gives an alarm or event to the operators for identifying and analyzing it. The SCADA in substations automatically controls the circuit breakers and switches for exceeding parameter limits, thereby continuous inspection of network status and parameters are performed regularly without a line worker.

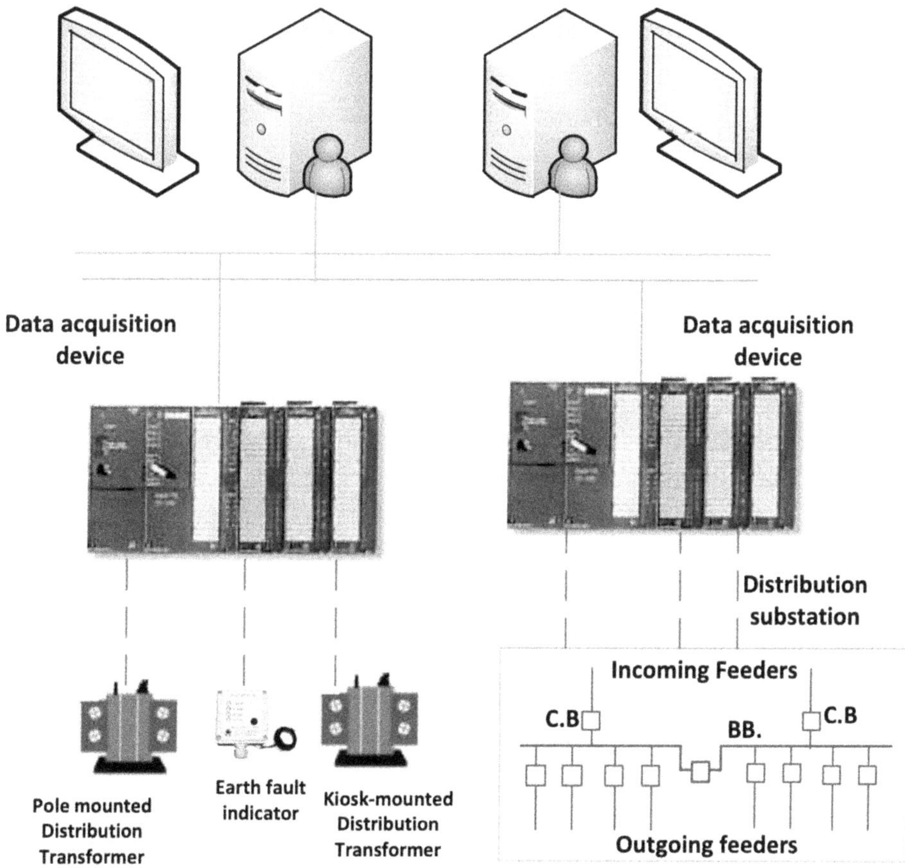

Figure 7.8 SCADA for Power Distribution System.

Some of the SCADA functions in power distribution system are given as follows:

- Improving power quality by maintaining power factor and harmonics contents.
- Limiting peak power demand.
- Continuous monitoring and controlling of various electrical components.
- Alarming to enable operators to fix outage problems.
- Historian data and viewing from remote locations.
- Quick response to customer service interruptions.
- Motorized control integration for control centers.
- Power Control i.e. tie-line control, peak shaving and load sharing.
- Load Shedding in response to frequency variations

SCADA in Hybrid Power System: Introducing renewable energy sources, such as solar photovoltaic, fuel cell, and wind energy has an excellent potential of contribution to power generation. Combining two or more different of renewable energy sources will form the system called "Hybrid Power system". A schematic diagram of a hybrid renewable power system is shown in Figure-7.9. The reliability of the HPS mainly relies on the dynamic behavior of the renewable energy sources. Hence, it is important to analyze the dynamic characteristics of these units in real time for long-time periods. A major challenge lies in the development of the real-time control scheme for the HPS. In order to test the HPS controllers, it requires, a controller that can interface with a hardware simulator and the inputs and outputs can be processed at the real time. All the Modules should be modeled in the PLC environment. PLC and SCADA system communicate with each other over a dedicated protocol which can be transmitted over serial port or Ethernet. The control structure of HPS is shown in Figure 7.9. Since the monitoring of PV cell and Fuel cell had done in PLC and is interfaced to SCADA/Energy management systems. So that the operator could be able to observe the parameters easily and control the parameters according to the changes in system requirements. The data of HPS is collected through PLC and connected to control room through communication protocol. The PLC is interfaced to energy management control unit to get control actions according to the respective microgrid loads and environmental conditions. The entire system is connected to SCADA for supervision and control.

The control room contains various I/O consoles such as engineering console and operator console. The Engineering console is responsible for adding new points or new IED devices to the system.

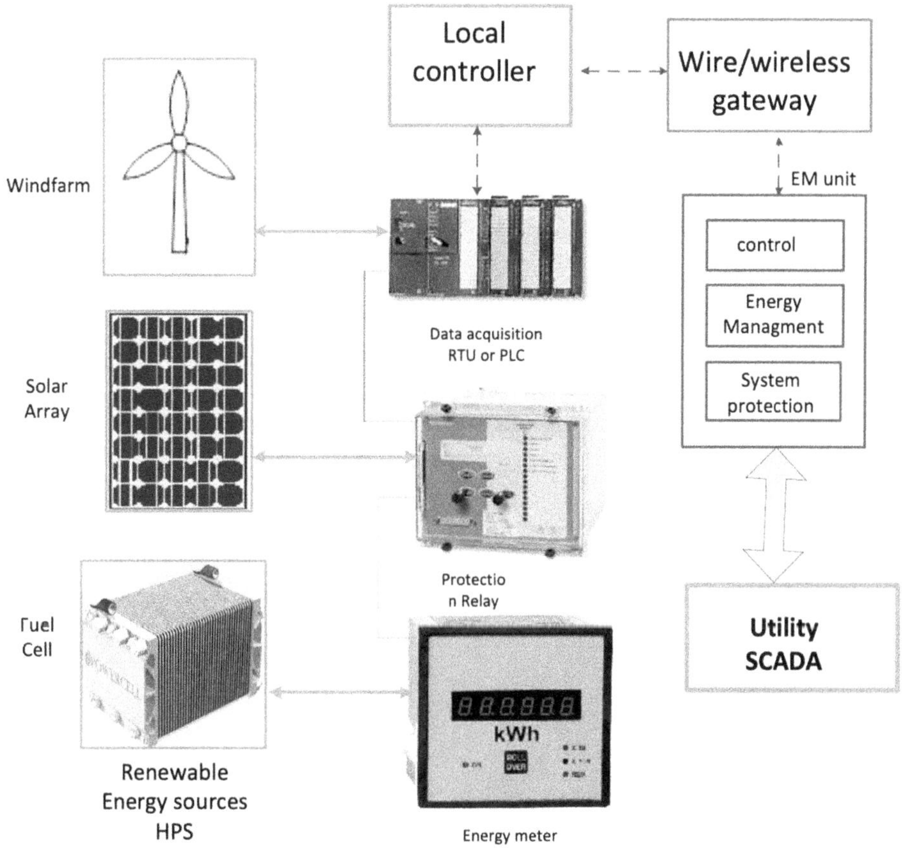

Figure 7.9 Schematic diagram of SCADA integrated Hybrid Power System.

7.3 Advanced Metering Infrastructure (AMI)

AMI is not a single technology; rather, it is a configured infrastructure that integrates a number of technologies to achieve its goals. The infrastructure includes smart meters, communication networks in different levels of the infrastructure hierarchy, Meter Data Management Systems (MDMS), and means to integrate the collected data into software application platforms and interfaces.

As shown in Figure 7.10, the customer is equipped with an advanced solid state electronic meter that collects time-based data. These meters can transmit the collected data through commonly available fixed networks, such as Broadband overpower Line (BPL), Power Line Communications, Fixed Radio Frequency, as well as public networks such as landline, cellular and paging. The metered consumption data are received by the AMI host system. Subsequently, it is sent to a MDMS that manages data storage and analysis and provides the information in a useful form to the utility service provider. AMI enables a two-way communication; therefore, communication or issuance of command or price signal from the utility to the meter or load controlling devices are also possible.

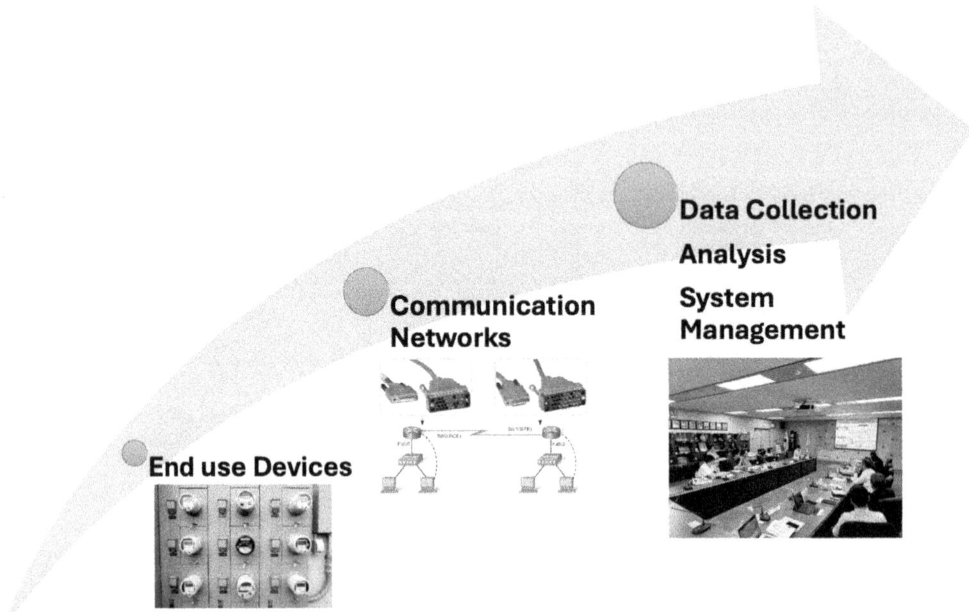

Figure 7.10 Schematic of Advanced Metering Infrastructure.

7.3.1 Subsystems of AMI

AMI is not limited to electricity distribution. AMI is applicable to gas and water networks also. Although the infrastructures for metering different forms of energy are very similar in several aspects, they still differ in some traits. Electric meters are typically fed from the same electric feed that they are monitoring. This is not the case for gas and water meters. Flow meters are typically powered by stored energy, i.e., batteries; therefore, have utilization constraints. These constraints are more evident in communication since power is needed for transmitting and receiving signals. Meters also have embedded controllers to manage the metering sensor, a display unit, and a communication module which is generally a wireless transceiver. Technical aspects of AMI are wide and vast. Hence, here the issues associated with utilization of AMI in electrical Smart Grids are discussed.

Smart Devices: End user devices are comprised of state-of-the-art electronic hardware and software capable of data collection or measurement in desired time intervals and with time stamping. These devices have an established communication with remote data center and are capable of transmission of such information to various parties in required time slots set by system administrator. Unlike Automatic Meter Reading (AMR), communication in AMI is bidirectional; therefore, smart devices or load controlling devices can accept command signals and act accordingly. At the consumer level, a smart device is a meter that communicates consumption data to both the user and the service provider. The In-Home Displays (IHD) illustrate the smart devices' data to consumers, making them aware of their energy usage. Utility pricing information supplied by the service provider enables load controlling devices to regulate consumption based on pre-set user criteria and directives. If Distributed Energy Resources (DER) and storages are available, the system can come up with an optimized solution in terms of share of each source to meet the demand.

From the measurement point of view, smart meters have three distinct categories (in broad): electrical, fluid, and thermal. There are also a number of sensors or devices that measure factors like humidity, temperature and light which affect the consumption. Home automation systems deal with the proper selection, placement and utilization of various sensors within the home premises. Smart meters have two functions: measurement and communication, and therefore each meter has two sub-systems: metrology and communication. The metrology part varies depending on a number of factors including region, measured phenomenon, required accuracy, level of data security, application. There are also multiple factors, including security and encryption, which define the suitable communication method. There are a number of essential functionalities meters should irrespective of quantity they measure. These functionalities include:

- *Quantitative measurement:* Ability to accurately measure the quantity of the medium.
- *Control and calibration:* Ability to compensate the small variations in the system.
- *Communication:* Capable of sending stored data and receiving operational commands
- *Power management:* Ability to maintain its functionality in the event of a primary source failure.
- *Display:* A display is also needed as demand management at customer end is not possible without the customer's knowledge of the real time consumption.
- *Synchronization:* Timing synchronization is critical feature for reliable transmission of data to central hub or other collector systems for data analysis and billing.

Hence, the key features of smart electricity meters can be summarized as follows:

- Time-based pricing.
- Providing consumption data for consumer and utility.

- Net metering.
- Failure and outage notification.
- Remote command (turn on/off) operations.
- Load limiting for Demand Side Management.
- Power quality monitoring.
- Energy theft detection.
- Communication with other intelligent devices.
- Optimizing power consumption to protect environment.

Communication Network: Smart meters should be able to send the collected information for analysis and to receive operational commands from operation center through a communication system. Considering the number of users and smart meters at each center, a highly reliable communication network is required for transferring the high volume of data. Hence, the following factors are to be considered for the design and selection of an appropriate communication network:

- Amount of data transfer.
- Data access restriction.
- Confidentiality of sensitive data.
- Representing complete information of customer's consumption.
- Presentation grid status.
- Authenticity of data and precision in communication with target device.
- Cost effectiveness.
- Ability to host modern features beyond AMI requirements.
- Support for future expansion.

Various topologies and architectures can be used for communication in Smart Grids. The most practiced architecture is to collect the data from groups of meters in local data concentrators and then transmit the data using a backhaul channel to central command where the servers, data storing and processing facilities as well as management and billing applications reside. As different types of architectures and networks are available for realization of AMI, there are various mediums and communication technologies for this purpose as well. Some of them include:

- Power Line Carrier (PLC).
- Broadband over Power Lines (BPL).
- Copper or optical fiber.
- Cellular.
- Wi-Max.

- Bluetooth.
- General Packet Radio Service (GPRS).
- Internet.
- Satellite.
- Peer-to-Peer.
- Zigbee.

At AMI level, devices within the premises of the house communicate with each other as well as the utility network through smart meters. This network, in short, is called as Home Area Networks (HAN). The Home Area Networks (HAN) communicates with the utility provider, forming another network that could be called as utility network. HANs connect smart meters, smart devices within the home premises, energy storage and generation (solar, wind, etc.), electric vehicles as well as IHD and controllers together. Since their data flow is instantaneous rather than continuous, HANs required bandwidth vary from 10 to 100 Kbps for each device, depending on the task. As the distances among nodes are short, low power transmission wireless technologies can be dominant solutions for HANs. These technologies include 2.4 GHz WiFi, 802.11 wireless networking protocol, ZigBee and HomePlug. Zigbee is based on the wireless IEEE 802.15.4 standard and is technologically similar to Bluetooth. Home Plug, on the other hand, transmits data over the electrical wiring existing at the home. There is still no unique standard or practice for in-home communication in the market; however, Zigbee, and to lesser extent Home plug and ZWave, are the dominant solutions. Advantages of Zigbee include providing wireless communication, low power consumption, flexibility and economic efficiency. The main disadvantage of Zigbee is the low bandwidth. In commercial buildings, a wired technology named BACnet is the prominent communication protocol. Recently, a wireless version of BACnet has become available using short range wireless networks such as Zigbee.

Figure 7.11 depicts four levels of utility network namely, core backbone, backhaul distribution, access points and Home Area Network (HAN). The smart meters typically act as the access points. HANs will connect to the access points in their immediate above layers. The information will then be taken from access points to aggregation points through backhaul distribution. Although aggregation points are usually local substations, they could be communication towers too. The requirement for this network is the same as HANs. however, network topology is important in this regard. If data from each appliance is to be transferred to aggregation point, then a higher bandwidth is needed. Backup power is not required for smart meters as they are not considered critical. However, backup power is needed at aggregation points. Currently, PLC addresses the communication needs between in-home system and aggregation points. The Advantages of PLCs are their low cost and expansion and penetration in utility provider's territory. Their disadvantages, however, include the low bandwidth of up to 20 Kbps, and data distortion around transformers which necessitates bypassing transformer points using other techniques. PLC is more or less the prominent practice in current market due to the aforementioned advantages and also because this grid is already up and running, minimizing the deployment cost.

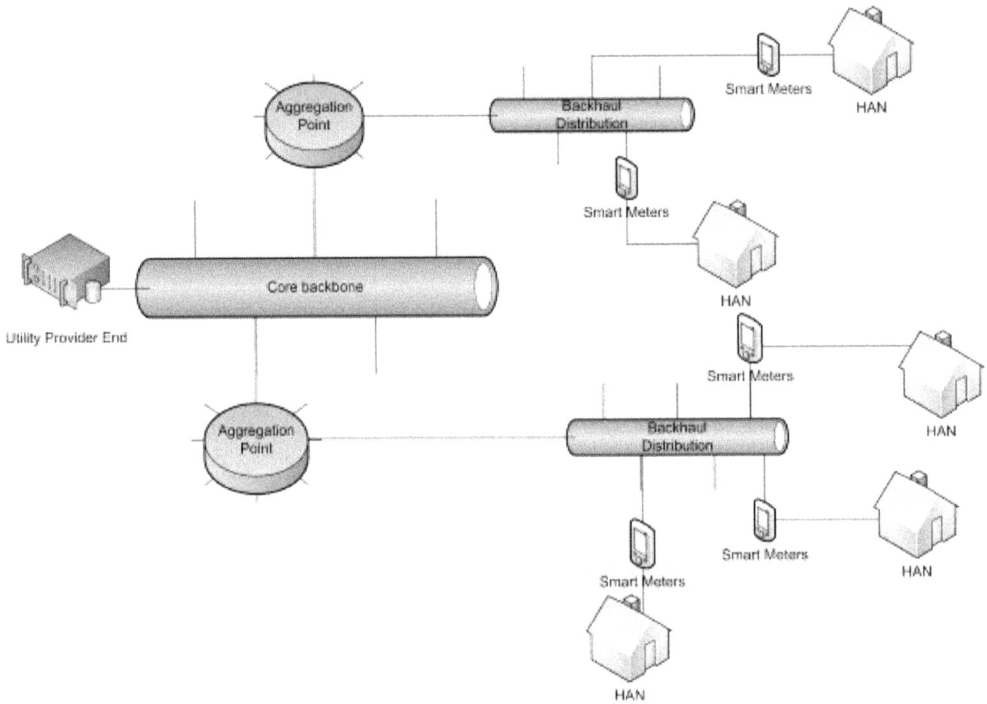

Figure 7.11 Overview of utility network.

PLC is specifically valuable in remote locations where the number of nodes (consumers) is relatively low and no wireless (cellular, GPRS) coverage is available.

The availability of reliable wireless technologies in urban areas, led to utilization of Mesh networks. In Mesh networks, in order to propagate information to the end point, each node is responsible for collecting its own data, as well as relaying the information by other nodes in the network. The wireless mesh networks are mainly owned and operated by utility companies. These networks are capable of supporting up to 900 MHz through unlicensed radio spectrums. As the demand for bandwidth increases, broad band technologies such as IEEE 802.16e, mobile WiMAX and broadband PLC are going to play a key role in newer installations. Today, the Long-Term Evolution (LTE) standard for wireless technology is believed to answer the market's demand. LTE enables high speed, high-capacity wireless communication with good Quality of Service (QoS) as well as low latency. These characteristics make

LTE suitable for critical applications in SG as well as for Neighborhood Area Networks, Wide Area Networks, substation automation and many more.

Table 7.1 Comparison of some of the available communication technologies.

	LTE-A	3G (HSPA+)	PLC	802.22
Latency (ms)	<5	<50	<10	<20
Data rate (Mbps) Download/upload	1000/ 500	28/11	3/3	18/18
Range (km)	100	10	5	100
Main disadvantage	-	Limited number of supported connections	Alternate technology (bypass) needed at transformers	No QoS due to faulty spectrum sensing

Cost and spectrum are the two main factors that prevent utility companies from adopting private LTE networks. On the other hand, they are hesitant to rely on public LTE networks. For many years, utility companies have used specific DR applications that utilize private communication networks by default. Utilities feel that the higher resilience against natural disasters, the ability to maintain service throughout a utility's service territory, avoidance of prioritization of other services when recovering from outages and the cost of service make the private networks a superior option over commercial services.

Data Management System: At the utility provider end, a system for storing and analyzing the data for billing purposes is needed. It should also handle demand response, consumption profile and real time actions to changes and emergencies in the grid. Modules of DMS are as follows:

- Meter Data Management System (MDMS).
- Consumer Information System (CIS), billing system, and the utility website.
- Outage Management System (OMS).
- Enterprise Resource Planning (ERP), power quality management and load forecasting systems.
- Mobile Workforce Management (MWM).
- Geographic Information System (GIS).
- Transformer Load Management (TLM).

MDMS could be considered the central module of the management system with the analytical tools required for communication with other modules incorporated within it. It also has the responsibility to perform validation, editing and estimation on the AMI data to ensure accurate and complete flow of information from customer to the management modules under possible interruptions at lower layers. In existing AMIs with data collection intervals of 15 min, the collected data is huge and in the order of terabytes referred to as "Big Data". Managing and analyzing such Big Data requires special tools. The sources of Big data in Smart Grids are as follows:

- Collection of consumption data at given frequency.
- Collection of data from real-time control of the system (up to 30 samples per second per sensor).
- Third-party systems connected to the grid, e.g. storages, Distributed Energy Resources or electric vehicles.
- Communication between central command and smart components in the network including updating firmware.

Different vendors have different definitions for MDMS and design their system based on their specific concept. Regardless of features or complexity, all MDMS suites should be able to address three demands: *improvement and optimization of operation of utility grids, improvement and optimization of utility management, and enabling customer engagement.*

Data analytics is to use all the available data from inside and outside of the grid, connect them together with available data analysis and data mining techniques, and extract useful information for decision makings. From an infrastructure and hardware point of view, the following components are necessary for such a system

Data center infrastructure: The building hosting the system and all related auxiliary systems, i.e., backup power, ventilation, etc.

Servers: Hardware needed for data handling.

Storage system: All hardware needed for storing data and connecting with other hardware in the system.

Data base system: Software needed for data analysis.

Virtualization systems: allow more efficient use of discrete storage and computing resources.

Since the collected data contains critical personal as well as business information, the storage facilities should be disaster proof, and all required back up and contingency plans for different scenarios should be carefully designed for them. The cost associated with such provisions is huge. Virtualization and cloud computing have been suggested as a solution for this problem. Virtualization allows all available resources to be merged together in order to improve the efficiency and return on investment. However, it requires additional technology and complexity. Cloud computing enables access to virtual resources in different locations and yet it brings serious concern regarding the security of data. Cloud computing can also be problematic since different regulations and laws apply to the data collected in different locations. Cloud computing however, reduces the cost of special purpose data centers as it uses the capacity of different service providers.

7.3.2 Security Challenges

As the number of smart meters increase exponentially, security issues associated with SG and AMI grow substantially from within the system as well as from outside. Some of the challenges include:

- Detailed information of customers' consumption is critical as it can reveal their lifestyle.
- Transmission of data over long distance as well as storing the data in various places for re-transmission or analysis can also create vulnerabilities in terms of data theft or manipulation.
- The price signal and commands received at consumer end are also potential areas for cyber and physical attack for the purpose of espionage, damaging infrastructure or power theft.
- Furthermore, consumer's peace of mind is critical in the success of smart meters and expansion of AMI.
- If consumers believe that their personal data is used against their will, or experience poor service or power quality due to external manipulation of the system by unauthorized parties or hackers, then they most likely resist the implementation of AMI.
- Potential health hazards and higher bills after installation of such smart meters will also affect the consumers' decision.

The governments need to take these issues seriously and work on procedures to guarantee customers' privacy of information. The governments also need to launch campaigns to increase the public's awareness on smart meters and to address their legitimate concerns regarding health and cost issues. Utility companies as well as installation technicians have to play an important role in this concern.

7.4 Wide-Area Monitoring System (WAMS)

Most of the present electrical grids operate closer to their stability limits because of ever expanding power demands, aging infrastructure, complex power transfers among regions, and challenging renewable integration. All these trends present an important challenge to the reliability and stability of the grid. In this scenario of complexities, carrying out monitoring, protection on real time basis and responding to contingencies are critical.

SCADA systems are widely used as situational awareness technology however they provide only the steady state view of dynamically changing power system. Wide area measurement systems (WAMS) have come forward as a prominent technology option to improve the visibility and situational awareness in the modern electrical grids. Synchro phasor technology acts as the heart of WAMS system that has enabled state measurement in WAMS compared to state estimation in conventional SCADA systems. WAMS measurements are more accurate and faster compared to their SCADA systems. The faster and more accurate synchro phasor measurements enable accurate and faster analysis of current grid situation almost in real time which in turn provides operators with options to carry out preventive measures and time to act. WAMS thus addresses not only the immediate reliability concerns but also operational issues by conducting real-time dynamic analysis, identifying and calculating security margins and indices, facilitating early detection, prediction of emergency states and initiating restorative actions to avoid instability. It also helps in postmortem analysis of disturbances grid.

7.4.1 Working of WAMS

In WAMS synchronized measurements are collected from the Phasor Measurement Units (PMUs) and this data is sent through communication networks, which is received and concentrated at a decision and control support system called as Phasor Data Concentrators (PDC) that determines appropriate preventive, corrective and protective measures. The decisions made by the support system helps operators at control centers to take smarter operator control actions. These actions are translated into feedback signals that are sent through communication networks to exploit the controllability and protection resources of the power system.

PMU and PDC thus act as backbones of any WAMS system. PMU is a function or logical device that provides synchro phasor (angle and magnitude), system frequency and rate of change of frequency measurements based on the data collected from one or more primary sensors like current (CTs) and potential (PTs) transformers. PMUs may optionally provide information such as calculated real (MW) and reactive (MVAr) powers, sampled measurements and Boolean status words. PDC, a function or logical device, works as a node in a communication network where synchro phasor data from a number of PMUs and/or PDCs is collected, time aligned, aggregated and sent out as a single stream to the higher level PDCs and/or applications. PDC optionally has to execute real-time wide-area protection and control applications. With the increasing number of PMUs installed in WAMS, there is a need of an efficient architecture of data collection and management for the efficient utilization of the data collected from PMUs.

7.4.2 Classification of WAMS Architectures

WAMS architectures can be classified as Centralized, Distributed and Decentralized architectures. The discriminating factors among these types are information or dataflow between the location of data acquisition, the location of decision making and the location where action based on decision is performed. Different types of WAMS architecture are discussed in detail as follows.

Centralized WAMS Architecture: In a centralized WAMS architecture, PMU data acquisition, data analysis and enactment of remedial action is performed at central location. Figure 7.12 encapsulates the centralized WAMS architecture. PMUs from various substations send the phasor data to Central PDC where time alignment and data concentration activities take place. The concentrated data is used for analytics and visualization. The remedial actions coming out of analysis are passed on to primary devices.

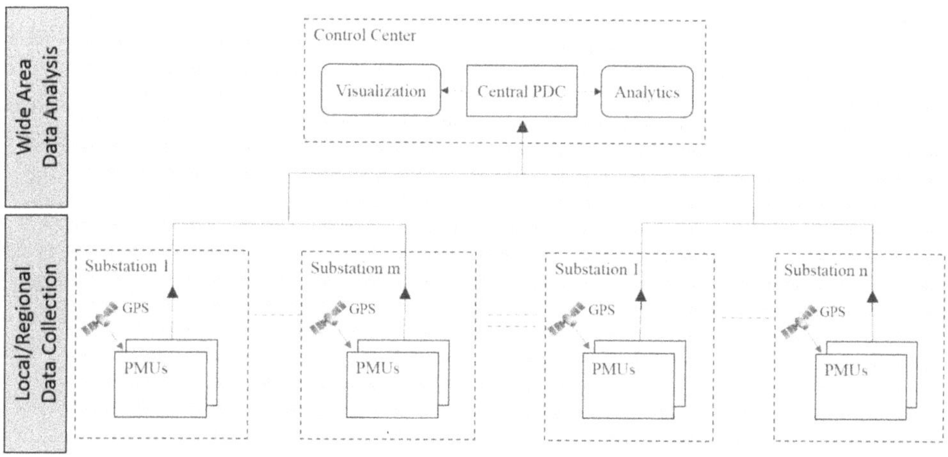

Figure 7.12 Centralized WAMS Architecture.

Decentralized WAMS Architecture: In decentralized WAMS architecture, the wide area monitoring is split into multiple small areas and PDCs control the small areas locally using local data. The local controllers are connected to each other if there is a need to solve larger area problem. Figure 7.13 depicts the decentralized WAMS architecture. PMUs within a local area such as substation or particular region send phasor data to respective local PDC for processing. Local PDCs analyze the data to take any remedial action to protect or control respective local assets. Although all distributed local PDCs are connected to each other for data exchange in order to monitor larger area, this is not efficient solution for monitoring wider area. Coordinated concentrated-data acquisition from local PDCs and their analysis for large area monitoring is often challenging and does not meet the goal most of the times.

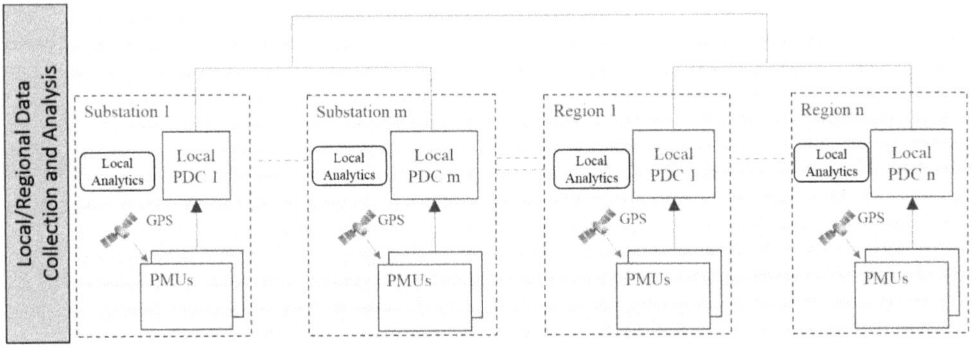

Figure 7.13 Decentralized WAMS architecture.

Distributed WAMS Architecture: Distributed WAMS architecture can be mapped between centralized and decentralized architectures. It includes local as well as central controllers. It can be thought of as centralized control with decentralized execution stage. Figure 7.14 describes the distributed WAMs architecture. It comprises of local PDC situated at substation or region level and master PDC located at central control station. PMUs within a local area such as substation or particular region send phasor data to respective local PDC. All local PDCs are connected to master PDC at central control station. The difference is in the flow of information as local PDC may process the PMU data locally, supervised and controlled by master PDC.

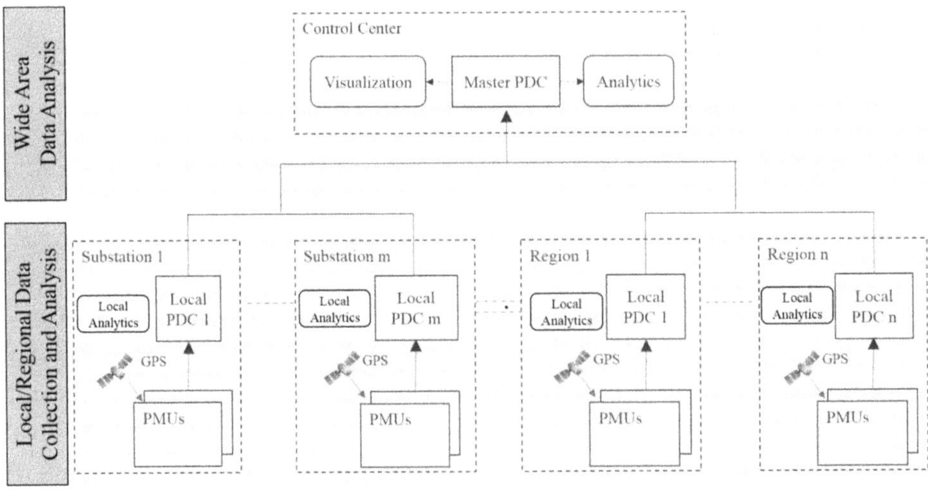

Figure 7.14 Distributed WAMs architecture.

7.4.3 Components of WAMS

Phasor Measurement Unit (PMU): The PMUs are intellectual computing devices to address many power quality problems around the world and they are developed from the symmetrical components distance relay. PMUs are becoming an integral part in many of the applications of the power system from load flow analysis, state estimation for analyzing the blackout causes. PMUs deploy the measurement of voltage and current that is synchronized along with a clocking signal obtained from the global positioning system (GPS). These devices can calculate the magnitudes of current and voltage, phase angles, real power flow, reactive power flow, etc.

Phasor Data Concentrators (PDC): These devices at the highest level of the hierarchy in the WAMS. They have a typical function to synchronize and gather data from several PMUs, reject bad data, align the time stamps, error checking, and create a coherent record of simultaneously recorded data from the observed parts of the power system at every time instant independent of when the data was received. Similar to the PMU, the PDC time needs to be synchronized. Most of the PDCs have been custom built by researchers or manufacturers of the PMU. The PDC can retrieve the data by itself. The PDCs also monitors power quality indices such as voltage sag, frequency, active power and reactive power. The data stream between the PMUs and PDCs is defined by a standard developed for measurement and transfer of synchro phasors. PDC has option to execute real-time wide area protection and control applications

Communication System: A reliable and high-speed communication infrastructure that enables secure sharing of data play an important role in the design of WAMS. It is also crucial to secure WAMS in order to ensure the availability, integrity, and confidentiality of the data transmitted over the network. The communication infrastructure should therefore be able to support different quality of services (QOS) classes for traffic and should be able to prioritize one class over another. An interface device such as modem or router is required between the PMU communication link and the communication system. Many utilities today use multi-protocol label switching, virtual private network, or Frame Relay technology for WANs.

To transmit the data stream between the PMUs and PDCs, there is a prerequisite of communication media. The choice of the communication medium depends up on the bandwidth, data transfer rate, and latency. The permissible time delay during data transfer between PMU and PDC is 20 ms, PDC to Super PDC is 40 ms, and from control center to relay at substation is 50 ms. Internet protocols are increasingly used for PMU communications, and they are mostly used over Ethernets. Transport control protocol (TCP) rearranges the packets of data in the specified order and retransmits corrupted or lost data.

7.4.4 Synchrophasor Technology

Among various devices placed in the power system, for improving the security, reliability and efficiency, the important device is the Phasor Measurement Unit (PMU) which closely observes the critical situations such as, unavailability of supply, load variations, contingencies or faulty conditions. The monitoring of the system in real time can be achieved with the help of Global Positioning System (GPS) enabled time stamping. PMU is a device that produces synchronized phasor, frequency and Rate Of Change Of Frequency (ROCOF) estimated from voltage and/or current signals and a time synchronizing signal. The device may also perform as digital fault recorder (DFR) by recording the power system waveforms.

Traditional technology for monitoring the system state is Supervisory Control And Data Acquisition (SCADA), which provides slow static measurement with low resolution. Hence, it fails to observe the real time dynamics of the system. Because of incomplete measurements, the system may go through the contingencies leading towards less secure system. To overcome aforementioned problem, fast monitoring device, that is, PMU is placed in the system, which is sufficiently able for real time voltage and current phasor measurements along with frequency and ROCOF monitoring for operating the system in safe mode. The key advantage of PMUs in WAMs is that these provide actual state of the system.

A brief comparison of SCADA and PMU given below justifies the better fitness of PMUs in the modern grid.

- SCADA observes steady state measurements, whereas PMU captures dynamic or transient measurements.
- SCADA can monitor analog measurements, while PMU can monitor digital measurements.
- PMU can measure the magnitude along with phase angle, whereas SCADA can only provide magnitude of the phasor.
- SCADA doesn't have time synchronization feature and provides local area monitoring, whereas PMU has the feature of time synchronized through GPS and hence can provide wide area monitoring of the system.
- SCADA system acquires the data in every 2-4 seconds, whereas PMU is able to obtain 30 samples per second to 120 samples per second. Such a high sampling rate helps in fast fault detection as well as detailed analysis of blackout like events.

The real time measurements obtained from PMUs encourage great monitoring of the system by detecting the exact problem and taking corrective actions during improper operation of the system. Hence the appropriate PMU placement increases the efficiency of the system. However, as the number of PMUs increases, the system becomes more expensive. Moreover, for obtaining completely observable system, the placement of PMUs at each bus is not mandatory.

Synchrophasor measurements: Phasor is defined as a complex equivalent of a sinusoidal wave quantity such that the complex modulus is the cosine wave amplitude, and the complex angle is the cosine phase angle. The representation of a sinusoidal signal is given as

$$x(t) = X_m \cos(\omega t + \phi) \quad .....(7.1)$$

Equation (7.1) can be expressed as a phasor given as Equation (7.2)

$$X = \left(\frac{X_m}{\sqrt{2}}\right) e^{j\phi} = \left(\frac{X_m}{\sqrt{2}}\right)(\cos\phi + j\sin\phi) = X_{re} + jX_{im} \quad .....(7.2)$$

where $\left(\frac{X_m}{\sqrt{2}}\right)$ represents the signal's rms value and subscript X_{re} and X_{im} represent real and imaginary parts of X. ϕ is the phase angle dependent on time scale especially at t=0 (Figure 7.15)

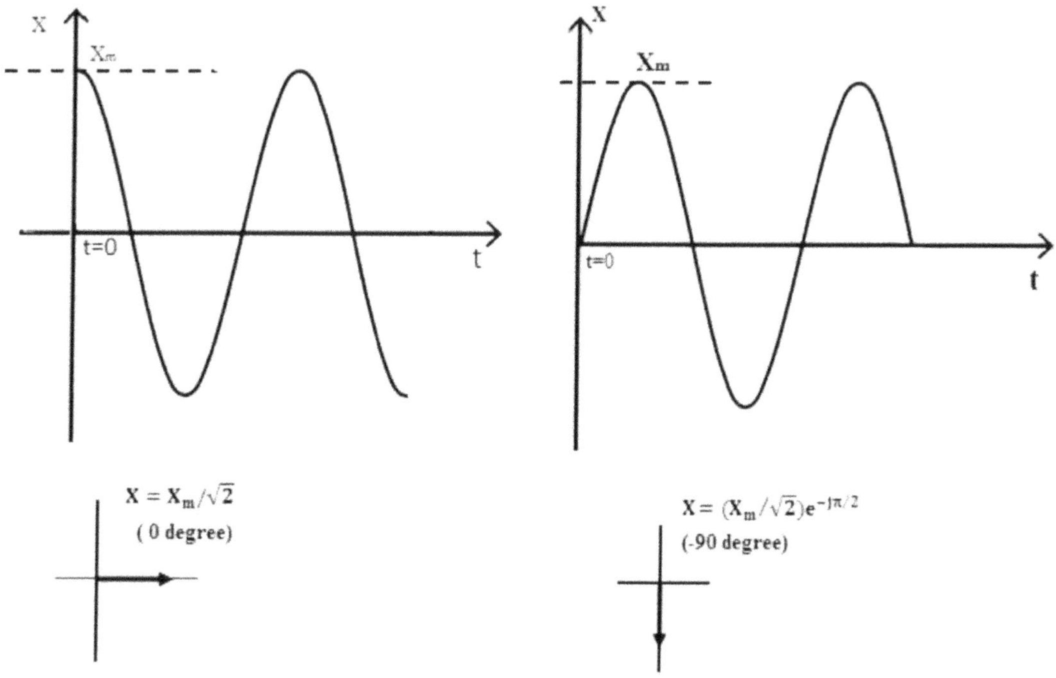

Figure 7.15 Representation of the Synchrophasors.

As defined in IEEE standard C37.118.1-2011, in Equation (7.1) signal $x(t)$ is the synchrophasor expression of the value X; in Equation (7.2), where ϕ is the instantaneous phase angle relative to cosine function at the nominal system frequency synchronized to Coordinated Universal Time, (UTC).

History of Synchrophasor Technology: The origin of the synchronized phasor measurement technology is computer relaying. The computer relaying first time launched in 1960s. Many algorithms were developed for computer relaying during 1970s-1980s. The most important outcome of which was symmetrical component distance relaying for protection of high voltage transmission lines. The major advantage of this algorithm is that for identification of the fault location of any type of fault, only one equation. has to be solved. With introduction of GPS satellite system time synchronization of the sampling clocks became feasible. Then the development of first prototype PMU was done by Virginia Tech research team in 1988. With many innovations in PMU, Macrodyne Co. developed commercial PMU along with numerous inbuilt features. They also developed phasor data concentrator which collected the data from the different PMUs. This data collection was time stamped, and the output was provided to the application software as per the requirement. The beginning of field installation by Virginia Tech team was done in three utilities in US which were Bonneville Power Administration, American Electric Power and New York Power Authority. After that Macrodyne started manufacturing these and next numerous field installations were done by the Macrodyne units.

Working Principle: Figure 7.16 presents the functional block diagram of the synchrophasor measurement system that is divided into seven blocks. In block-1 (CT/ PT) converts the three-phase current and voltage to the appropriate level required for A/D converter in block-3. At block-2, the harmonics and various interferences, from the signal are eliminated as they pass through an anti-aliasing filter (with cut-off frequencies as per Nyquist criterion) before reaching block-3. The combination of these three blocks is termed as data collection block-A. GPS block-5 contains the crystal oscillator, which provides sampling pulses to data collection block. The sampling clock and GPS clock are phase locked. CPU block-4 contains digital signal processing or microprocessor to evaluate the power frequency components of the inputs. The phasor measurements must be time stamped with GPS clock signal. Then communication block-6 gets time stamped measurements through the communication link through appropriate modems to phasor data concentrator. All the blocks of the PMU get power supply from power supply block-7.

7.5 Architecture of Communications Network in Smart Grid

IEEE 2030-2011 standard is being widely accepted as the industry's first guideline regarding its architecture and interoperability even though it is very difficult to reach into a consensus about the architecture and scope of a highly cross-functional infrastructure like the smart grid. The standard provides the smart grid interoperability reference model that uses a systems-level approach to provide guidance on interoperability among various components of communications, power systems, and information technology platforms in the smart grid. The communication technology perspective of the model provides a broad set of communication networks to interconnect the smart grid generation, transmission, and distribution and customer domains. Together they form an end-to-end smart grid communications network. A representative model of the end-to-end smart grid communications architecture based on the IEEE 2030 standard is presented in Figure 7.16

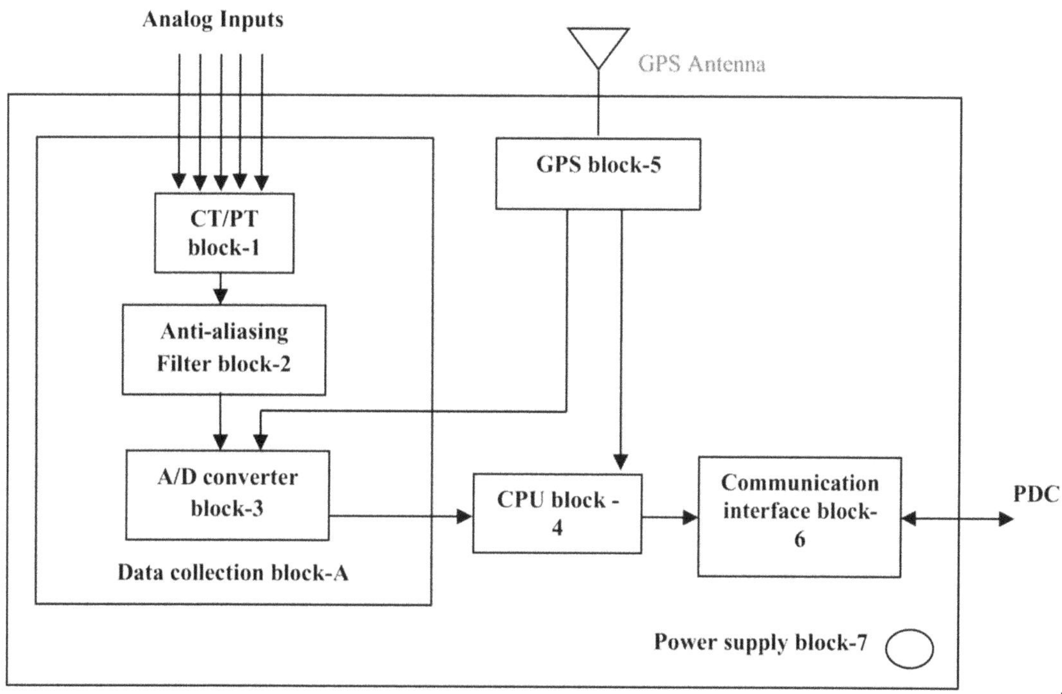

Figure 7.16 Functional block diagram of Synchrophasor measurement system.

The power generation and transmission domains are generally comprised of large substations that already have legacy communication infrastructures in place. A backbone network connects these substations with utility control center and 3rd party networks using mostly wired infrastructures such as Digital Subscriber Line (DSL), fiber and cables. The power distribution domain contains substations, feeders and end-user loads over a vast geographic area. A wide area communications network connects these utility infrastructures with the utility control center to enable grid-wise monitoring and control applications. In addition, it provides the 'last mile' connectivity to the customer premises to enable end-user applications such as metering and billing, demand management and load control. A backhaul network connects it with the utility's backbone network.

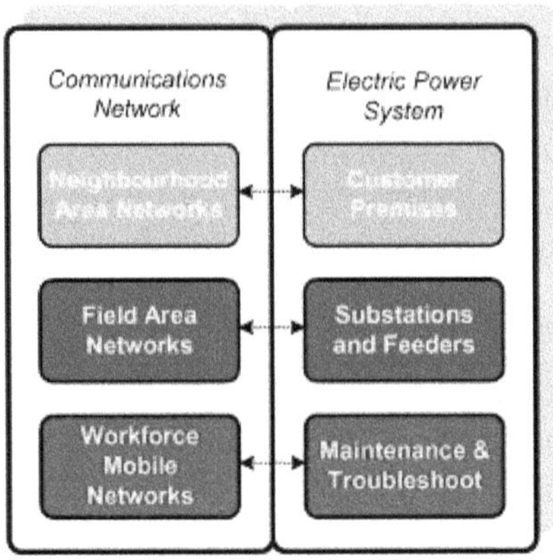

Figure 7.17 Logical architecture of the smart grid communications network.

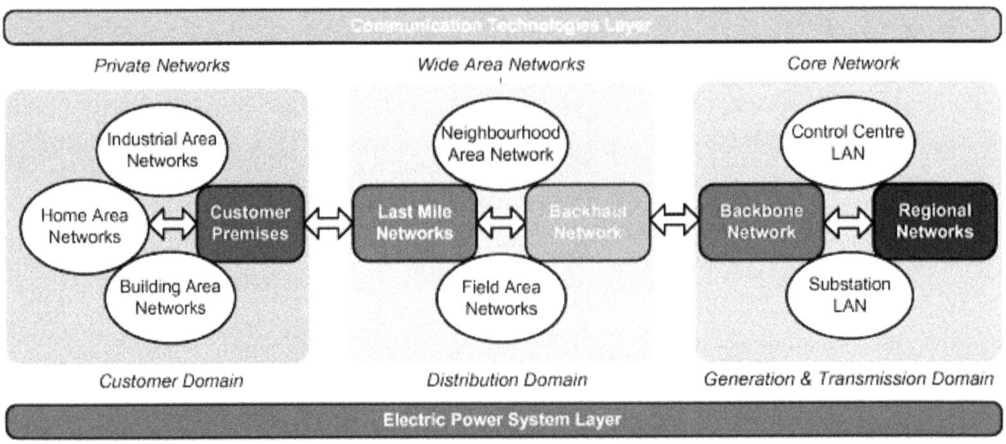

Figure 7.18 Smart Grid communications architecture as per the IEEE 2030 standard.

The IEEE 2030 standard specifies a logical architecture for the smart grid communications network comprising of the following three subnetworks – (1) Neighborhood Area Network for the customer premises, (2) Field Area Network for the utility infrastructures and (3) Workforce Mobile Network for the utility workforce as depicted in Figure 7.18. A brief description of these logical subnetworks is given in the following subsections.

7.5.1 Neighbor Area Networks (NANs)

A NAN is the logical representation of an Advanced Metering Infrastructure (AMI) system that connects customer premises with the utility control center. The basic constituent of an NAN/AMI network is the smart meter that performs a variety of intelligent metering tasks such as consumption metering, power quality monitoring and optionally perform load control activities. A smart meter can also act as an Energy Services Interface (ESI) that allows the private networks, for example, the Home Area Network (HAN) to exchange information with the AMI system. It supports a number of advanced applications such as remote load control, monitoring and control of distributed energy resources and electric vehicles, in-home display of customer usage and reading of non-energy meters (e.g., water and gas). Figure 7.19 illustrates the types of connectivity within a smart grid NAN. The end-points of a NAN can either be a standalone smart meter or a Data Aggregation Point (DAP) that collects data from/to a group of smart meters and then sends the aggregated information to the Meter Data Management System (MDMS) via a backbone network.

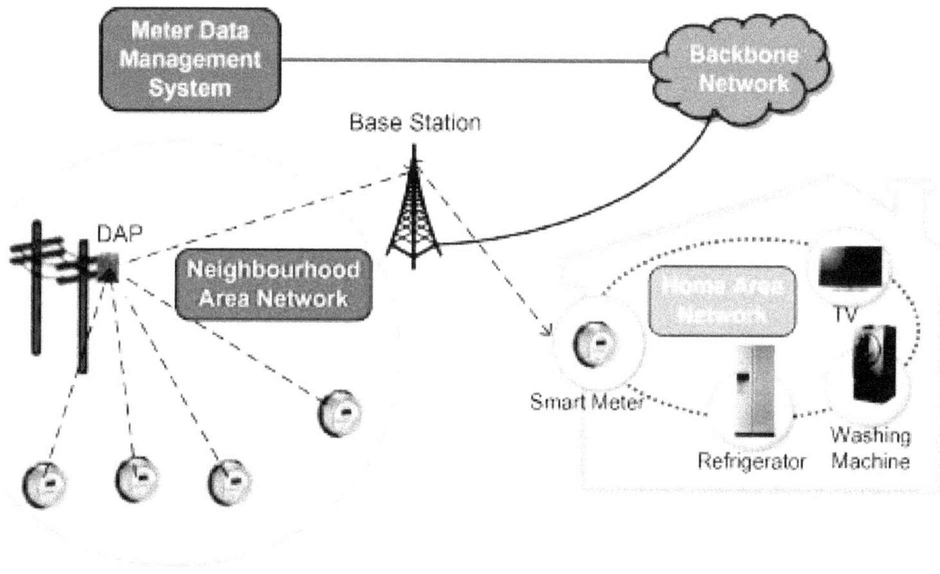

Figure 7.19 Connectivity within a Neighborhood Area Network.

7.5.2 Field Area Networks (FANs)

The FAN is responsible for facilitating information exchange between the utility control center and the distribution substation and feeder equipment for various monitoring, control and protection applications. The distribution substations convert the High Voltage (HV) electricity into a low voltage one as required by the homes and businesses. In addition to voltage transformation, they also isolate faults and are used as a point of voltage regulation. In a smart grid environment, they will be equipped with advanced monitoring and control equipment such as remote terminal units (RTUs), Phasor Measurement Units (PMUs) and IEDs to perform various substation automation functions.

The distribution feeders include transmission lines, cable poles and tower to provide electricity connection to the customer premises. They also act as the point of common coupling for the DER/microgrids that are connected with the distribution side of the grid. In the smart grid, there would be a number of sensor and actor networks overlaid along the distribution feeders for distribution supervision and monitoring applications.

7.5.3 Workforce Mobile Networks (WMNs)

The WMN is used by the utility workforce to provide dispatch, maintenance, and normal day-to-day operations. Typical requirements of a WMN include broadband connectivity to employees, including virtual private network (VPN), VoIP, and geographic information system (GIS) based applications for logistics or asset management. Moreover, in-vehicle applications and fleet telematics such as location-based services (LBSs) with global positioning system (GPS) based tracking and navigation, and automatic vehicle locating (AVL) are also expected to be integrated with the WMN system.

The WMN should be able to access both NAN and FAN via the wide area network to collect customer and equipment information and status. The communication requirements of a WMN are similar to that of standard telecommunication services such as voice, video, and internet applications. However, wide coverage and mobility support are typically required for many of its applications

References

1. M. M. Ahmed and W. L. Soo, "Supervisory Control and Data Acquisition System (SCADA) based customized Remote Terminal Unit (RTU) for distribution automation system," *2008 IEEE 2nd International Power and Energy Conference*, Johor Bahru, Malaysia, 2008, pp. 1655-1660, doi: 10.1109/PECON.2008.4762744
2. https://automationforum.co/different-types-of-scada-system-architecture/#htoc-architecture-of-scada-system
3. Shahzad A.,Musa S., Aborijilah A. & Irfan M. (2014), The SCADA Review: System Components, Architecture, Protocols and Future Security Trends. American Journal of Applied Sciences, 11(8), 1418-1425.

4. Geeta Yadav, Kolin Paul, Architecture and security of SCADA systems: A review, International Journal of Critical Infrastructure Protection, Volume 34, 2021, 100433, ISSN 1874-5482.
5. Sayed, Khairy & A.Gabbar, Hossam. (2017). SCADA and smart energy grid control automation. 10.1016/B978-0-12-805343-0.00018-8.
6. Momoh JA, Smart grid design for efficient and flexible power networks operation and control In Power Systems Conference and Exposition, PSCE '09. IEEE/PES; 2009. p. 1–8.
7. National Energy Technology Laboratory for the U.S. Department of Energy. Advanced metering infrastructure, NETL modern grid strategy; 2008.
8. Electric Power Research Institute (EPRI). Advanced metering infrastructure (AMI); 2007.
9. Depuru SSSR, Wang L, Devabhaktuni V. Smart meters for power grid: challenges, issues, advantages and status. Renewable and sustainable energy reviews; 2011. p. 2736–42.
10. US Department of Energy. Communications requirements of smart grid technologies; October 5, 2010.
11. Deign J, Salazar CM. Data management and analytics for utilities. FC Business Intelligence Ltd.; 2013.
12. Anderson D, Zhao C, Hauser C, Venkatasubramanian V, Bakken D, Bose A. A virtual smart grid. IEEE power & energy magazine; December 13, 2011. p. 49–7.
13. Cohen R. Is cloud computing really cheaper? Forbes; 2012
14. IEEE, IEEE 2030-2011, IEEE Guide for Smart Grid Interoperability of Energy Technology and Information Technology Operation with the Electric Power System, End-use Applications, and Loads, September 2011.
15. WiMAX Forum, WiMAX Applications for Utilities, October 2008.
16. Shahraeini, Mohammad & Javidi, Hossein. (2012), Wide Area Measurement Systems. 10.5772/35466.
17. Lee, K. Y. Perakis, M. Sevcik, D. R. Santoso, N. I. Lausterer, G. K. & Samad, T. (2000). Intelligent Distributed Simulation and Control of Power Plant, IEEE Transactions on Energy Conversion, vol. 15, no. 1, pp. 116-123
18. Cai, J. Y.; Huang, Z.; Hauer, J. & Martin, K. (2005). Current status and experience of WAMS implementation in North America, Proceeding IEEE/Power Eng. Soc. Transmission and Distribution Conference Exhibition, Aug. 2005, pp. 1–7.
19. Cassel, W. R. (1993). DISTRIBUTION MANAGEMENT SYSTEMS: FUNCTIONS AND PAYBACK, IEEE Transactions on Power Systems, Vol. 8, NO. 3, pp. 796-801
20. Clarke G.; & Reynders D. (2004). Practical Modern SCADA Protocols: DNP3, 60870.5 and Related Systems. Oxford: Elsevier EPG & CERTS, (2006). Phasor Technology and Real Time Dynamics Monitoring System (RTDMS) frequently asked question (FAQs), Electric Power Group and CERTS

21. Eshraghnia, R.; Modir Shanehchi, M. H.; & Rajabi Mashhadi, H. (2006). Generation maintenance scheduling in power market based on genetic algorithm, Proceeding 2006 IEEE Power Systems Conference and Exposition, 2006, pp. 1814 – 1819.
22. "IEEE Standard for synchrophasor Measurements for power systems", *IEEE C37.118.1-2011*.
23. "IEEE Standard for synchrophasor Data Transfer for power systems", *IEEE C37.118.2-2011*.
24. Rahul Gore and Mallikarjun Kande, "Platform Analysis in Embedded Systems", *2nd International Conference on Emerging Technology Trends in Electronics Communication and Networking (ET2ECN)*, 2014.
25. M. Kanabar, M.G. Adamiak and J. Rodrigues, "Optimizing Wide Area Measurement System Architectures with Advancements in Phasor Data Concentrators (PDCs)", *IEEE Power and Energy Society General Meeting (PES)*, 2013.
26. M.G. Adamiak, M. Kanabar, M.D. Zadeh and J. Rodrigues, "Design and implementation of a synchrophasor data concentrator", *IEEE PES Conference on Innovative Smart Grid Technologies-Middle East*, 2011.
27. Nageswara Rao, A., Vijaya Priya, P., Kowsalya, M., & Gnanadass, R. (2017). Wide area monitoring for energy system: a review. *International Journal of Ambient Energy*, *40*(5), 537–553.
28. O. Muhayimana and P. Toman, "Overview on Synchrophasors Technology and Trend of their Application in Electrical Network Protection," *2022 22nd International Scientific Conference on Electric Power Engineering (EPE)*, Kouty nad Desnou, Czech Republic, 2022, pp. 1-6, doi: 10.1109/EPE54603.2022.9814107.
29. S. Almasabi and J. Mitra, "An overview of synchrophasors and their applications in smart grids", *International Conference on Intelligent Control Power and Instrumentation (ICICPI) IEEE 2016*.
30. M. Mohanty, R. Kant, A. Kumar, D. Sahu and S. Choudhury, "A Brief Review on Synchro Phasor Technology and Phasor Measurement Unit", 2020.
31. S. Nuthalapati, Power System Grid Operation Using Synchrophasor Technology, Cham:Springer International Publishing, 2018.
32. M. N. bin Mohd Nasir, A. Sabo and N. I. A. Wahab, "A Review on Synchrophasor Technology for Power System Monitoring", *in 2019 IEEE Student Conference on Research and Development (SCOReD)*, pp. 58-62, 2019

CHAPTER

8

Smart Grid Protection

8.1 Introduction

Failures in the power system can occur due to various reasons, such as accidental damage to cables, weather-related events, damage to power poles, and other unforeseen events. During fault conditions, abnormally high energy can flow through critical equipment, leading to failures and disruptions in electricity supply. Therefore, it is vital to protect these equipments and ensure reliable operation for the entire electricity grid. Protection devices such as fuses, and protective relays are widely used to identify and isolate faulty areas from operational networks. These devices operate at a pre-defined protection setting, known as conventional protection, which is obtained from a power system study on the static electrical network.

Power networks, especially distribution networks, have been undergoing substantial changes since the application of new technologies. Technology development in the beginning of the 21st century has opened up new horizons for automated, efficient and reliable power grids. New technologies, while enhancing the capability of the electrical networks and providing efficient solutions to the challenges of future networks, can also introduce new challenges that should be investigated and mitigated. Distributed Generation (DG) is one of such new technologies that improves the operation of power grids. The rate of integration of DG units to the distribution level to meet the increasing power demand grows as a reasonable replacement for costly network expansion. While this integration brings many advantages to consumers and power grids, it also becomes a challenge from protection and control perspective. Integration of DG units cause challenges and concerns, regarding how a fault should be detected and isolated in active distribution networks. The major effects of DG integration include change in the direction of fault current flow, increase or decrease of fault current magnitude, the blindness of protection, sympathy feeder trip, nuisance trip of the circuit interrupting devices, and disruption of coordination between protective devices (PDs). This chapter presents various challenges in smart grid protection followed by the proposed protection schemes based on Adaptive techniques, Wide Area Monitoring Systems (WAMS) and Multi Agent Systems (MAS) to mitigate these challenges.

8.2 Challenges in Smart Grid Protection

Electrical networks are subjected to short circuit faults that can damage equipment and cause hazardous conditions to utility crews and consumers. The faults should be detected and cleared as fast as possible in order to minimize the damage and restore the supply at the earliest. Usually, a three-phase fault poses the highest threat in terms of stability of the grid. Under certain circumstances, the other fault would also cause the highest fault current. For instance, the current magnitude of a phase-to-ground fault near a solidly grounded generator is drastically higher than a three-phase one. The fault current level depends on the network topology, grounding arrangements, as well as the number of in-service generators. In addition to the pick-up settings of Protective Device (PDs), coordination between some types of PDs depends on the fault current level. Thus, anything that can affect them, should be considered in the fault calculation.

8.2.1 Changes in Short Circuit Levels

In a grid with numerous high-capacity sources, the addition of small DG units does not change the short circuit current magnitude. On the other hand, the integration of many small-and medium-sized DG units or high-capacity DG units should be considered in the fault current calculations since these certainly impact the fault current level. The contribution of a DG unit to a fault current is depends on the type of DG unit, which can be categorized as synchronous generators, induction generators and inverter-based units. Based on several studies it has been observed that the contribution of inverter-based DG units to the fault current would not exceed 200% of their rated current. The contribution of different types of DG units to the fault current is summarized in Table-8.1

Table 8.1 Fault current contribution of different types of DGs.

DG Type	Contribution to Fault Current
Inverter based	1-2 times the inverter rated current; Duration: from a ½ a cycle to several cycles depending on the control method
Separately Excited Synchronous Generator	5–10 times the generator rated current for the sub transient and transient cycles, 2–4 times the generator rated current under steady-state
Induction Generator or Self-excited Generator	5–10 times the generator rated current for the first few cycles, negligible after 5–10 cycles

It can be understood that regardless of its type, will contribute to the fault current even if this contribution is negligible, the integration of DG units at the distribution level would cause the short circuit current in other parts of the network to change.

Since many DG units can be installed at the distribution level, the effect of DG units on the recloser-fuse sets has been investigated in several studies. The connection of a DG unit after recloser as shown in Figure 8.1(a) can affect the fault current flowing through the recloser. If a fault is developed on the tapped line and the capacity of the DG is significant, then the fault current through the fuse could be greater than the fault current through the recloser. According to Figure-8.1(b) this condition will worsen with several DG units connected to downstream feeders.

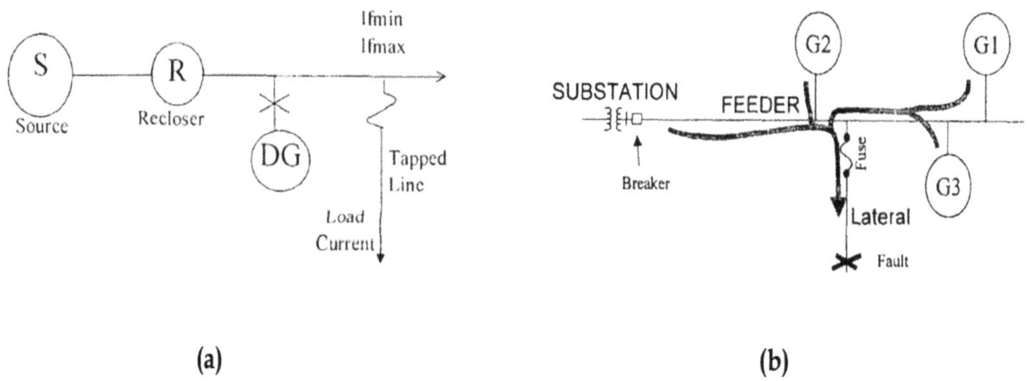

Figure 8.1 (a) Effect on the current of recloser and fuse; (b) Recloser–fuse set with several DG units connected after recloser.

In addition, the distance between the interconnected DG unit and the fault location would affect the fault current seen by recloser and fuse. Higher change in fault level will occur if the DG unit happens to be a synchronous generator. On the other hand, inverter-based DG units have the least effect on the short circuit current.

Generally, several distribution feeders are connected in parallel to sub-transmission substations. In this case, any DG unit on one feeder would affect the fault current through the upstream circuit breaker (CB)/recloser as well as fuses on parallel feeders. For instance, as illustrated in Figure 8.2 for a fault on feeder 2, the fault current seen by CB1/Fuse1 will be greater than the short circuit current through the main CB on the grid side due to contribution of DG1. In other words, the introduction of the DG unit to the feeder 1 will lead to the decrease of the fault current through upstream CB and increase of the fault current through CB1/Fuse1.

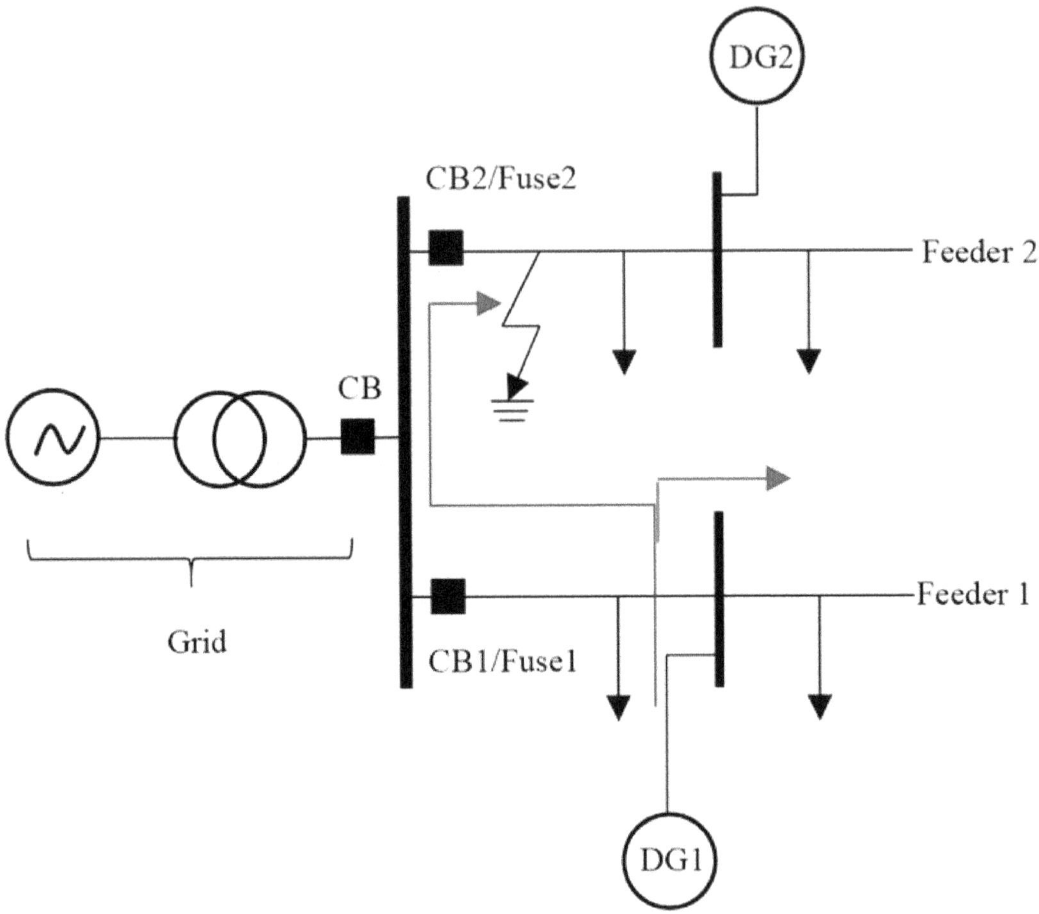

Figure 8.2 Parallel distribution feeders with their interconnected DG units.

The fault current seen by CB2/Fuse2 in this case depends on whether there is a DG on feeder 2 or not. In the case of a small size DG2 on feeder 2, the short circuit current through CB2/Fuse2 is substantial. On the other hand, a larger DG2 leads to a small fault current flowing through CB2/Fuse2.

Several DGs Connected to Feeder: Figure 8.3 presents a generalized circuit to calculate the contribution of each DG to the CB/fuse short circuit current. Here, PD is the corresponding CB or fuse of the feeder. I_1 indicates the contribution of the DGs on adjacent feeders and the grid side. I_2 is the contribution of all the PVs between the PD and the fault. and I_3 represents the contribution of all the DGs after the fault location.

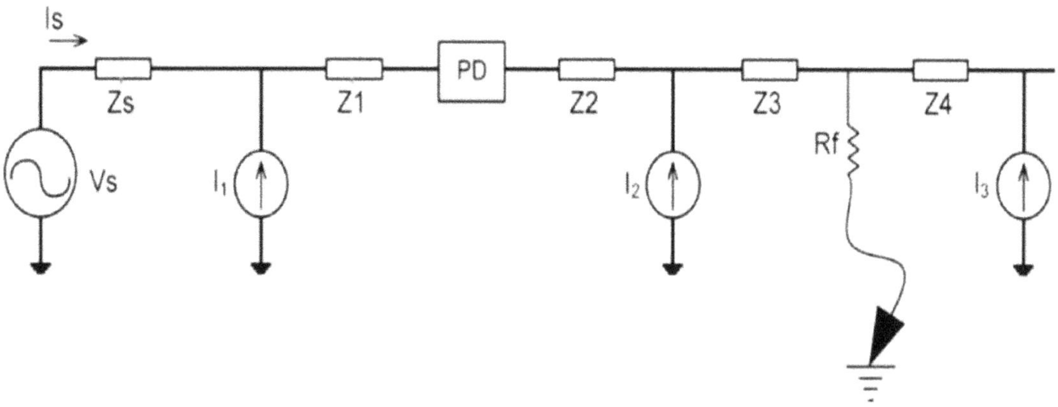

Figure 8.3 Contribution of different PV sources to the fault current.

The fault current seen by the PD can be approximately given by

$$I_{PD} = I_s + \frac{Z_S}{Z_S+Z_1+Z_2+Z_3+R_f} I_1 - \frac{Z_S}{Z_S+Z_1+Z_2+Z_3+R_f} I_2 - \frac{Z_S}{Z_S+Z_1+Z_2+Z_3+R_f} I_3 \qquad .....(8.1)$$

Depending on the size of DGs at each part of the feeder, the fault current would increase or decrease. In addition, thc duration of thc contribution of each DG to the fault current is related to their distance from the fault. The farther the DG is, the longer its contribution to the fault will be. The reason is that the terminal voltage of a remote PV source drops lower compared to the one close to the fault. Since the short circuit current of the PV source does not exceed twice its rated current, the contribution of the PV sources will be smaller than the contribution from the grid. However, any level of contribution to the fault changes the fault current seen by CBs and fuses which leads issues in protection.

It has been observed that the sizes of connected DGs and the location of the fault have a significant impact on the fault current. For a fault on one feeder, the short circuit current in the adjacent feeder is higher when more DGs on the adjacent feeder come into service. The higher fault current on the neighboring feeder happens when the fault is developed farther from the beginning of the other feeder except for the phase-to-ground fault cases.

8.2.2 Reversed Power Flow

In the conventional grid, the power flows from the transmission to the downstream loads on the distribution side, which makes the flow of power unilateral. The integration of DG units to the distribution level of the power network would cause the power flow direction to change depending on the size of the DG unit. Figure 8.2 also presents an example for reverse power flow (ignoring the fault and DG2). Without DG1, the power only flows from the grid to the loads on feeders 1 and 2 while the presence of DG1 would change the direction of power flow in part of feeder 1. The size of the DG should be large enough to provide the required power for loads on feeder 2. If the reverse current flowing through CB1/Fuse1 goes beyond the current setting of CB1 or rated current of Fuse1, CB1/Fuse1 would operate, which is to be avoided.

8.2.3 False Tripping of Protective Devices (PDs)

False tripping, also known as a nuisance or unwanted tripping, is a condition in which an inappropriate PD operates or backup PD functions sooner than the primary one. False tripping can cause a healthy feeder to become disconnected or an unnecessary outage of a distribution feeder to occur (this type of false tripping is also known as sympathetic tripping). For example, in Figures 8.1(a) and 8.1(b) the fuse on the lateral feeder should only clear permanent faults. All temporary faults, which occur most of the time, should be cleared by the upstream recloser on the substation side. Locating the DG units at any point after the recloser will increase the fault current through the fuse and therefore, the fuse would blow up for a temporary fault and give rise to an unnecessary feeder disconnection. Another example of unwanted tripping is illustrated in Figure 18.2, in which a fault on feeder 2 may cause the CB or the fuse located on the adjacent healthy feeder to operate incorrectly. In this case, before DG integration, the fault on feeder 2 is only fed by the grid while after DG integration, the fault current can also flow through feeder 1. If the capacity of connected DGs is significant, the fault current through CB1/Fuse1 can exceed the setting of the CB or the rated current of the fuse and prompt incorrect operation of CB1/Fuse1.

It has been established by several researchers that the interconnection of DG units to downstream feeders in distribution networks would cause false tripping of PDs on the healthy feeders. In looped distribution networks or networks with multiple distribution feeders connected to a substation, the situation can get complicated and even deteriorate because different PDs can be falsely affected, and numerous healthy feeders may be taken out of service due to DG units on these healthy feeders.

8.2.4 Blinding of Protection

Blinding of protection happens when a PD, which must operate and clear the fault, is blocked. For instance, when the fault current seen by an overcurrent PD (OCPD) falls beyond its setting due to changes in network configuration, the PD cannot detect the fault and, consequently, it does not trip.

Most of the faults in transmission and distribution networks are temporary faults and, therefore, the faulty feeders do not need to be taken out. Reclosers, in this case, operate two or three times and close the feeder breaker. If the fault is a self-cleared one, the reclosing action will be successful. On the other hand, when a permanent fault is developed on the lateral feeder, the recloser will be blocked after two or three recloses and other PDs will function and remove the fault. Some of the case studies reported in the literature are as follows:

Case-1: Figure 8.1(a) presents one of the most common blinding of protection cases. When a temporary fault occurs on the tapped line, the fault current through the recloser drops due to the presence of DG after the recloser and as a result, the recloser would not clear the fault before the fuse melts. This problem is depicted in Figures 18.1(b) and 18.2. The size of the DG unit, location of the fault, network impedance and fault resistance can affect the fault current through an OCPD.

Case-2: According to equation (8.1) of Figure 8.3, if the sizes of DGs (after the PD and before the fault) are significant, their contribution to the fault will grow, resulting in a decrease of the fault current seen by PD which may lead to blinding of OCPD.

Case-3: In this case, the effect of DG impedance on location of the fault/OCPD blinding is demonstrated for a medium voltage (MV) network with a DG unit as shown in the Figure 8.4(a). Figure 8.4(b) presents its Thevenin equivalent circuit. An OCPD is installed at the beginning of the feeder before the DG unit.

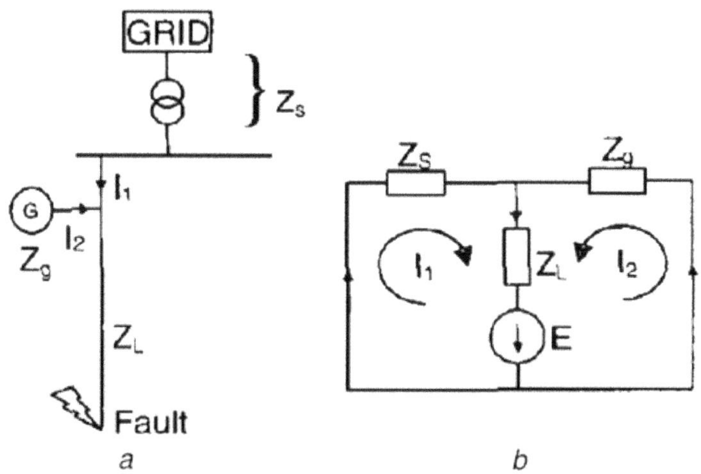

Figure 8.4 MV network with DG unit; (a): Single-line diagram, (b): Thevenin equivalent circuit.

If I_k and I_1 are the fault currents seen by OCPD for a feeder fault in the absence of DG and in the presence of the DG, then the ratio between these two currents can be given by equation (8.2).

$$\frac{I_k}{I_1} = \frac{Z_g(Z_s+Z_L)}{Z_s(Z_g+Z_L)+Z_L Z_g} \quad .....(8.2)$$

where Z_s is the aggregate of the impedances on the grid side, $Z_g$shows the impedance of the DG and $Z_L$represents the impedance of the feeder to the fault point.

It can be seen from the equation (8.2) that the impedance of the DG Z_g can affect the fault current in addition to the fault location. In other words, the connection of the DG will result in a reduction of the fault current seen by the relay at the beginning of the feeder. If the grid impedance is very high and the impedance of the feeder to the fault point is negligible (e.g. fault occurs at the very beginning of the feeder), the fault current seen by the feeder relay before and after the connection of the DG will be almost equal. However, if the size of the DG is high and the impedance to the fault point is significant, the fault current seen by the relay in the presence of the DG unit will decrease significantly.

Note: The voltage magnitude and angle of the DG unit in Figure 8.4 is assumed to be equal to the grid voltage magnitude and angle during the fault for the sake of simplicity. In reality, these voltages are different, and they have an impact on the fault current measured by the feeder relay. However, this effect can be neglected comparing to the effect of impedances.

Case-4: Distance relay calculates the impedance based on the measurement of the voltage and current at the installation point, and therefore any change in the voltage and/or the current can change the measured impedance. For instance, as shown in Figure 8.5 a fault at point F_2 is in the second zone of relay R_1 before interconnection of the DG unit.

After DG integration, the fault current I_{F1} decreases and relay may see the fault in its third zone meaning the DG unit gives rise to the under-reach of the distance relay or R_1 cannot detect the fault at all. The same problem can occur for the fault at point F_1. This fault is in the first zone of relay R_1 before the DG connection. However, the presence of the DG unit will cause the fault current through this relay, I'_{F1} to reduce, resulting in the under-reach phenomenon in distance relayR_1 .

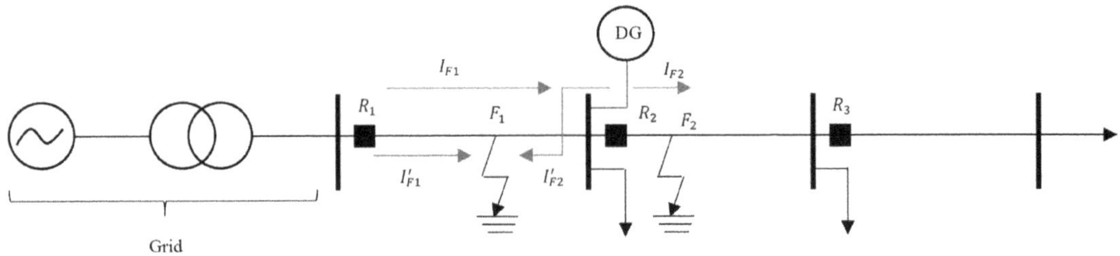

Figure 8.5 Effect of DG units on distance relays.

8.2.5 Nuisance Islanding

Nuisance islanding, also known as unwanted islanding or unwanted/nuisance tripping of production units. This refers to a condition in which a part of distribution system unintentionally becomes isolated from the rest of the power system.

In conventional distribution networks, when the distribution network gets disconnected from the rest of the system for any reason, the network is completely de-energized, and therefore no problem arises in terms of personnel safety and protection systems. However, active distribution networks, with several DG units, would still be energized even though it is isolated from the main generation source. It is hazards for maintenance crews. Moreover, since the protection system of distribution networks is not designed to deal with the islanding conditions. the short circuit level in the isolated part becomes low due to a reduction in generation. If a fault occurs in the isolated part, the protection system might not detect it during islanding mode and the fault remains energized. Consequently, in addition to the destructive damage to equipment, people will also be exposed to fatal electrical hazards.

8.2.6 Unsynchronized Reclosing

Most of the faults throughout power systems are temporary and self-clearing. Therefore, the application of reclosers for momentary disconnects is prevalent now-a-days, especially with respect to distribution grids to improve the power quality of the networks. When a fault happens, the PDs detect and initiate a trip to CBs, and thus the fault feeding of will be interrupted. After the reclose interval period (dead time), the recloser closes back the CB. The reclose interval time is usually between 0.2 and 0.5 s depending on the network conditions. This procedure is typically implemented two to four times. For temporary faults, the reclosing function will restore power to consumers, while for permanent faults, the recloser will be blocked after two to four reclosing attempts and consumers at the distribution level will be exposed to sustained interruption until maintenance crews detect and eliminate the root cause of the fault.

The introduction of DG units to distribution networks means continuous feed to the fault even though the CB at the head of the feeder is tripped by its corresponding PD. The DG will keep injecting the current to the fault, and hence the fault is not cleared and the arc at the fault location is not fully dissipated. Consequently, the recloser procedure will fail. This condition is called unsynchronized/out-of-phase reclosing

The damaging of rotating DGs, deterioration of the equipment insulation, reduction of apparatus service life and increase in maintenance costs are some impacts of out-of-phase reclosing. Additionally, power quality issues, such as prolonged power outages, is another undesirable effect of reclosing failure. However, inverter-based DG units would be less vulnerable to unsynchronized reclosing because they are equipped with current surge protection.

8.2.7 Loss of Main Power

If the main grid is disconnected for any reason, a part of the system will be in islanding mode. DG units in the islanded part can cause safety problems. Moreover, the DG units contributing to the fault current may not trigger the OCPDs in the islanded part because the low fault current magnitude.

8.3 Adaptive Protection Schemes

Failures in the electrical network can occur due to various reasons, such as accidental damage to cables, weather-related events, damage to power poles, and other unforeseen events. During fault conditions, abnormally high energy can flow through critical equipment, leading to failures and disruptions in electricity supply. Therefore, it is vital to protect these equipments and ensure reliable operation for the entire electricity grid. Protection devices such as fuses, and protective relays are widely used to identify and isolate faulty areas from operational networks. These devices operate at a pre-defined protection setting, known as conventional protection, which is obtained steady state analysis of power system network.

With the rise in renewable energy integration, the load current supplied from the grid varies, leading to a mismatch between the existing protective relay settings and the actual network conditions, necessitating a reassessment of the settings which can no longer accurately reflect the network state. This calls for a dynamic or adaptive protection system for varying current and voltage due to penetration of renewable intermittent sources, changing load demand and network topology variation. Conventional relay fails to detect these variances, which poses challenges to its reliability and stability. This section presents some of the adaptive protection schemes reported in the literature to mitigate the challenges due to the formation of the grid.

Scheme-1: Figure 8.6 presents an adaptive protection scheme where a central computer responsible for processing calculations in the distribution system. The power system model's relay status is monitored by the substation computers, which communicate with the central computer to provide relay data. Time delay setting (TDS) of the relay is updated based on the changes in the line current in the network.

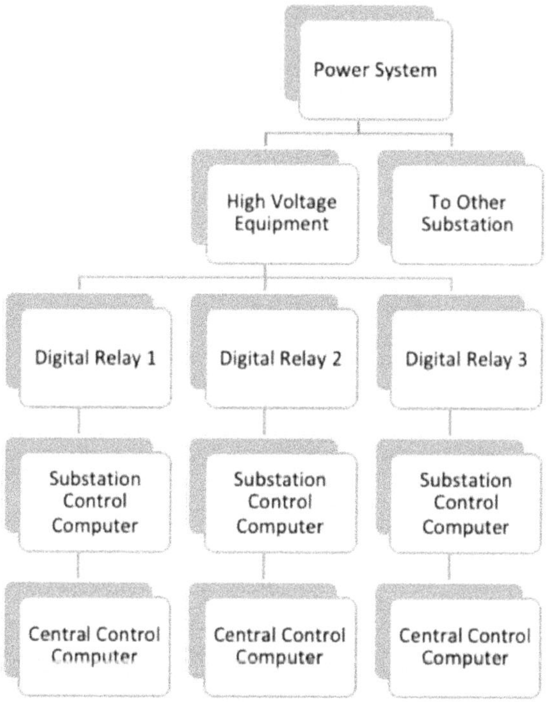

Figure 8.6 An adaptive protection scheme based on central computer monitoring.

Network status checks are done by detecting the status of all relays in the power system. The substation computers monitor the relay status and communicate with the central computer. It updates the power system model based on the network relay status and recalculates the protection settings.

Scheme-2: In this adaptive method overcurrent relay settings are updated using two techniques i.e., change in load current and insertion or removal of distributed generation. In this scheme pick-up current setting of the relay is adjusted periodically based on the load current in the distribution network as per the equation given by equation (8.3).

$$I(Pickup) = 1.5 * I(Line) \qquad \text{.....(8.3)}$$

when a DG is connected to the network or disconnected from the network, the pick-up current and TDS of the relays are updated per the equations (8.4) and (8.5)

$$I(Pickup: New) = I(Pickup: Old) \pm \Delta I(Line) \qquad \text{.....(8.4)}$$

$$\frac{TDS\ new}{TDS} = \frac{\left(0.0515+0.114(M^{0.02}-1)\left(M'^{0.02}-1\right)\right.}{\left(0.0515+0.114\left(M'^{0.02}-1\right)(M^{0.02}-1)\right.} \qquad \text{.....(8.5)}$$

The flowchart in Figure 8.7 presents the proposed algorithm

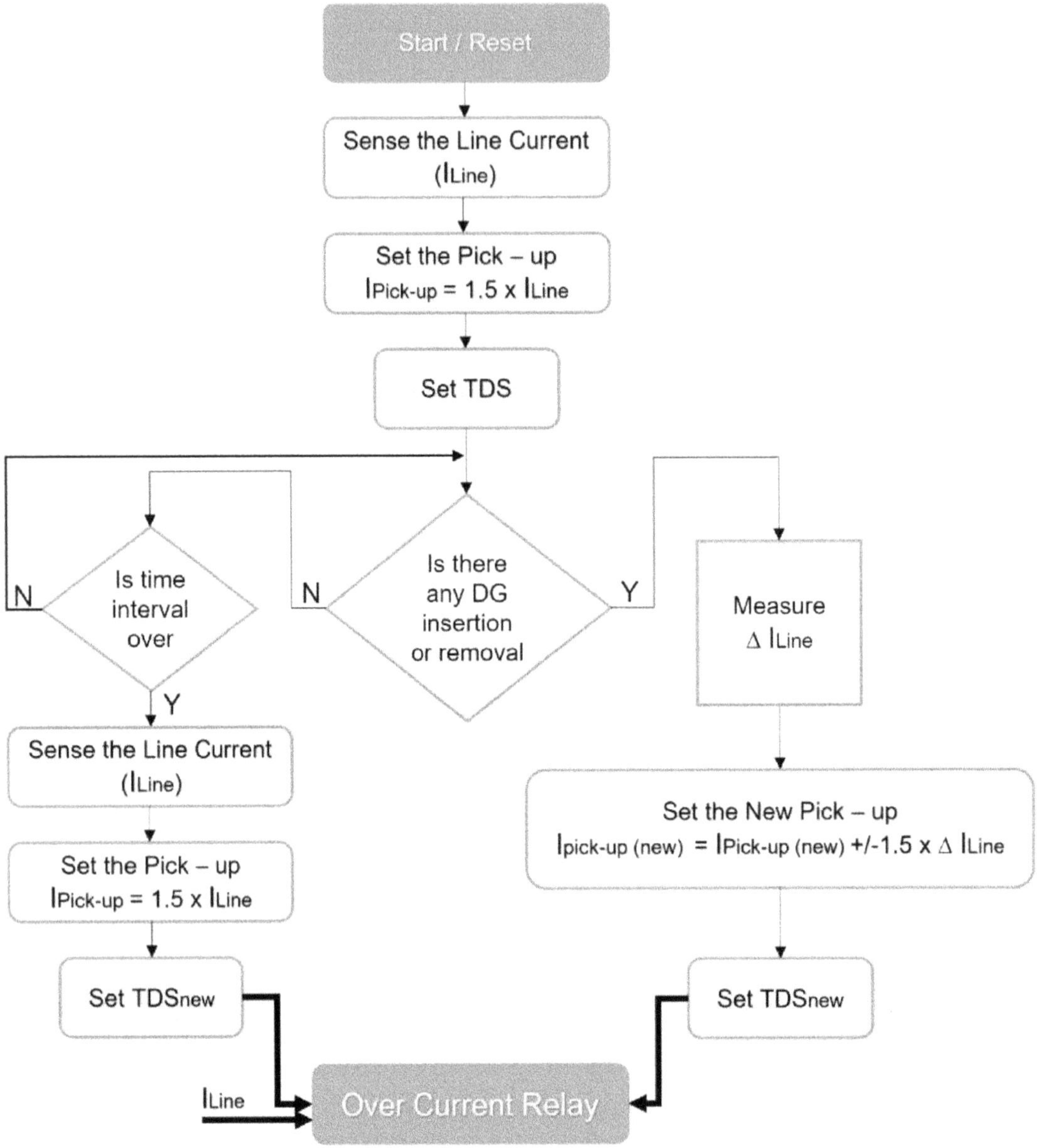

Figure 8.7 Flowchart for the proposed algorithm.

Scheme-3: Figure 8.8 depicts a scheme in which the connections and data flow within the PLC system for an adaptive algorithm deployed in a distribution network with DG integration. The protective relay receives data from the network, particularly the system voltage (U) and current (I) and transfers it to the PLC. The PLC then utilizes these data to calculate new relay settings based on predefined algorithms. PLC sends the data to the processor, which commands the switching devices to operate. The structure comprises a graphic terminal for display and operation.

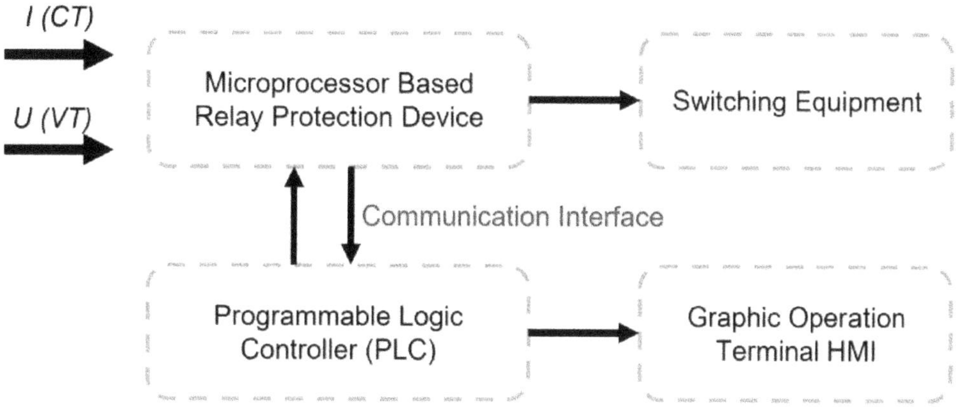

Figure 8.8 Structural diagram for proposed adaptive protection scheme.

However, it is important to note that even though the PLCs are effective in performing certain tasks, they are not usually used as a power system analysis tool, especially in the context of large distribution networks. Figure 8.8 highlights that the PLC values are derived from separate analysis results. Therefore, to achieve a more comprehensive analysis, it would be beneficial to integrate the PLC system with other power system analysis tools or software. Integrating the PLC with additional analysis capabilities, such as dedicated power system analysis software or hardware, can enhance the system's ability to handle complex analysis tasks. This integration would provide a more holistic approach to power system analysis within a large distribution network.

Scheme-4: In this proposed adaptive approach setting of the relays are updated based on a voltage threshold limit. The threshold limits are set at 0.95 and 1.05 per unit. In Figure 8.9 the process involved in the relay setting changing is depicted.

When the measured system voltage falls outside these threshold limits, adaptive settings are employed. Fourier transforms are used to calculate RMS value of the system voltages. These techniques enable the identification and analysis of the system voltage characteristics, facilitating the decision-making process for relay setting adjustments.

Scheme-5: The proposed methodology follows a three-stage process, as illustrated in Figure 8.10. In the first stage, an offline model is developed to understand the behavior and characteristics of the distribution network. This offline model serves as a foundation for subsequent analysis and protection scheme design. In the second stage, protection zones are established based on the network's topology. These protection zones define specific areas within the network where fault detection and isolation are performed. By carefully designing the protection zones, it becomes possible to address the challenges posed by DGs and their impact on fault currents. In the final stage, real-time monitoring of variables such as current and angles is carried out to continuously assess the network's condition. If any violation of the protection zone occurs, a trip command is issued to isolate the fault and protect the network components.

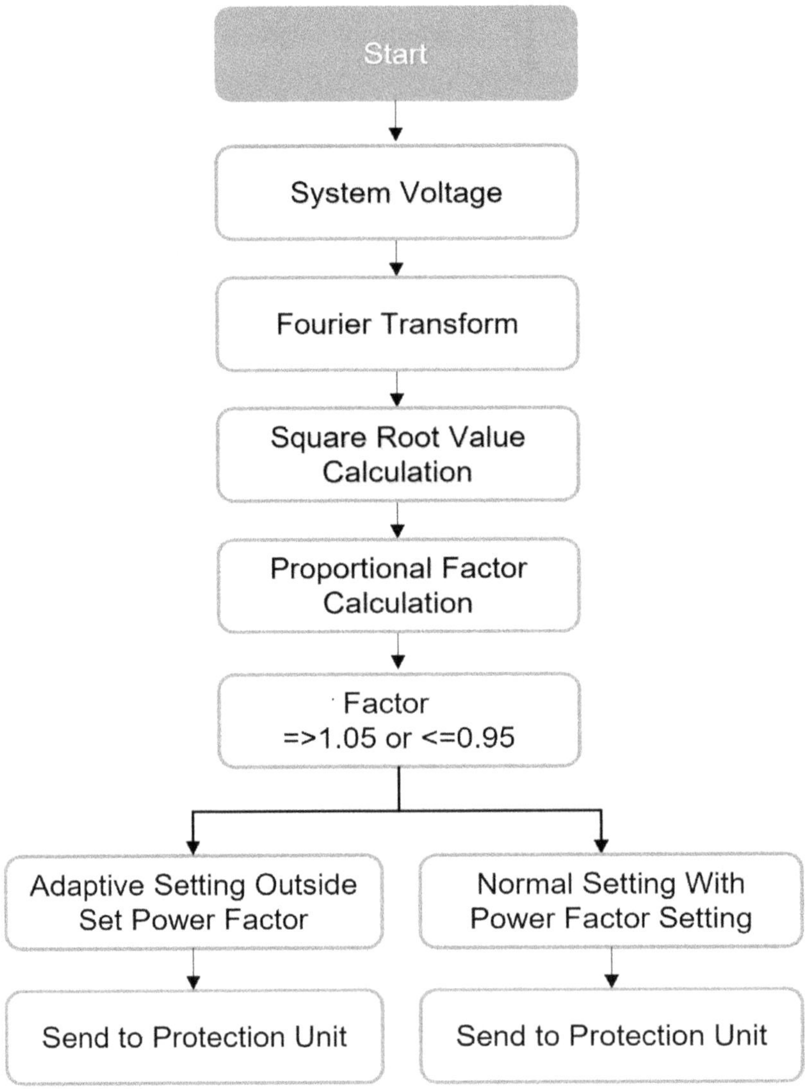

Figure 8.9 Flow chart for proposed adaptive algorithm.

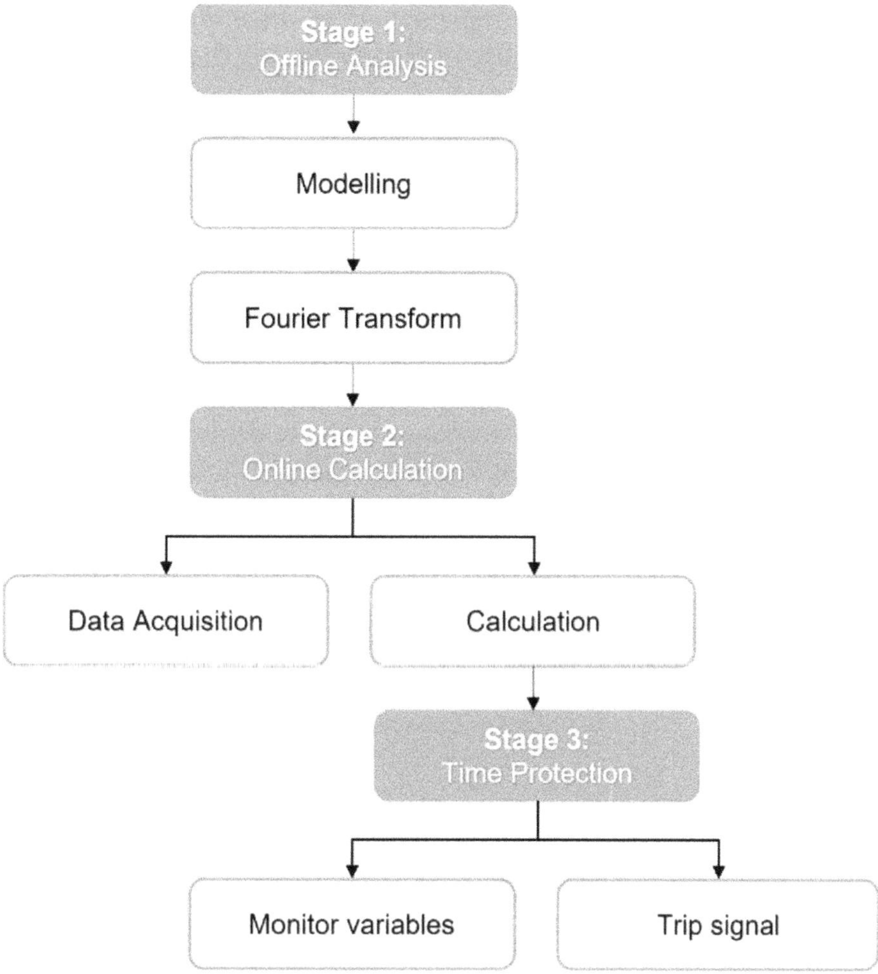

Figure 8.10 The proposed three stages adaptive protection scheme.

8.4 WAMS based Protection or Wide-Area Protection Systems (WAPS)

The most important device in the Wide-Area Protection And Control (WAPAC) is the phasor measurement unit (PMU). PMUs are deployed at various locations in the smart grid to provide real-time monitoring for measuring the voltage, current, angle, frequency etc. PMUs are used for protection and monitoring purposes. Wide-area monitoring system (WAMS) uses device sensors deployed everywhere on the grid in combination with global positioning system (GPS) satellites for accurate time stamping of measurements. The incorporated sensors will interact with the communication media. The information and communication technology (ICT) is highly applied in smart grid to utilize and gather data information about the conduct of suppliers and consumers, in digital method to ameliorate, and enhance the efficiency, economics and sustainability of the electricity. The smart grid considers the integration of electrical and communication infrastructures with advanced process automation and information technologies within the existing electrical network.

A WAMPAC can be achieved in the PMU by making it to act as a server in addition to its data collection functions. Figure 8.11 shows the WAMPAC system. Some modules respond faster to isolate the faulted portion. The PMU is designed to be located in a critical location on the power system. This location can be substations, generation buses or at interconnection points. The PMU sends the data through communication media to PDC. The data is exported in the shape of the stream and then correlated and normalized. The response of the wide-area control architecture is achieved through the control scheme. On a geographic wide area, the static and dynamic data can be collected and monitored in a centralized control center which termed as a wide-area measurement and control.

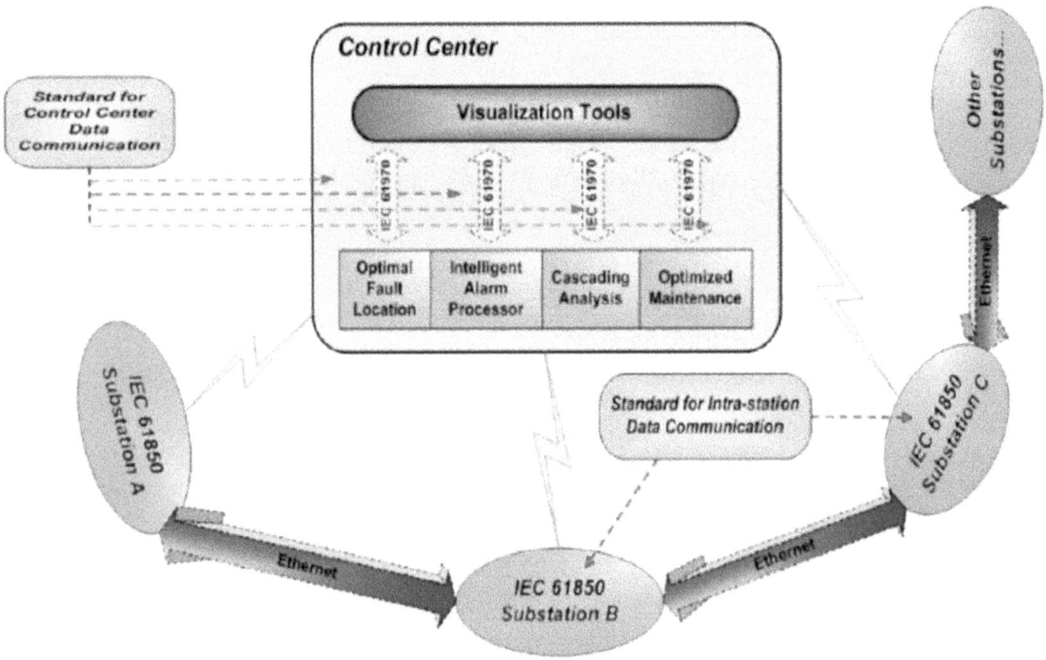

Figure 8.11 WAMPAC system.

In a power system equipped with WAMPAC systems, high-speed communication networks and synchrophasor data can be leveraged to update the relay setting. Moreover, new advancements and IoT-enabled relays have led to the large-scale production and installation of computer-based relays that can communicate with each other and the control center using various technologies and communication links (e.g., protocols specified in IEC 61850). Consequently, several new wide-area schemes can be added to the protection systems to improve power system performance. Some of the new wide-area schemes are briefed as follows.

Remote Access Service: It represents one of the main WAPS functions, which is defined as detecting an abnormal state of a system from a pre-defined set of possible events. The goal is to respond to detected abnormalities with rapid, corrective, and automated actions, such as load/generator shedding, generation rescheduling, and line tripping. Utilization of such actions should limit the fault consequences, prevent cascading events, and guarantee safety measures of the power system

Fault Localization*:* Electrical faults can occur in either transmission lines or substations. Despite the relatively simple fault localization in the substations, identifying the location of the fault in transmission lines is a challenging task due to their widespread and extended structure. It is of significant importance to precisely detect the fault location since it enables the Power System Operator to fix or remove faulty parts and prepare the line for normal operation

Backup Protection*:* In a power system, each relay's setting and location are coordinated so that each one protects on its own zone of the power system. These zones are designed deliberately to have overlaps, where some of the relays operate as backups to ensure no fault remains undetected. In realistic systems, however, due to under-stressed operation conditions, such a local backup protection scheme may mal-operate. As a result, wide-area backup protection schemes, which make use of PMU data and communication between the relays/control centers, can be deployed. This scheme ensures that the fault is cleared rapidly, and backup protection is provided for essential sections of the system followed by the malfunction of any relay.

Emergency Protection*:* The power system is subjected to various operating conditions, such as faults and overloading. The occurrence of these disturbances can make the power system vulnerable to cascading outages or instability. The existence of wide-area infrastructure provides the ability to predict such conditions and perform emergency protection schemes to save some parts of the system and prevent cascading outages.

Short Circuit Current Calculation: In the protection scheme, relays measure the current and then, according to their setting, which is often a time-current inverse curve, decide whether the fault has occurred or not. Thus, the current that passes through the fault and, consequently, its clearing time depend on power system impedance and topology. Since the power system topology can change over time, the wide-area structure provides the ability to rapidly calculate the short circuit current based on the new power system topology and update the relay setting accordingly.

The following example illustrates the application of WAPS. The Figure 8.12 depicts a power system where two generators at Buses 1 and 3 supply the load located at Bus 2. Assuming an event when, Line 1 (L1) is disconnected from the system due to a fault. In a system without a WAPS structure, the output of the generators remains unchanged, and the sum of power transferred through Lines L1 and L2 now passes through Line L2. Such as considerable increase in the power transfer may trigger Relay 2, disconnect the generator from the system, and put the system under immense stress. However, WAPS applications can observe the status of CBs and change the generation of G2 so that the power that passes through L2 remains below a critical level.

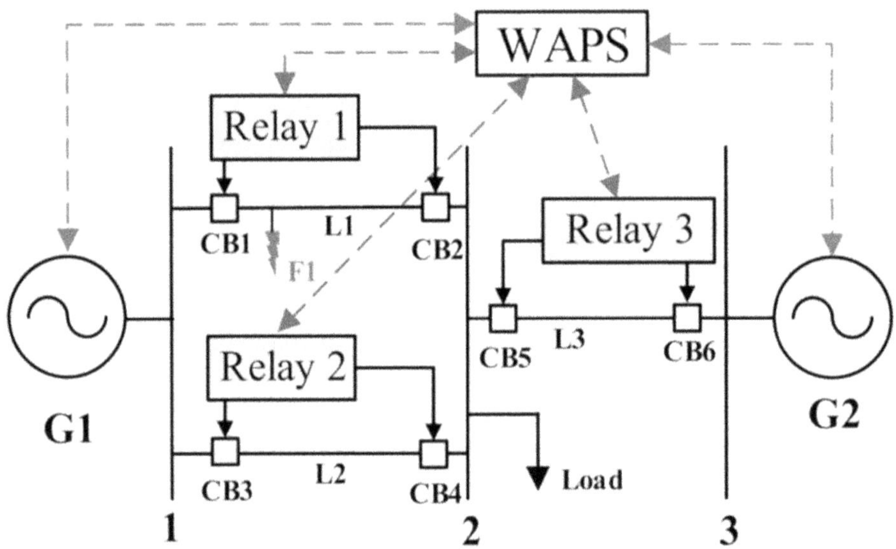

Figure 8.12 Illustrative example of a wide-area protection system.

8.5 Multi Agent Systems - Smart Grid Protection

To make the smart grid resilient, an advanced and intelligent analytical technique like agent methods that can handle the vast and variety of data must be used. Agents are intelligent computer systems, either software or hardware (or a combination of both), which can act autonomously or semi-autonomously, and have the ability to make a critical decision by using local or global information to achieve its design objectives. The system of agents working together to make a collective goal is known as Multi Agent Systems (MAS). The MAS is composed of several smaller programs or hardware (agents) which are easier to build and maintain, as well as can respond to their environmental changes in an adaptive way and speed. The MAS is beneficial in solving complex problems as it uses the divide and conquer technique. It is applied in various disciplines where the distributed computing and control are prevalent to construct robust, flexible, and extensible systems. The MAS is not an alternative or competitor for Artificial Intelligence (AI) techniques. The AI techniques can be implemented inside the agents to provide the agents with better computational and decision-making capability. Each agent in the MAS can be heterogeneous or homogeneous based on whether they differ or be identical in goals, actions, domain knowledge, sensor inputs, and outputs. The communication between agents can be direct or indirect through the central directory. MAS method is widely used in the areas of power system protection, optimal power flow, cybersecurity, voltage control, smart home, energy market and etc.

8.5.1 Agents and MAS

Agents are social elements and can decide by communicating, negotiating, and collaborating with their environment and other agents as shown in Figure 8.13. They are also characterized by autonomy which means independent from control calls from other systems or agents. Hence, as an agent can be included or removed without affecting the functionality of the remaining system. These abilities of agents make them very suitable for distributed applications. Agents can also be categorized as reactive agents if they respond to changes in their environment and cognitive agent if they take initiatives to achieve their goals. Cognitive agents have higher communication capability and more intelligence. In contrast, reactive agents lack higher-level decision-making capability and are mostly used in applications that require fast responses. Agents can be reflective based on whether they depend on their explicit knowledge or models of the environment. They can be also intelligent and learning if they can improve their performance from the previous experience of their action and environmental interaction.

An agent-based paradigm is an intrinsic extension to the object-orientation, where the agents have control over its execution. Agents can further implement the BDI (Belief-Desire-Intention) paradigm where an agent can have a set of beliefs about its environment and itself, as well as a computational state it wants to maintain (desires) and achieve (intentions). Based on this paradigm, reactive, deliberative, and hybrid agent architectures are proposed in as described in Table 8.2.

Table 8.2 Agents architecture.

Agents architecture	Descriptions
Reactive agent	The agents depend totally on sensory data for their decision without referring to the history. In addition, they base their decisions on the present and they simply respond to their environment. Moreover, the action to be performed is specified upon possible sensory input.
Deliberative agent	The goals and plans are explicitly represented.
Hybrid agent (reactive and deliberative)	They are autonomous, reactive and proactive agents. They can also exhibit social behavior.

The internal structure of an agent varies based on the application of the agents. A generic intelligent agent consists of an input interface, output interface, decision mechanism, and communication structure. A typical agent structure contains sensors, actuators, and brain. The sensors are used as an input interface to any change that occurs in the environment. The actuators are used as an output interface to the environment and transfer the decision inside the agent. The brain is used for decision-making. The internal structure of the agent for power system application is proposed and contains Device Unit (DU), Decision Making Unit (DMU), and Communication Unit (CU). The CUs are used to exchange information with other agents; DMU serves computing, controlling, and decision-making activities, and DU represents different elements of power systems like loads, DERs, etc. which affect the decisions of DMU as shown in Figure 8.13.

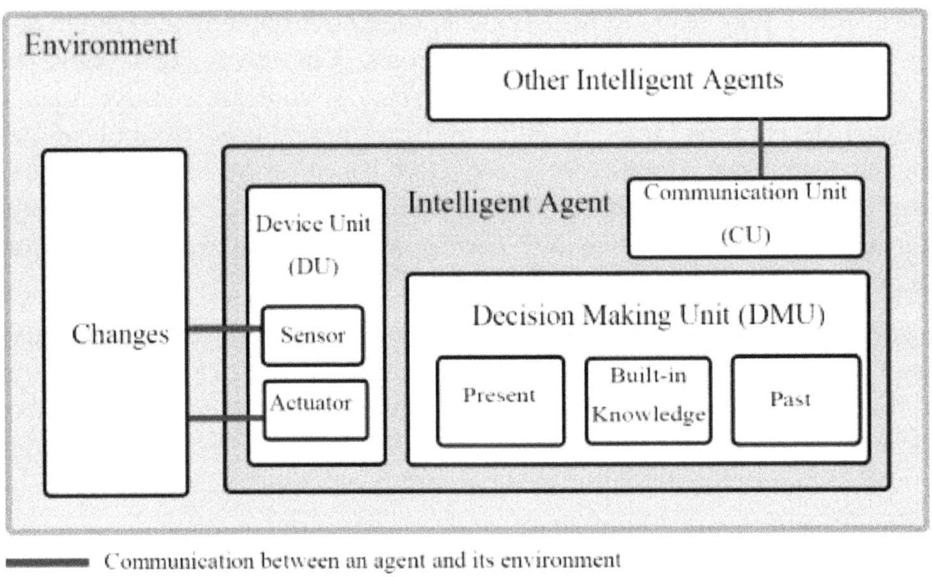

Figure 8.13 The internal structure of the intelligent agent.

8.5.2 MAS Architecture for the Smart Grid

The MAS architectures can be classified as centralized, distributed, hierarchical and hybrid from a reliability and communication perspective. In centralized MAS, a central agent manages non-communicative and homogeneous agents by master–slave relationship. This centralized approach can bring the best solution to the problems, but it is prone to a single point of failure. The disadvantages include need to handle big centralized data, heavy computing processes, and time consuming operations. MAS-2 represents the structure of this architecture is demonstrated in Figure 8.14. MAS-2 has four agents, i.e., agent-1 can be implemented with a cognitive agent, and agents 2–4 can be implemented with reactive agents.

On the other hand the distributed MAS consists of local communicative agents in a single control layer. In Figure 8.14, MAS-1 demonstrates distributed MAS and contains agents 1–3. Each one of the local agent coordinates and communicates with its neighbours to learn about their collective domain. Distributed MAS can help to build a robust and reliable control system. The distributed MAS architecture is suitable for a distributed control approach of the future power grid compared to the hierarchical and centralized MAS architectures.

In hierarchical MAS architecture, as demonstrated in Figure 8.14 as a MAS-3, agents 1–3 can have authority over the agents 4–7 in terms of actions and decisions. The hierarchical MAS can have two-level or more levels of hierarchy based on the application and the need.

In some cases, the central control agent may be used for decision making with distributed agents, which leads to hybrid MAS architecture, as illustrated as by MAS-4 in Figure 8.14.

Not only agents, but MAS can also communicate and interact to form MAS Society (MASS). The green bold line in Figure 8.14, indicates the communication between MAS inside the MASS, and the blue line indicates the communication between agents inside the MAS. Thus, the collection of agents form MAS, and the group of MAS can build a MASS. These ideas are used to propose a smart grid MAS architecture as follows.

The MAS-based technology is becoming very promising in satisfying the smart grid's computational, decision making, and control requirements. The smart grid, and the MAS technique share common characteristics like interoperability, maintainability, upgradeability, scalability, security, and flexibility. Both the smart grid and MAS techniques are also characterized by modularity and cooperation.

The smart grid actors are categorized into bulk generation, transmission, distribution, customer, markets, operations, and service providers domains that cooperate and interact. These seven domains can be further modularized into individual components and sub-components. Similarly, a single unit of the MAS is called an agent. The MAS can be grouped to form the MAS Society (MASS).

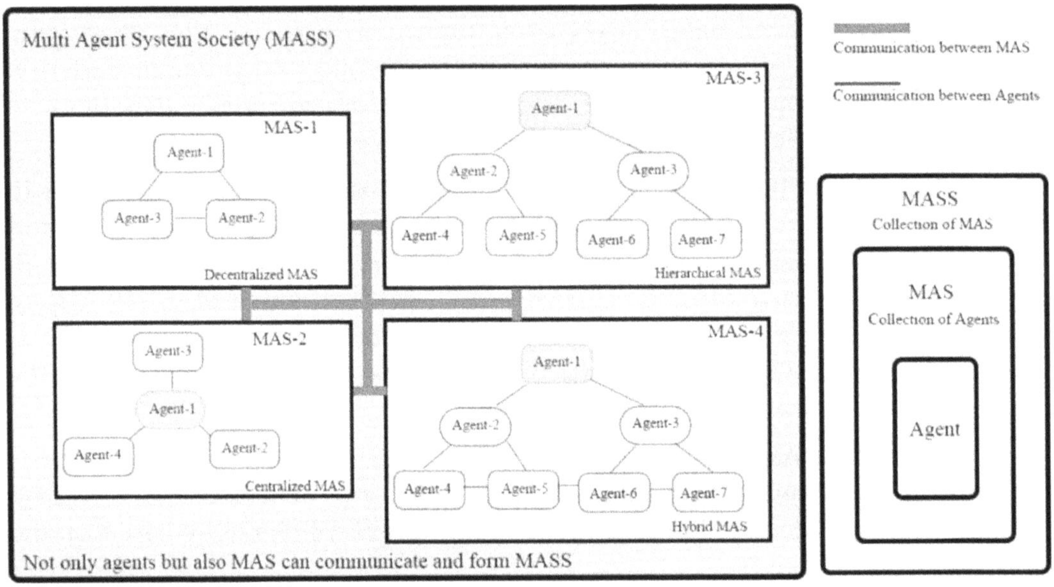

Figure 8.14 The generic distributed, centralized, hierarchical and hybrid MAS architectures and Formation of MASS.

The smart grid modularization proposed can be mapped to a multiagent modularization structure. Thus, each domain in the smart grid can be considered as a MASS, and the sub-components can be further represented as a MAS and agents. Thus, the smart grid can be modeled as the collection of MASS. These ideas of mapping the smart grid to the MAS architecture are further illustrated in Figure 8.15. The domains can be represented by a group of MAS which form MASS. For example, the transmission domain can be represented as a group of Control and Operation MAS (COMAS), FLISR MAS (FLISRMAS), Wide-area Monitoring and Measurement MAS (WMAS), Asset Management and Stabilization MAS (ASMAS), etc. Forming Transmission MASS (TMASS) is shown in Figure 8.16.

Figure 8.15 Smart Grid as group of MASS.

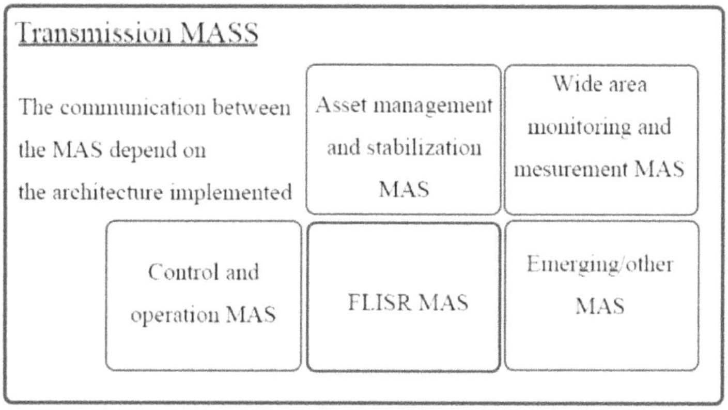

Figure 8.16 Transmission MASS.

Similarly, the remaining domains can be represented by Bulk Generation MASS (BGMASS), Market MASS (MMASS), Operation MASS (OMASS), and Service Providers MASS (SPMASS), as shown in Figure 8.16. The blue line in Figure 8.15 indicates the interconnection between the MASS. The MAS implemented in each MASS and their interconnection architecture can be specific to the particular domain requirement. It has to be also flexible enough to the possible expandability in each domain. Likewise, the agents implemented in each MAS and their interconnection architecture have to be flexible, expandable, and have the capacity to meet each domain's computational requirement.

8.5.3 MAS Based Smart Grid Protection

The dependency of fault current in a smart grid on the operation mode of the grid comprising topological change, bidirectional power flow, the type, quantity, location, power rating, control type, dynamic behavior, and intermittency of DGs, makes it necessary to deploy new methods such as MAS for smart grid protection. A generic MAS architecture for power system protection may consists of mainly coordination agents, configuration agents, and relay agents as shown in Figure 8.17.

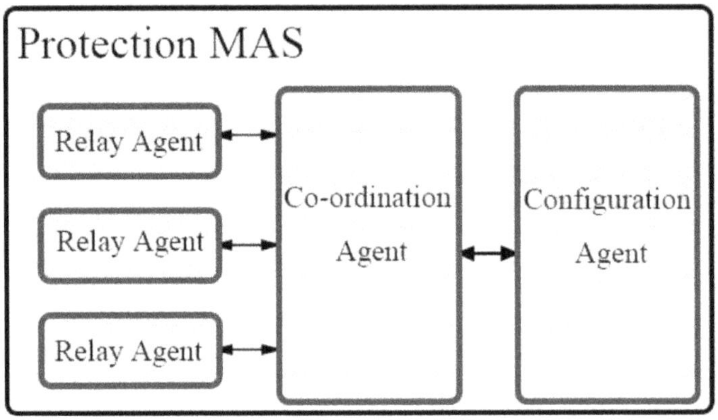

Figure 8.17 The generic MAS architecture for power system protection.

The coordination agent identifies the fault location by using the protection types implemented. Then it decides whether the relay has to operate or not. The protection technique which is used to detect the occurrence of a fault can be, overcurrent-based, sequence elements-based, distance-based, or differential current-based. The configuration agent decides which relay must operate. Even though similar MAS architectures are reported in many kinds of research, it is challenging to define power system protection strictly under one of the MAS architectures discussed in the previous sections. This is because the protection system's primary goal is assumed to clear the fault, and there is no generally accepted hierarchical classification for the protection system duties compared to the other control duties of power systems.

References

1. CIRED.: 'Dispersed generation; preliminary report of CIRED'. Int. Conf. on Electricity Distribution, Working Group WG04, Brussels, Belgium, June 1999.
2. International Energy Agency (IEA).: 'Distributed *generation in liberalized electricity market*' (IEA Publications, 2002)
3. IEEE.: 'IEEE standard for interconnecting distributed resources with electric power systems', IEEE Std 1547–2003, July 28 2003, pp. 1–28

4. Gulachenski, E.M., Kern, E.C.Jr., Feero, W.E et al.: 'Photovoltaic Generation Effects on Distribution Feeders (No. EPRI-EL-6754-Vol. 2)', Electric Power Research Inst., Palo Alto, CA (United States); New England Power Service Co., Westborough, MA (United States), Ascension Technology, Lincoln Center, MA (United States); Electric Research and Management, Inc., State College, PA (United States); Worcester Polytechnic Inst., MA (United States), September 1991.

5. Kroposki, B., Vaughn, A.: 'DG Power Quality, Protection, and Reliability Case Studies Report', National Renewable Energy Laboratory, August 2003

6. Katiraei, F. et al.: 'Investigation of solar PV inverters current contributions during faults on distribution and transmission systems interruption capacity'. Proc. 39th Annual Western Protective Relay Conf., Spokane, WA, 2012

7. Behnke, M., Ellis, A.: 'Contribution of photovoltaic power generation systems to AC short circuits – a survey of current modeling practices and challenges'. IEEE 39th Photovoltaic Specialists Conf. (PVSC), Tampa, FL, 2013, pp. 3128–3133

8. Turcotte, D., Katiraei, F.: 'Fault contribution of grid-connected inverters'. IEEE Electrical Power & Energy Conf. (EPEC), Montreal, QC, 2009, pp. 1–5.

9. Ball, G.J., Blackburn, D.C., Gish, W. et al.: 'Static power converters of 500 kW or less serving as the relay interface package for non-conventional generators', *IEEE Trans. Power Deliv.*, 1994, 9, (3), pp. 1325–133.

10. Firouz, Y., Farhadkhani, S., Lobry, J. et al.: 'Numerical comparison of the effects of different types of distributed generation units on overcurrent protection systems in MV distribution grids', *Renew. Energy*, 2014, 69, pp. 271–283

11. Barker, P.P., deMello, R.W.: 'Determining the impact of distributed generation on power systems: part I. Radial distribution systems'. IEEE Power Engineering Society Summer Meeting, Seattle, WA, 2000, vol. 3, pp. 1645–1656

12. George, S.P., Ashok, S.: 'Multiagent based adaptive relaying for distribution network with distributed generation'. Int. Conf. on Energy, Power and Environment: Towards Sustainable Growth (ICEPE), Shillong, 2015, pp. 1–6

13. deBritto, T.M., Morais, D.R., Marin, M.A. et al.: 'Distributed generation impacts on the coordination of protection systems in distribution networks'. IEEE/PES Transmission and Distribution Conf. and Exposition: Latin America, 2004, pp. 623–628

14. Driesen, J., Vermeyen, P., Belmans, R.: 'Protection issues in microgrids with multiple distributed generation units'. Power Conversion Conf. (PCC '07), Nagoya, 2007, pp. 646–653

15. Dugan, R.C., McDermott, T.E.: 'Operating conflicts for distributed generation on distribution systems'. Rural Electric Power Conf., Little Rock, AR, 2001, pp. A3/1–A3/6

16. S. K. Salman and I. M. Rida, "Investigating the impact of embedded generation on relay settings of utilities electrical feeders," Power Delivery, IEEE Transactions on, vol. 16, no. 2, pp. 246-251, Apr. 2001.

17. S. Horowitz, D. Novosel, V. Madani, and M. Adamiak, "System-wide Protection," Power and Energy Magazine, IEEE, vol. 6, no. 5, pp. 34-42, 2008.

18. D. Tholomier, D. Paraiso, and A. Apostolov, "Adaptive protection of transmission lines," in Power Systems Conference, 2009. PSC '09., 2009, pp. 1-14.

19. S. Teimourzadeh, F. Aminifar, M. Davarpanah, M. Shahidehpour, Adaptive Protection for Preserving Microgrid Security, IEEe Trans. Smart. Grid., 10 (1) (2019), pp. 592-600

20. J.V.B.U.K. Iskakov, K.K. Smagulova, Modeling of operating modes of the protection relays in isolated neutral system, Electrotechnica, 10 (2016), pp. 45-49

21. M.S. Sachdev, B. Chattopadhyay, T.S. Sidhu, A new approach to distribution system protection-adaptive relaying, *1993 Fifth International Conference on Developments in Power System Protection*, 30 March-2 April 1993 (1993), pp. 165-168

22. S. Mitra, G. Misra, P. Chattopadhyay, A Smart Approach for Development of an Adaptive Overcurrent Relay, *2018 Fifth International Conference on Emerging Applications of Information Technology (EAIT)*, 12-13 Jan. 2018 (2018), pp. 1-4

23. L. Hengwei, J.M. Guerrero, J. Chenxi, T. Zhenghua, J.C. Vasquez, L. Chengxi, Adaptive overcurrent protection for microgrids in extensive distribution systems, *IECON 2016 - 42nd Annual Conference of the IEEE Industrial Electronics Society*, 23-26 Oct. 2016 (2016), pp. 4042-4047,

24. Kantamneni A., Brown L.E., Parker G., Weaver W.W. Survey of multi-agent systems for microgrid control Eng Appl Artif Intell, 45 (2015), pp. 192-203

25. Nair A.S., Hossen T., Campion M., Selvaraj D.F., Goveas N., Kaabouch N., Ranganat han P. Multi-agent systems for resource allocation and scheduling in a smart grid, Technol Econ Smart Grids Sustain, Energy, 3 (1) (2018),

26. McArthur S.D., Davidson E.M., Catterson V.M., Dimeas A.L., Hatziargyriou N.D., Pon ci F., Funabashi T. Multi-agent systems for power engineering applications - Part I: Concepts, approaches, and technical challenges IEEE Trans Power Syst, 22 (4) (2007), pp. 1743-1752.

27. Huang Q., Jing S., Li J., Cai D., Wu J., Zhen W. Smart substation: State of the art and future development, IEEE Trans Power Deliv, 32 (2) (2017), pp. 1098-1105

28. NIST special publication 1108R2 NIST framework and roadmap for smart grid interoperability standards, NIST special publication 1108R2 NIST framework and roadmap for smart grid interoperability standards Nist Special Publication (2012), pp. 1-90

29. Tuballa M.L., Abundo M.L. A review of the development of Smart Grid technologies, Renew Sustain Energy Rev, 59 (2016), pp. 710-725,

30. Zhu Y., Song S., Wang D.Multiagents-based wide area protection with best-effort adaptive strategy, Int J Electr Power Energy Syst, 31 (2–3) (2009), pp. 94-99

31. Wan H., Wong K.P., Chung C.Y. Multi-agent application in protection coordination of power system with distributed generations. IEEE power and energy society 2008 general meeting: Conversion and delivery of electrical energy in the 21st century, PES (2008)

32. Abbaspour E., Fani B., Heydarian-Forushani E. A bi-level multi agent based protection scheme for distribution networks with distributed generation, Int J Electr Power Energy Syst, 112 (December 2018) (2019), pp. 209-220

33. Ghorbani J., Choudhry M.A., Feliachi A. Fault location and isolation using multi agent systems in power distribution systems with distributed generation sources. 2013 IEEE PES innovative smart grid technologies conference, ISGT 2013, IEEE (2013), pp. 1-6

34. Davidson E.M., McArthur S.D., McDonald J.R., Cumming T., Watt I. Applying multi-agent system technology in practice: Automated management and analysis of SCADA and digital fault recorder data. 2006 IEEE power engineering society general meeting, PES, Vol. 21, 1424404932 (2006), pp. 559-567,

35. Kantamneni A., Brown L.E., Parker G., Weaver W.W. Survey of multi-agent systems for microgrid control Eng Appl Artif Intell, 45 (2015), pp. 192-203,

36. Fani B., Abbaspour E., Karami-Horestani A. A fault-clearing algorithm supporting the MAS-based protection schemes, Int J Electr Power Energy Syst, 103 (February) (2018), pp. 257-266

CHAPTER

9

Role of Power Electronics in Smart Grid

9.1 Introduction

The power electronics constituted crucial components of the energy control applications for several years. The power circuits including HVDC controllers, FACTS, static synchronous compensator (STATCOM), inverters, and converters are implemented to control and facilitate the transmission and distribution of the power generated. Furthermore, the controllable power electronics are developed to precisely adjust amplitudes of voltage, current, and frequency while increasing the quality of the power supplied to the consumers. The latest implementations of power electronics are assisted with monitoring, communication, real time simulation such as OPAL-RT and RTDS and several other autonomous features that facilitate to install smart grid and power line communication (PLC) infrastructures. The PLC allows obtaining the signalling feature over high voltage power systems where it can be used as a communication medium for remote sites.

The conventional grid structure is being widely studied in order to move to a more flexible, intelligent, and more than just an electrical carrier with communication features that will meet the involvements of this century. The recent trends in grid structure are widely based on the integration of small DG sources and distributed storage systems where the energy is generated by energy sources such as wind, solar, fuel cell, and biomass. The smarter and distributed generation structure of electrical grid that is defined as SG brings the intelligent communication technologies together where wired or wireless options can be easily installed. Thus, the integration of renewable energy sources and microgrids (MGs) to the utility grid should be handled in a logical way. The DC and AC MGs tend to supply the generated power under diverse conditions supported by the quality operation of the power electronic converters. Several control methods are implemented in order to eliminate disruptive effects on voltage and current waveforms, harmonic contents, and to increase the quality of power transmitted to utility grid. The DC power, control, which is particularly involved in solar plant or fuel cell based generation, is performed in DC-DC converters. On the other hand, parallel inverter control that is required in both AC and DC power conversion is mostly focused on the power flow control.

Droop control is one of the most recently studied methods that focus on active and reactive power flow control as per frequency and amplitude values. The parallel inverter operation that is essential in converting HVDC to HVAC is coped with several control loops together to increase the reliability and flexibility of the generation sides. Besides these methods, the harmonic current share or output impedance conditioning is based on the latest control methods that are implemented to operate with nonlinear load profiles. Both the grid-tied and islanded DG structures deploy properly controlled converters. While the power of a dispatchable DG system is controlled by power control strategies such as droop control or others mentioned above, the generated power of undispatchable DG units as solar plants or wind farms is controlled in a predefined optimum operation range.

On the other hand, the DG system transmits the generated power to both loads and utility grid in grid-tied operation. The reference voltage of grid and active power control should be combined in this situation. The optimized control of power flow between the DG and the grid can be implemented by using agent-based strategy that sets a communication infrastructure through generators and loads. However, this strategy cannot operate in parallel with droop control since it avoids interacting with all sides at the point of common coupling (PCC). The smart transformer (ST) that is configured with tap changing feature and located at the PCC copes with the inadequacy of agents.

The ST manages the power flow between grid and DG section, and can control the power transfer rates and directions instantly. In this way, the monitoring section of grid needs to observe only the ST instead of entire DG elements reducing the overall installation and operation costs. The ST adjusts the DG side voltage in order to control the power flow between grid and MG. The primary and secondary side controls of ST manage the voltage-based control of power regarding the demands directed by the loads to utility grid. The following sections introduce specific applications power electronic converters in SG and MG structures.

9.2 Power Flow in Smart Grid

A complete smart grid can be mapped as shown in Figure 9.1 with power generation, transmission and distribution (T&D), and customer sections. The diagram illustrates bidirectional electricity and communication flows. The generators-section consists of several sources including renewables. The generation section may be assisted by energy storage systems (ESS) to ensure the sustainability of the system. The T&D section consist of substations and control centre mechanism in which the generated and consumed electricity rates are observed instantly incorporating with wired or wireless sensors together. The control environment, which contains switches, meters, technical personnel, is now being managed owing to remote control and telemetry features of smart grid. To achieve this development, utility grid is equipped with remotely activated switches, automatic circuit breakers, and reclosers, which involve power electronics.

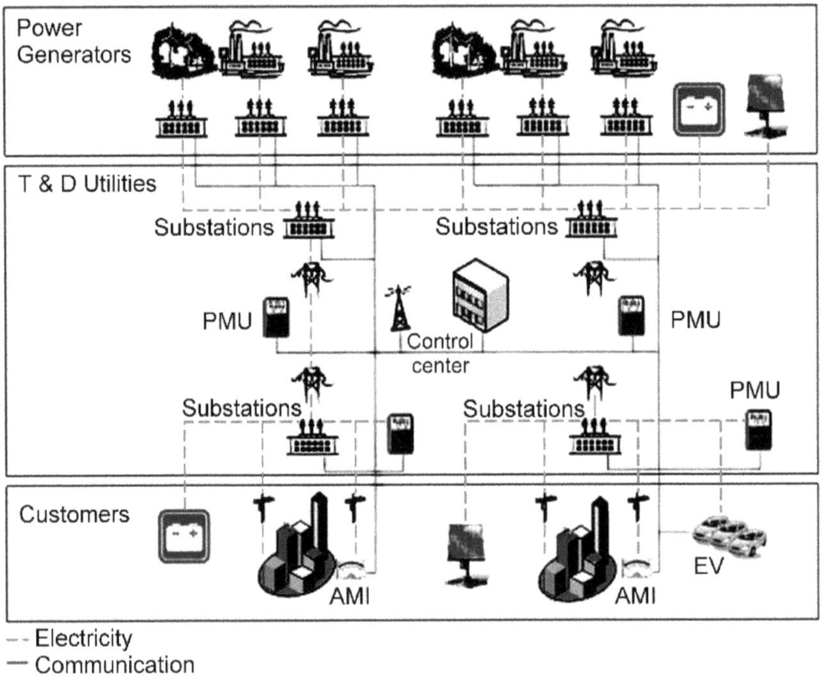

Figure 9.1 Power generation, control, and measurement diagram across the distribution network and consumers.

Besides the control centre in substations, phase measurement units (PMU) are also integrated to the system to measure the phase values of the supplied power to consumers. The T&D section includes wide power control infrastructure such as remote monitoring and remote control of power circuits as inverters and converters in MG scale besides switches and breakers. The control elements of power electronics in a smart grid cover the power factor controllers, sensors, and other components of a SCADA-like system. The last section of the diagram consists of customers in where the advanced metering infrastructure (AMI) is installed to each node in order to perform monitoring and billing requirements. The AMI infrastructure can be connected to control centre either using wireless communication standards or wired communication methods over power lines. The PLC technologies are recently involved for several applications in smart grid that cover a wide range from domestic energy control to Internet service supplying known as broadband power lines (BPL). All the electricity and communication background are bidirectionally connected end-to-end between power generating section and customers over control centre.

Figure 9.2 presents another smart grid control diagram with PLC infrastructure. The power control sections of the smart grid can be classified according to voltage levels as high voltage (HV) in where the primary power generation plants are located, medium voltage (MV) covering the DG plants and T&D centre, and low voltage (LV) section that includes the consumers with communication infrastructures.

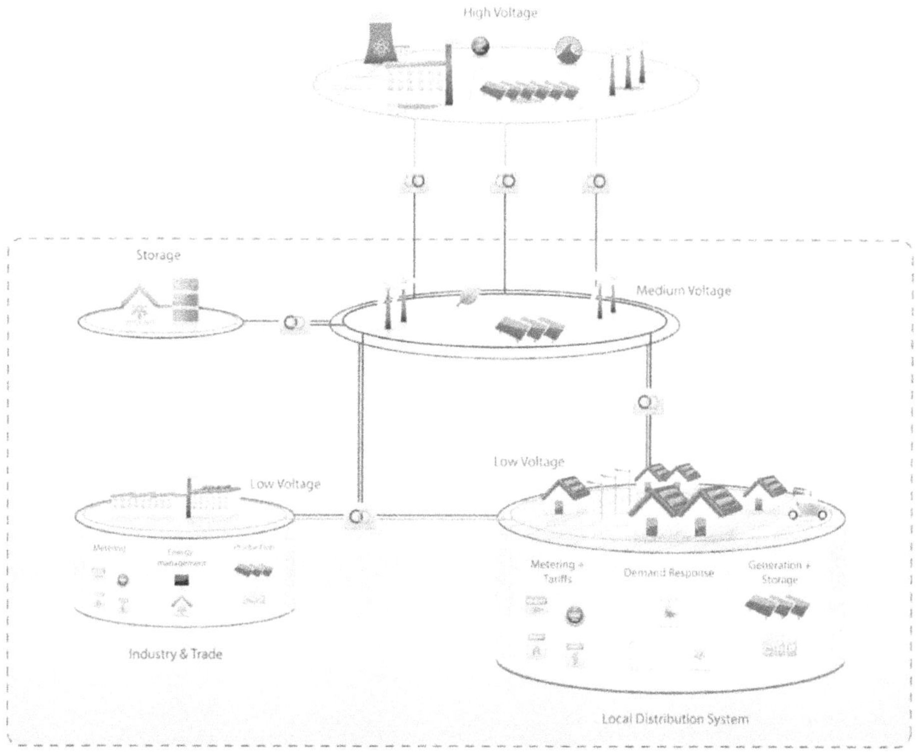

Figure 9.2 PLC based smart grid scenario.

The transformer centre decreases voltage level across the sections in order to supply to customers enabling the interconnection of DG sources to utility grid. The MV section consists of storage facilities to ensure the sustainability of the renewable energy, while the LV section includes individual RESs such as solar, wind, and fuel cell. The bidirectional communication between HV and LV sections is performed by BPL technology where metering, energy management, demand response, and storage controls are instantly managed. The local distribution system manages the metering, the tariff, the demand response, the generation and the storage control of each customer regarding to power control and communication strategies.

The improvements achieved in smart grid infrastructure rapidly increased the usage of RESs in where the domestic and industrial applications are easily implemented owing to convenient management features for generation and storage processes. A novel term, namely future renewable electric energy delivery and management (FREEDM) system building a utility grid by integrating distributable and scalable RESs to utility grid is proposed. The FREEDM system is intended to install an alternative interface for PV converters that holds a 400 V dc bus.

It is obviously known that the power electronic converters are more important for power control, energy management and efficiency issues as well as interfacing several circuits across RESs and utility grid. The interfacing requirements and integration capabilities of sources define the power electronics to be used in control level. The transient response of the sources has a key role in device control and management operations besides overall efficiency. The transient response of the systems of these types increased by integrating the hybrid generation systems together. Furthermore, battery or capacitor banks increase the dynamic performance of ESS. The Fuel Cell and Electrolyser Cell based systems that are similar for PV applications require highly efficient DC–DC converter structures with a wide range of input voltage variations and high current output. Recently implemented power switches made from silicon carbide (SiC), gallium nitride (GaN), and several magnetic materials can increase the power control stability of these systems.

The power control of a HVDC link in Kii-Channel uses thyristors of rating of 8 kV–3.5 kA. Whereas line frequency commutated converters of gate turn off thyristor (GTO) and gate commutation thyristor (GCT) with ratings of 6 kV–4 kA and 6 kV–6 kA, respectively, are also used in power control and interfacing of RESs to utility grid. Another widely used power device is manufactured with insulated gate bipolar transistors (IGBT) that have rating of 4.5 kV–1.2 kA used in series connection in each arms of converter systems. Besides its high-speed switching capacity owing to SiC structure itself, another advanced power electronics semiconductor injection enhanced gate transistor (IEGT) provides rating of 4.5 kV–900 A. The Static Var Generator (SVG) is the recent large power application of GTCs with the capacity of controlling 80 MVA power and 77 kV bus voltages in Kanzaki substation.

The SVG is commutated with pulse width modulation (PWM) switching signals that reduce the switching losses without any AC filter requirement. The ESS systems are also involved to compensate the voltage sags in addition to their contribution in load supplying. The recent advances in battery technology allowed the implementation of sodium-sulphur batteries, namely NAS battery, and a pioneer application is performed with voltage compensation of 3 MW during 10 sec. wherein the switching response from grid to battery is less than 5 ms that is considered as a great increment of transient response in ESS.

Micro Grids (MGs) contribute in the advances of the smart grid which consist of several DGs, power converters, ESS and communication systems together. The MG can be operated with conventional centralized control of utility grid or decentralized control of islanded mode wherein both of the control mechanisms increases the efficiency and stability. The power control of a MG is related to voltage and frequency (*V–f*) regulation, active and reactive power (*P–Q*) flow control, synchronization, and energy management between DG sources and utility grid. The hierarchical control strategy of MGs is proposed with three control stages that are named as primary, secondary and tertiary control levels. The primary control level includes the voltage and current control loops, droop voltage controls, and impedance controls. The decentralized control operations that deal with *P–f* and *Q–V* droop controls are assumed in the primary control level with its autonomous power share processes. Since the primary level control is completely focused on main power control processes, it does not involve communication infrastructures in where the secondary level control does . The major advantages of droop control are great flexibility, high reliability, and not requiring any communication. The primary control namely field level control produces the deviation values of amplitude and frequency.

The secondary level control is based on first communication systems which are based on lower bandwidth to measure specific values in MG and featured control loops. The secondary level control is also responsible for management of islanded and grid-tied operation by making decision to shift operation mode according to demands and requirements. MG Central Controller (MGCC) that provides data feedback to DG system to control the MG instantly performs the relaying system as presented in Figure-9.3. The MGCC is designed to compensate *V–f* variations, voltage unbalances, and harmonic contents caused by input and output impedances of primary control generated in the MG. Further, the *Q–V* droop control that is affected by the unstable voltage of MG is also implemented by an external control loop in the secondary level control. It is extensively studied to perform droop control in the primary level control by improving virtual inductors to manage the *P–Q* coupling, *Q* sharing algorithms, and frequency droop control. The tertiary control operates at the grid level and calculates the demanded active and reactive power where the calculation results are transferred to management level that consists of secondary control loops.

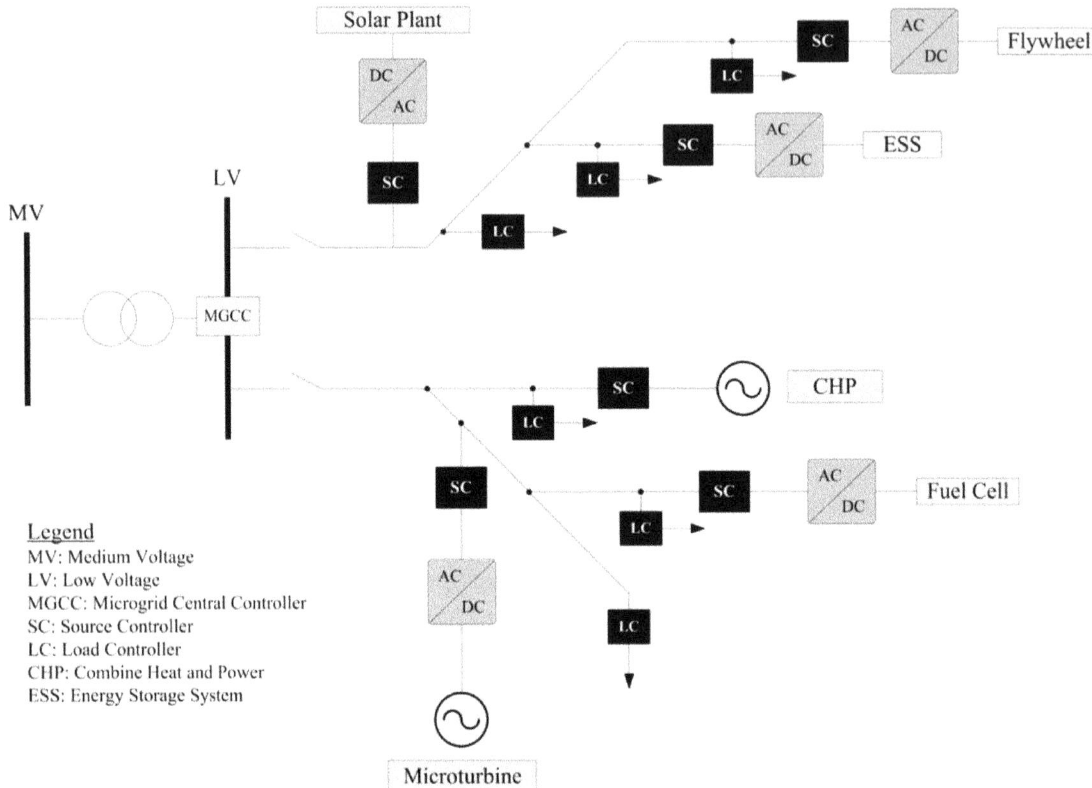

Figure 9.3 A complete Micro Grid structure with DG sources.

Since the voltage along the MG is not stable, and the secondary level control loops employed in MGCC cannot perform completely managed reactive power sharing, the pricing control and economical optimization of the MG are carried out in tertiary level control by including related algorithms to MGCC. Tertiary level control provides a better communication possibility between MG and utility due to its centralized control level, while the primary level architecture is based on decentralized control.

It can be observed from Figure-9.3 that the MGCC is installed in MV–LV substation wherein the primary and secondary level controls are performed owing to its functions. The tertiary level control is performed in each source and load nodes by using source controller (SC) and load controller (LC) those carry out the communication requirements between MGCC and utility grid by operating each DG unit. The SC controls the P–Q generation and sharing levels at each individual DG unit, while LC interfaces load connections to DG sources and utility grid.

The individual inverter performs the V–f control in islanded operation mode or P–Q control in grid connected modes. On the other hand, the parallel inverter operation is essential in P–f droop control, Q–V droop control and power sharing controls. The smart grid studies should consider the transportation utilities such as electrical vehicles (EV) and plug-in hybrid electric vehicles (PHEV) that are integrated to utility grid regarding to the recent advances. The increased integration of EVs to the utility grid causes several compensation problems which should be handled by similar control loops managed by MGCC and LC systems. One of the recent proposals for power control is related to resonant and isolated converters that focus on decreasing the voltage stresses on the switching devices either in inverters or in converters. The power converters used in SG applications including MG and RES are presented in the following sections.

9.3 AC-DC Converters in SG

The wind power generation systems (WPGS) can be classified into four ways i.e. fixed speed with squirrel cage induction generator (SCIG), variable speed with wounded rotor induction generator (WRIG), variable speed with double fed induction generator (DFIG), and variable speed with synchronous generator (SG) based on the generator used in the wind energy conversion. The SG used in variable speed WPGS is implemented with wounded rotor SG (WRSG) or permanent magnet SG (PMSG) in where the PMSG is the most widely used type. The multi-pole structure of a generator used in WPGS provides to construct gearless operation in where it is achieved by using a PMSG. Figure-9.4 illustrates the AC–DC converter types which are either in uncontrolled structure or in bidirectional structure used in a PMSG wind turbine.

The uncontrolled converter seen in Figure-9.4(a) is constituted with front-end rectifier that is built in three-phase full bridge topology. Whereas, the second block diagram given in Figure-9.4(b) illustrates bidirectional AC–DC converter that provides the power flow between wind turbine and utility grid. Furthermore, the mechanical load that affects the tip speed of the turbine can be controlled by the bidirectional control of converter where energy storage is managed by a capacitor between rotor side and grid side converters. In addition to this, matrix converters enable direct conversion without including any capacitor whether the DC-link allows to implement a method known as the low-voltage ride-through capability (LVRT) that specifically makes the generator able to operate through lower voltage periods for critical loads.

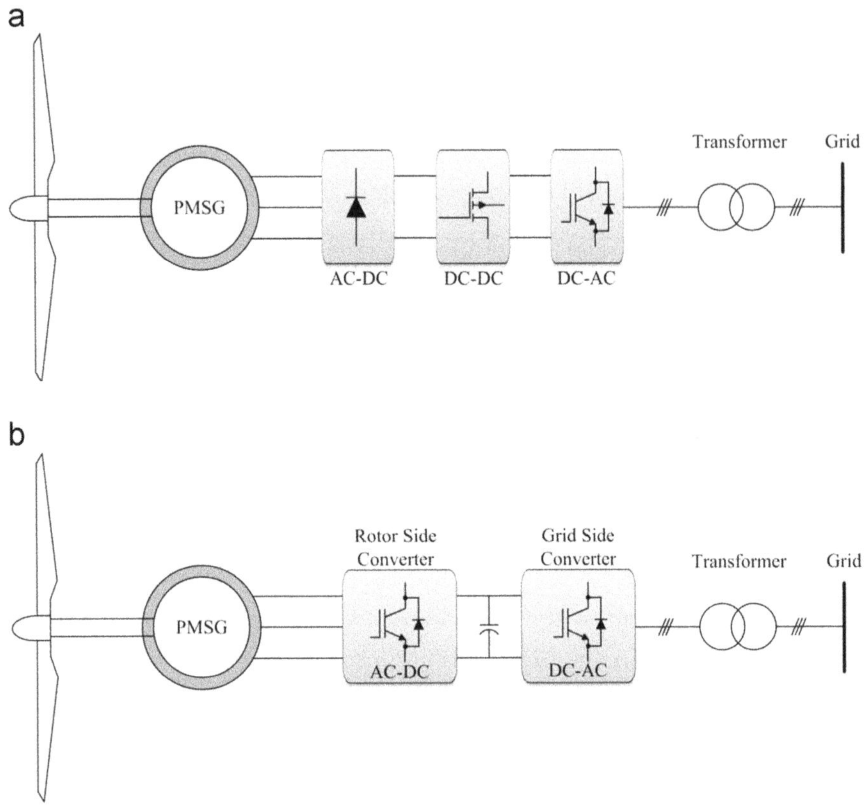

Figure 9.4 AC–DC converters in PMSG wind turbine: (a) uncontrolled converter and (b) bidirectional converter.

The DFIG-based WT consists of a wound rotor where the slip rings build the output connection of rotor circuit, and AC–DC converter controls the electrical frequency of the rotor, while the stator circuit is connected to the grid as shown in Figure-9.5. The AC–DC converter controls the mechanical speed of the WT by decoupling the rotor speed from frequency of the grid. The DFIG is the most recent and most widely preferred WT among others owing to its speed range, which varies up to 60% of nominal speed. The power converters used in DFIG are named as partial-scale since they are designed to control 30% of the rated power limited by the power of rotor circuit.

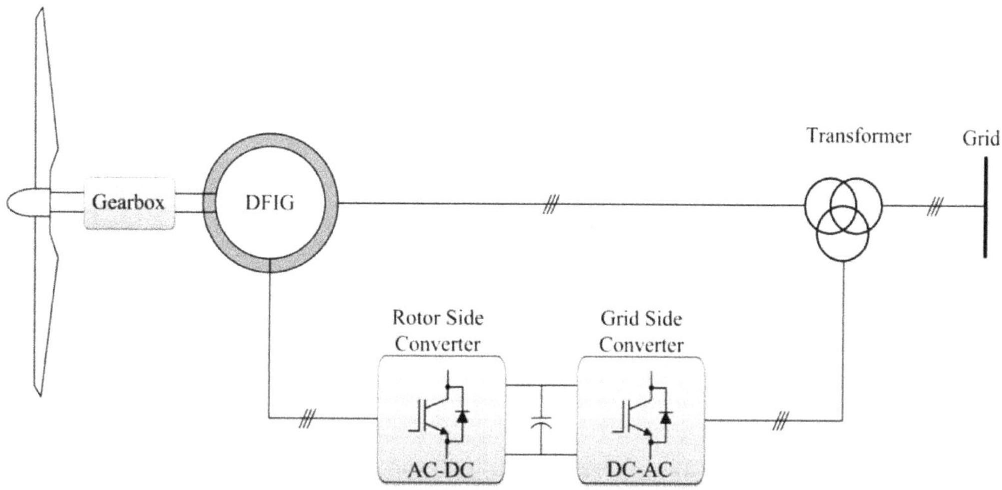

Figure 9.5 AC–DC converters in Doubly Fed Induction Generator wind turbine.

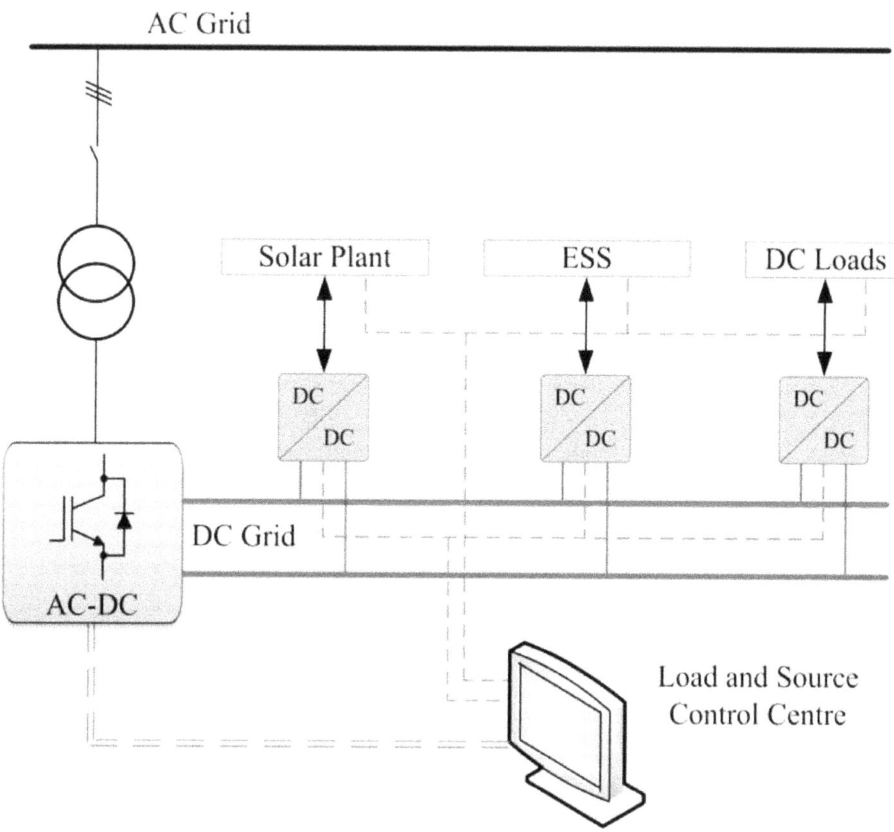

Figure 9.6 Block diagram of an islanded LVDC smart grid.

Another application area of the AC–DC converter is widely observed in DC islanding operations in AC utility grid. Apart from HVDC the DC grids are not only constructed for high voltage but also at the low voltage which is known as the LVDC grid. The power injection to the LVDC requires strict voltage regulation processes wherein the power control should also be performed on the load side in addition to source-side control. Thus, the communication infrastructures should be extended to load-side as done in source-side. The recent studies about LVDC grid are based on separating a bus from conventional utility grid over bidirectional AC–DC converter that provides a flexible way to interact through PV and several storage alternatives. An islanded LVDC smart grid shown in Figure-9.6 can be obtained separating either the bus between rotor side and grid side converters of a WPGS or from a conventional utility grid. The improved diagram enables to connect many small DG units along the bidirectional AC–DC converters, ESS units, and communication infrastructures together.

Another application is a Solid-state transformer (SST) which consists of power electronic devices (Figure 9.7) and provides multiple taps. Such a distribution transformer assists a novel AC–DC converter structure. The SST is implemented as a high frequency (HF) transformer that is connected to active converter bridges on primary and secondary sides to build a DC island from conventional utility grid. The SST makes the bidirectional communication owing to its structure enabling additional functionalities as the power control, the current demand control, and the ESS management opportunities.

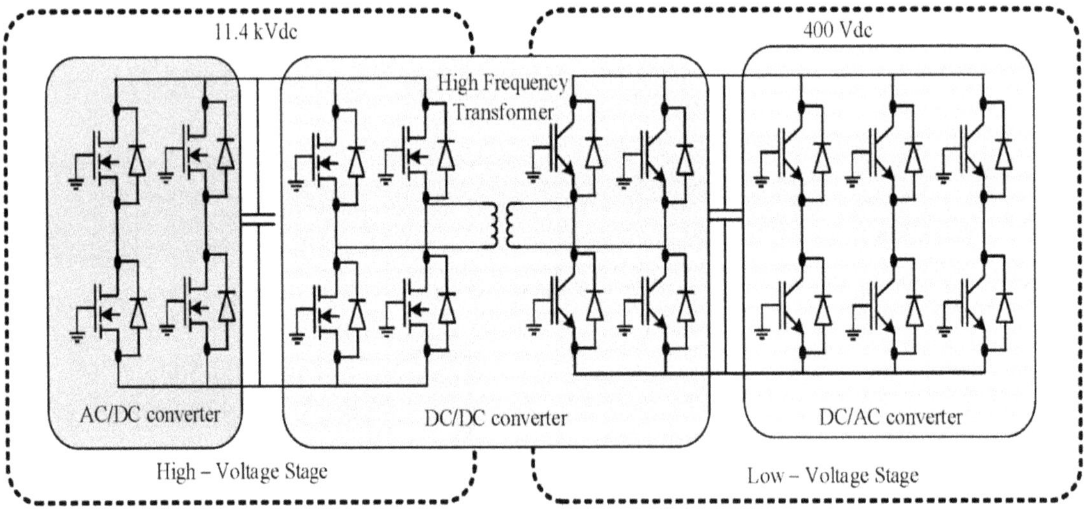

Figure 9.7 A 11.4kVdc/400Vdc Solid State Transformer Circuit.

9.4 DC-AC Converters in SG

The DC–AC converter, namely inverter, has a wide utilization area containing transportation, heavy-duty railways, electric vehicles, aircrafts besides industrial and power generation plants. The latest Tsukuba Express has PWM switched IGBT inverters in its substations wherein the DC-line voltage generated during breaking is converted to AC. Apart from the hard switching voltage source inverter (VSI) based on sinusoidal PWM (SPWM) or space vector PWM (SVM), the recently developed multilevel inverters (MLI) present a wide application area in medium and high voltage market promoting the multi-MW wind turbines and individual generators. The initial studies of MLIs are based on three-level hard-switching inverter. The following topologies are implemented owing to its lower switching losses, decreased common mode voltage, and harmonic elimination features. The most widely known topologies of MLI are named as diode clamped, flying capacitor, and cascaded H-bridge. The most popular usages of MLI cover motor drives are FACTS, STATCOMs, adjustable speed drives (ASD), and active filters. Figure 9.8 presents the most common MLI topologies in three-level structure in where Figure 9.8(c) depicts a five-level H-bridge that is cascaded by using three-level cells.

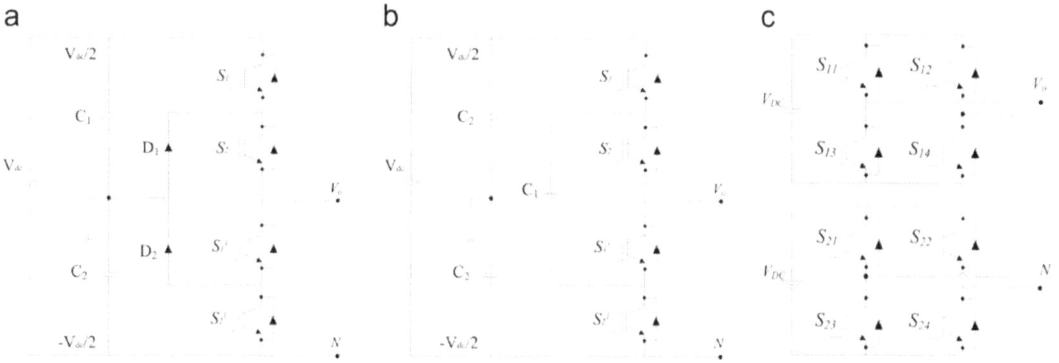

Figure 9.8 Multilevel inverter topologies: (a) diode clamped, (b) flying-capacitor, and (c) cascaded H-bridge.

A microgrid can include many inverters in different topologies to interface aforementioned DG units and loads. The most important point to figure out is to control the parallel inverters in the islanded operation to achieve the highest performance for power and voltage regulations. The parallel inverters require P–f droop control and Q–V droop control to manage the power flow, while the single inverter operation involves V–f and P–Q control in isolated and grid-tied connection, respectively.

The main disadvantage of parallel inverter operation is that the change occurred in voltage and reactive power due to the load deviation. One of the recent methods to cope with this issue is proposed, where the control scheme is based on droop control to govern the active power share of parallel inverters in DG units. In the proposed method, the reactive power of each inverter could be adjusted to an exact value owing to reactive power generation capacity of the DG units in the microgrid. The Q value of a VSI in the system was unregulated and could vary with the load.

A comprehensive block diagram of AC and DC smart grid with several DG units, VSIs generating AC bus, AC–DC converters comprising the DC bus by using diesel generator, WT, and flywheel is shown in Figure-9.9. The DC sources are interfaced to DC bus over DC–DC converters that are based on PWM controlled switches to set the output voltages to a desired value. The DC loads can be connected to DC smart grid over DC bus similar to AC loads interfaced in AC bus.

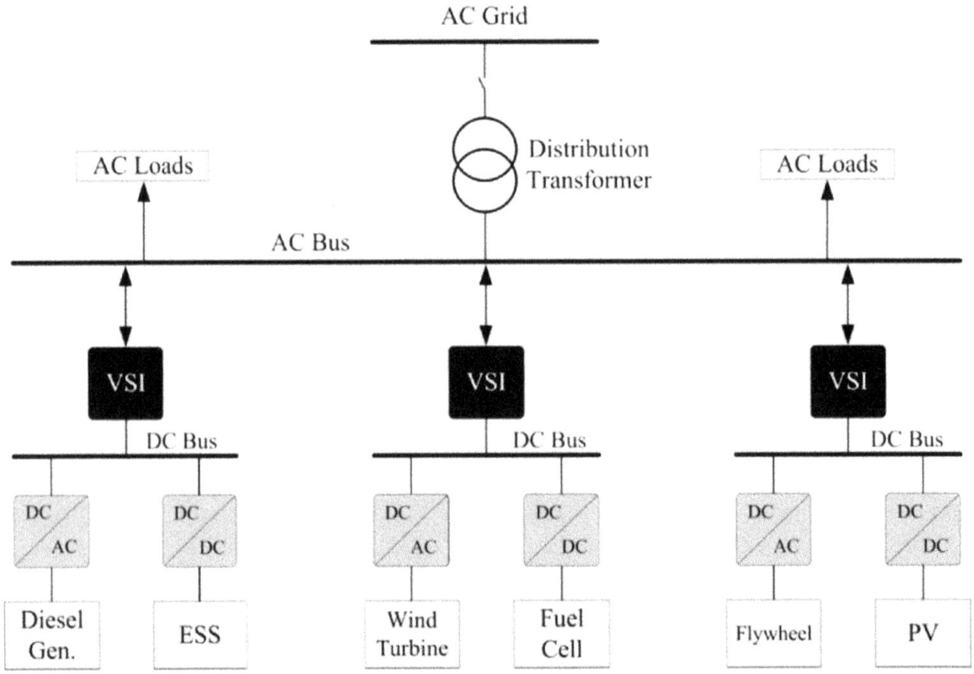

Figure 9.9 Block diagram of AC and DC smart grid managed by VSI.

9.5 DC-DC Converters in SG

The DC power supply in DG units is commonly obtained from PV panels, ultra-capacitors, fuel cells, and batteries. The photo voltaic effect of solar irradiation produces DC energy and AC power is derived from inverters tied to grid. The batteries or ESS are commonly used in the islanded configurations to sustain the energy in the periods where the irradiation is unavailable. On the other hand, the grid-tied system does not require ESS assistance since the generated energy is instantly transmitted to the utility grid. Several DC-Dc converter topologies in SG are the derivatives of traditional buck, boost, and buck-boost topologies. Bidirectional conversion is one of the extensively used DC–DC converter.

The DC–DC converter topologies have become an important area where a lot of improvement is observed in terms of circuit topology and control methods owing to widespread usage of EV and DG applications. The bidirectional power flow control involves cooperation of converter with an inverter and an ESS by using the capabilities provided by communication infrastructure. The implementation of these devices wherein the DC sources are used makes the smart grid a compact power source with enhanced power flow control. The smart grid including converters should manage the bidirectional power flow between ESS and converter on demand, besides implementing featured charging algorithms for each battery type as lead-acid, Ni–Cd, Li-ion, and super capacitors in ESS. Furthermore, the DC–DC converter should track and perform the necessary operations received from the control and management center.

The requirements for a dc-dc converter used in EVs are similar to a microgrid, supplied by several DG units. Thus, its topology of a converter should meet the requirements in bidirectional modes. The circuit diagrams of buck, boost and buck-boost converters are introduced in the succeeding sections. The design analysis of several other (topologies such as Cúk and Sepic) in terms of important parameters such as output voltage (V_O), minimum inductor value (L_{min}) for continuous conduction mode (CCM), CCM boundary load current (I_{LB}), minimum capacitor value for CCM (C_{min}), and minimum–maximum load current ($I_{Lmin\text{–}max}$) are given in Table-9.1. The design calculations can be easily done using these equations. V_S stands for the supply voltage, while T denotes the switching period while f is the switching frequency, and D is the duty cycle.

A DC–DC converter designed according to one of the topologies and criteria illustrated in Tabe-9.1 is not enough to operate in bidirectional mode itself. The converter requires a proper control algorithms. The control algorithms can be classified according to their loop either in feedback or in feed-forward structure. The feedback methods are based on tracking the output current or output voltage where the measured values are evaluated by a decisive algorithm such as PI, PID or soft computing methods to control the duty cycle of switching device. On the other hand, the feed-forward methods are responsible to control the actual state of power supply particularly used in PV or battery supplied converters. The renowned feed-forward method among others is named maximum power point tracking (MPPT) algorithm that generates the switching signals regarding to variations occurred in the supply voltage.

Table 9.1 Physical parameters, design criteria, and equations pf DC-DC converters.

	Buck	Boost	Buck–Boost
Output voltage	$V_O = \frac{t_{on}}{T} V_S = DV_S$ (6)	$V_O = \frac{V_S}{(1-D)}$ (11)	$V_O = \frac{D}{1-D} V_S$ (16)
Min. inductor for CCM	$L_{min} \geq \left(\frac{TR}{2}\right)(1-D)$ (7)	$L_{min} \geq \left(\frac{TR}{2}\right) D(1-D)^2$ (12)	$L_{min} \geq \left(\frac{TR}{2}\right)(1-D)^2$ (17)
Min. CCM load current	$I_{LB} = \frac{V_S TD(1-D)}{2L}$ (8)	$I_{LB} = \frac{V_O D(1-D)T}{2L}$ (13)	$I_{LB} = \frac{V_S D}{R(1-D)^2} - \frac{V_S DT}{2L}$ (18)
Min. capacitor for CCM	$C_{min} \geq \frac{V_o(1-D)}{8V_r L f^2}$ (9)	$C_{min} \geq \frac{V_o D}{V_r R f}$ (14)	$C_{min} \geq \frac{V_O D}{V_r R f}$ (19)
Min.–max. load current	$I_{L\ min,\ max} = \frac{V_o}{R} \pm \frac{V_O T(1-D)}{2L}$ $(+) \rightarrow I_{L\ max} \mid (-) \rightarrow I_{L\ min}$ (10)	$I_{L\ min,\ max} = \frac{V_S}{R(1-D)^2} \pm \frac{V_S DT}{2L}$ $(+) \rightarrow I_{Lmax} \mid (-) \rightarrow I_{L\ min}$ (15)	$I_{L\ min,\ max} = \frac{V_S D}{R(1-D)^2} \pm \frac{V_S DT}{2L}$ $(+) \rightarrow I_{L\ max} \mid (-) \rightarrow I_{L\ min}$ (20)

9.6 Fundamentals of Power Electronic Converters

This section presents topologies, working principle and performance of AC-DC, DC-AC, and DC-DC converters for the quick reference of the readers which helps to brush up background required for the topic covered in the previous sections. Basics of Pulse Width Modulation (PWM) are also included in one of the subsection with same intention.

9.6.1 AC-DC Converters

A rectifier is a circuit that converts an ac signal into a unidirectional signal. A rectifier is a type of ac–dc converter. A rectifier may also be considered as an absolute value converter. If v_s is an ac input voltage, the waveform of the output voltage v_0 would have the same shape, but the negative part will appear as a positive value. That is, $v_0 = |v_s|$. Depending on the type of input supply, the rectifiers are classified into two types: (1) single phase and (2) three phases. A single-phase rectifier can be either a half wave or a full wave. A single-phase half-wave rectifier is the simplest type, but it is not normally used in industrial applications. For the sake of simplicity, the diodes are ideal.

Performance Parameters: Although the output voltage of a rectifier in Figure-9.10(a) should ideally be a pure dc, the output of a practical rectifier contains harmonics or ripples as shown in Figure 9.10(b). A rectifier is a power processor that should give a dc output voltage with a minimum amount of harmonic contents. At the same time, it should maintain the input current as sinusoidal as possible and in phase with the input voltage so that the power factor is near unity. The power-processing quality of a rectifier requires the determination of harmonic contents of the input current, the output voltage, and the output current. We can use the Fourier series expansions to find the harmonic contents of voltages and currents. The performances of a rectifier are normally evaluated in terms of the parameters like average value of the output voltage (V_{dc}), average value of the output current(I_{dc}), output dc power($P_{dc} = V_{dc}I_{dc}$), rms value of the output current(I_{rms}), rms value of output voltage (V_{rms}), and output ac power ($P_{ac} = V_{rms}I_{rms}$).

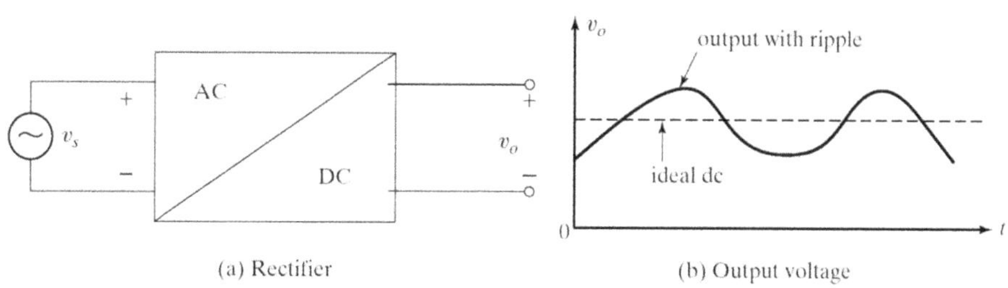

Figure 9.10 Input and output relationship of a rectifier.

The efficiency of rectification of a rectifier is defined as

$$\eta = \frac{P_{dc}}{P_{ac}} \quad .....(9.1)$$

It should be noted that η is not the power efficiency. It is the conversion efficiency which is a measure of the quality of the output waveform. For a pure dc output, the conversion efficiency would be unity. The output voltage can be considered as composed of two components i.e. the dc component and the ac component or ripple.

The effective (rms) value of the ac component of output voltage is

$$V_{ac} = \sqrt{V_{rms}^2 - V_{dc}^2} \quad .....(9.2)$$

The form factor, which is a measure of the shape of output voltage is

$$FF = \frac{V_{rms}}{V_{dc}} \quad .....(9.3)$$

The ripple factor, which is a measure of the ripple content, is defined as

$$RF = \frac{V_{ac}}{V_{dc}} \quad .....(9.4)$$

Substituting (9.2) and (9.3) in (9.4), the ripple factor can be expressed as

$$RF = \sqrt{\left(\frac{V_{rms}}{V_{dc}}\right)^2 - 1} = \sqrt{FF^2 - 1} \quad .....(9.5)$$

The transformer utilization factor is defined as

$$TUF = \frac{P_{dc}}{V_S I_S} \quad .....(9.6)$$

where V_S and I_S are the rms voltage and current of the transformer secondary, respectively. The input power can be determined approximately by equating input power with the output ac power. That is, the power factor is related by

$$PF = \frac{P_{ac}}{V_S I_S} \quad .....(9.7)$$

Crest factor (CF), which is a measure of the peak input current $I_{S(peak)}$ as compared with its rms value I_S, is often of interest to specify the peak current ratings of devices and components. CF of the input current is defined by

$$CF = \frac{I_{S(peak)}}{I_S} \quad .....(9.8)$$

Single Phase Full Converter: The circuit diagram of a single phase fully controlled bridge converter is shown in the Figure-9.11 with a highly inductive load and a dc source in the load circuit so that the load current is continuous and ripple free (constant load current operation). The fully controlled bridge converter consists of four thyristors T_1 T_2, T_3 andT_4 connected in the form of full wave bridge configuration. Each thyristor is controlled and turned on by its gating signal and naturally turns off when a reverse voltage appears across it. During the positive half cycle when the upper line of the transformer secondary winding is at a positive potential with respect to the lower end the thyristors $T_1$and T_2 are forward biased during the time interval $\omega t = 0$ to π . The thyristors T_1 and T_2 are triggered simultaneously $\omega t = \alpha$ $(0 \leq \alpha \leq \pi)$, the load is connected to the input supply through the conducting thyristors T_1 and T_2. The output voltage across the load follows the input supply voltage and hence output voltage $v_o = V_m sin\omega t$. Due to the inductive load T_1 and T_2 will continue to conduct beyond $\omega t = \pi$, even though the input voltage becomes negative. T_1 and T_2 conduct together during the time period π to $(\pi + \alpha)$, for a time duration of π radians. During the negative half cycle of input supply voltage for $\omega t = \pi$ to 2π the thyristors T_3 and T_4 are forward biased. T_3 and T_4 are triggered at $\omega t = (\pi + \alpha)$. As soon as the thyristors T_3 and T_4 are triggered a reverse voltage appears across the thyristors T_1 and T_2 and they naturally turn-off and the load current is transferred from T_1 and T_2 to the thyristors T_3 and T_4.The output voltage across the load follows the supply voltage $v_o = -V_m sin\omega t$ during the time period π to $(\pi + \alpha)$. In the next positive half cycle when T_1 and T_2 are triggered, T_3 and T_4 are reverse biased and they turn-off. The Fgure-9.12 shows the waveforms of the input supply voltage, the output load voltage, the constant load current with negligible ripple and the input supply current.

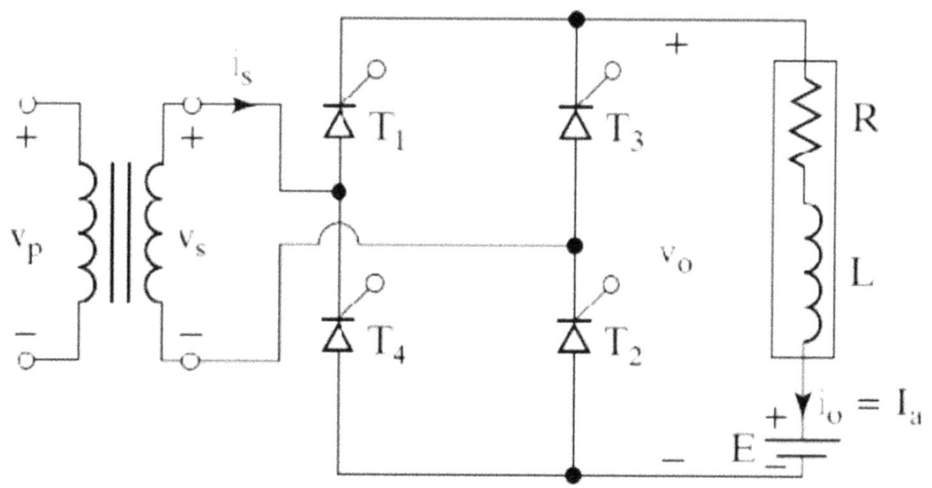

Figure 9.11 Single Phase fully controlled rectifier with RLE load.

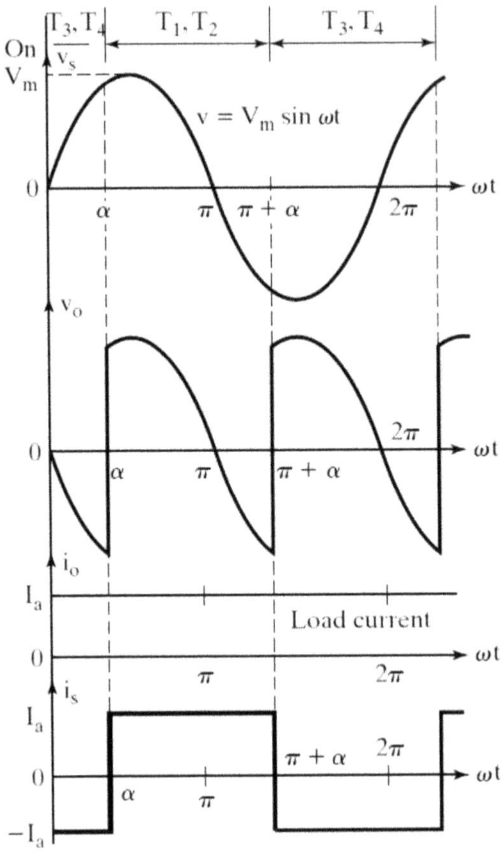

Figure 9.12 Input and output waveforms of the rectifier shown in Figure 9.11.

During the time period ωt = α to π , the input supply voltage v_S and the input supply current i_S are both positive and the power flows from the supply to the load. The converter operates in the rectification mode during ωt = α to π. During the time period ωt = α to (π+α), the input supply voltage v_S is negative and the input supply current i_S is positive and there will be reverse power flow from the load circuit to the input supply. The converter operates in the inversion mode during the time period ωt = α to (π+α) and the load energy is fed back to the input source.

Expression for DC output Voltage: The average (dc) output voltage can be determined by using the expression

$$V_{O(dc)} = V_{dc} = \frac{1}{2\pi}\left[\int_0^{2\pi} v_O . d(\omega t)\right] \quad .....(9.9)$$

$$V_{O(dc)} = V_{dc} = \frac{2}{2\pi}\left[\int_{\alpha}^{\pi+\alpha} V_m \sin\omega t.d(\omega t)\right] = V_{dc} = \frac{2V_m}{\pi}\cos\alpha \quad .....(9.10)$$

The dc output control characteristic can be obtained by plotting the average or dc output voltage V_{dc} versus the trigger angle α as shown in Figure-9.13.

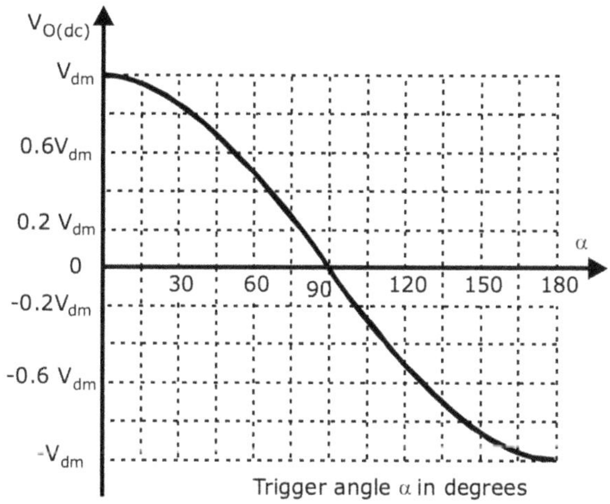

Figure 9.13 DC output control characteristic of full controller.

It can be noticed from the control characteristic that by varying the trigger angle α we can vary the output dc voltage across the load. Thus, it is possible to control the dc output voltage by changing the trigger angle α . For trigger angle α in the range of 0 to 90 degrees, V_{dc} is positive and the average dc load current I_{dc} is also positive. The average or dc output power P_{dc} is positive, hence the circuit operates as a *controlled rectifier* to convert ac supply voltage into dc output power which is fed to the load. For trigger angle α > 90, dc output voltage V_{dc} becomes negative, but the load current flows in the same positive direction i.e., I_{dc} is positive. Hence the output power becomes negative. This means that the power flows from the load circuit to the input ac source. This is referred to as line *commutated inverter operation.* During the inverter mode operation for 0α > 90 the load energy can be fed back from the load circuit to the input ac source.

The rms value of the output voltage is calculated as

$$V_{O(RMS)} = \sqrt{\frac{1}{2\pi}\left[\int_{0}^{2\pi} v_O^2.d(\omega t)\right]} = \frac{V_m}{\sqrt{2}} = V_S \quad .....(9.11)$$

From the waveform of the Figure-9.13 it can be derived that the average and RMS values of the thyristor current are $\frac{I_{dc}}{2}$ and $\frac{I_{dc}}{\sqrt{2}}$ respectively.

Three Phase Full Converter: Three phase full converter is a fully controlled bridge-controlled rectifier using six thyristors connected in the form of a full wave bridge configuration. All the six thyristors are controlled switches which are turned on at an appropriate times by applying suitable gate trigger signals. The three-phase full converter is extensively used in industrial power applications up to about 120 kW output power level, where two quadrant operation is required. Figure-9.14 shows a three-phase full converter supplying highly inductive load. This circuit is also known as three phase full wave bridge or as a six-pulse converter. The thyristors are triggered at an interval of π/3 radians. The frequency of output ripple voltage is 6fs and the filtering requirement is less than other converters. Where fs is the supply frequency.

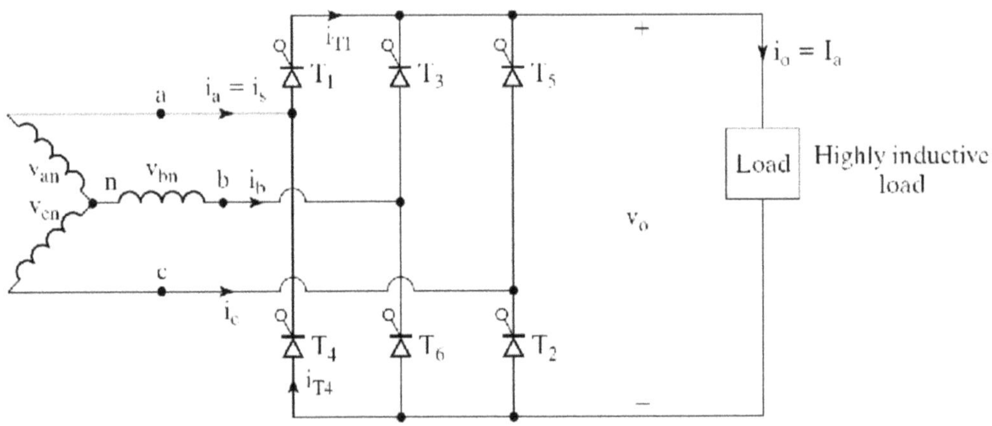

Figure 9.14 Three-phase full converter supplying highly inductive load.

We define three-line neutral voltages (3 phase voltages) as follows:

$$V_{an} = V_m \sin(wt);$$

$$V_{bn} = V_m \sin(wt - 120^o);$$

$$V_{cn} = V_m \sin(wt + 120^o);$$

The corresponding line voltages can be obtained as:

$$V_{ab} = V_{an} - V_{bn} = \sqrt{3}V_m \sin(wt + 30^o);$$

$$V_{bc} = V_{bn} - V_{cn} = \sqrt{3}V_m \sin(wt - 90^o);$$

$$V_{ca} = V_{cn} - V_{an} = \sqrt{3}V_m \sin(wt + 90^o);$$

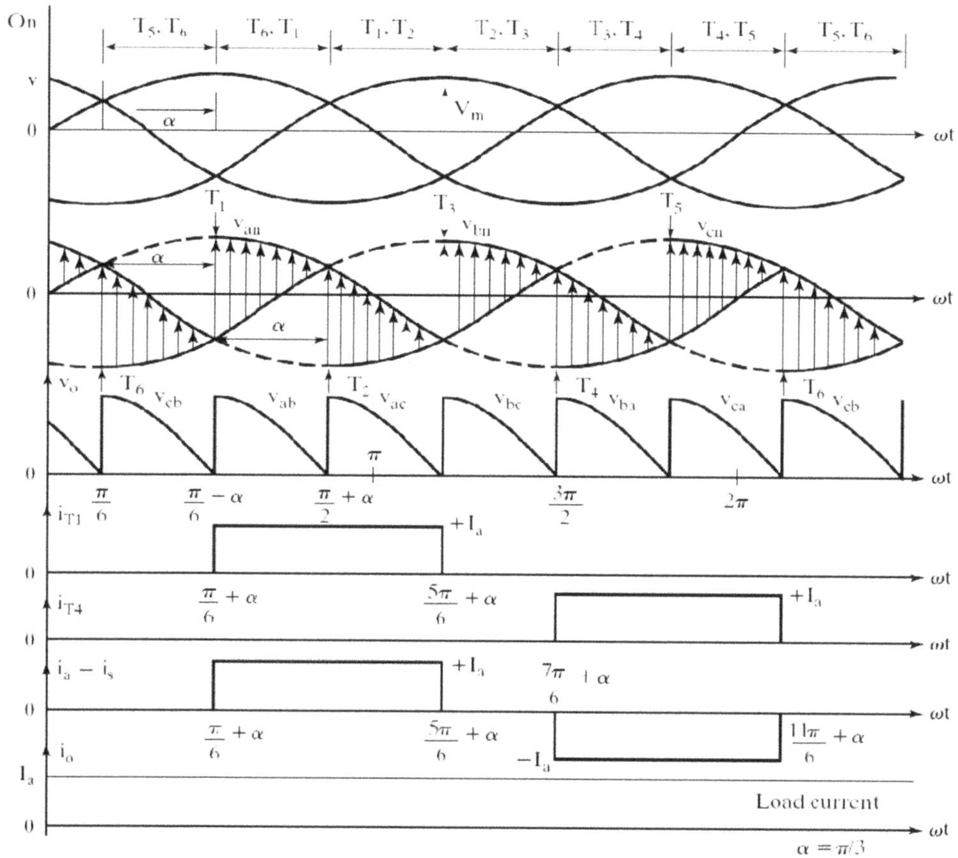

Figure 9.15 Various waveforms associated with 3-phase full converter.

where V_m is peak phase voltage. The three phase voltage waveforms can be seen in Figure 9.15 along with other waveforms. The thyristors are numbered in the circuit diagram corresponding to the order in which they are triggered. The trigger sequence (firing sequence) of the thyristors is 12, 23, 34, 45, 56, 61, 12, 23, and so on.

At $wt = (30^o + \alpha)$, thyristor T_6 is already conducting when the thyristor T_1 is turned on by applying the gating signal to the gate of T_1. During the time period $(30^o + \alpha)$ to $(90^o + \alpha)$ thyristors T_6 and T_1 conduct together and the line-to-line supply voltage V_{ab} appears across the load. At $wt = (90^o + \alpha)$, the thyristor T_2 is triggered and T_6 is reverse biased immediately and T_6 turns off due to natural commutation. During the period $(90^o + \alpha)$ to $(150^o + \alpha)$ thyristor T_1 and T_2 conduct together and the line-to-line supply voltage $V_{ac} = -V_{ca}$ appears across the load. The Figure-9.15 shows the waveforms of three phase input supply voltages, output voltage, the thyristor current through T_1 and T_4 the supply current through the line 'a'.

The output load voltage consists of 6 voltage pulses over a period of 2π radians, hence the average output voltage is calculated as

$$V_{O(dc)} = V_{dc} = \frac{6}{2\pi}\int_{\frac{\pi}{6}+\alpha}^{\frac{\pi}{2}+\alpha} v_O.d\omega t = \frac{3}{\pi}\int_{\frac{\pi}{6}+\alpha}^{\frac{\pi}{2}+\alpha} \sqrt{3}V_m \sin\left(\omega t + \frac{\pi}{6}\right).d\omega t = \frac{3\sqrt{3}V_m}{\pi}\cos\alpha \quad .....(9.12)$$

Similarly, rms value of the output voltage can be calculated as:

$$V_{O(rms)} = \left[\frac{6}{2\pi}\int_{\frac{\pi}{6}+\alpha}^{\frac{\pi}{2}+\alpha} v_O^2.d(\omega t)\right]^{\frac{1}{2}} = \left[\frac{6}{2\pi}\int_{\frac{\pi}{6}+\alpha}^{\frac{\pi}{2}+\alpha} v_{ab}^2.d(\omega t)\right]^{\frac{1}{2}} = \sqrt{3}V_m\left(\frac{1}{2} + \frac{3\sqrt{3}}{4\pi}\cos 2\alpha\right)^{\frac{1}{2}} \quad .....(9.13)$$

9.6.2 DC-AC Converters

Figures 9.16(a) & (b) show the typical power-circuit topologies of a single-phase and a three-phase voltage source inverter respectively. These topologies require only a single dc source and for medium output power applications; the preferred devices are n-channel IGBTs. 'Edc' is the input dc supply, and a large dc link capacitor (Cdc) is put across the supply terminals. Capacitors and switches are connected to dc bus using short leads to minimize the stray inductance between the capacitor and the inverter switches. Needless to say, that physical layout of positive and negative bus lines is also important to limit stray inductances. Q1, Q2, Q3 etc. are fast and controllable switches. D1, D2, D3 etc. are fast recovery diodes connected in anti-parallel with the switches. 'A', 'B' and 'C' are output terminals of the inverter that get connected to the ac load. A three-phase inverter has three load-phase terminals whereas a single-phase inverter has only one pair of load terminals. The current supplied by the dc bus to the inverter switches is referred as dc link current and has been shown as 'idc' in Figures-9.16(a) & (b). The magnitude of dc link current often changes in step (and sometimes its direction also changes) as the inverter switches are turned on and off. The step change in instantaneous dc link current occurs even if the ac load at the inverter output is drawing steady power. However, average magnitude of the dc link current remains positive if net power-flow is from dc bus to ac load. The net power-flow direction reverses if the ac load connected to the inverter is regenerating. Under regeneration, the mean magnitude of dc link current is negative.

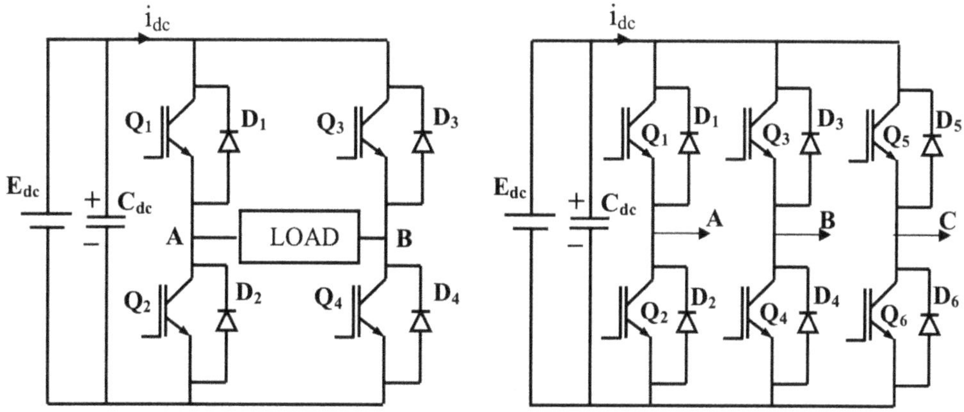

(a)Topology of a 1-phase VSI (b)Topology of a 3-phase VSI

Figure 9.16 Voltage Source Inverters.

For an ideal input (dc) supply, with no series impedance, the dc link capacitor does not have any role. However, a practical voltage supply may have considerable amount of output impedance. The supply line impedance, if not bypassed by a sufficiently large dc link capacitor, may cause considerable voltage spike at the dc bus during inverter operation. This may result in deterioration of output voltage quality; it may also cause malfunction of the inverter switches as the bus voltage appears across the non-conducting switches of the inverter. Also, in the absence of dc link capacitor, the series inductance of the supply line will prevent quick build up or fall of current through it and the circuit behaves differently from the ideal VSI where the dc voltage supply is supposed to allow rise and fall in current as per the demand of the inverter circuit.

The dc link capacitor should be put very close to the switches so that it provides a low impedance path to the high frequency component of the switch currents. The capacitor itself must be of good quality with very low equivalent series resistor (ESR) and equivalent series inductor (ESL). The length of leads that interconnect switches and diodes to the dc bus must also be minimum to avoid insertion of significant number of stray inductances in the circuit. The overall layout of the power circuit has a significant effect over the performance of the inverter circuit.

Single Phase Inverters: The full bridge circuit will have two pole-voltages (VAO and VBO), which are similar to the pole voltage VAO of the half bridge circuit. Both VAO and VBO of the full bridge circuit are square waves but they will, in general, have some phase difference. Figure-9.17 shows these pole voltages staggered in time by 't' seconds.

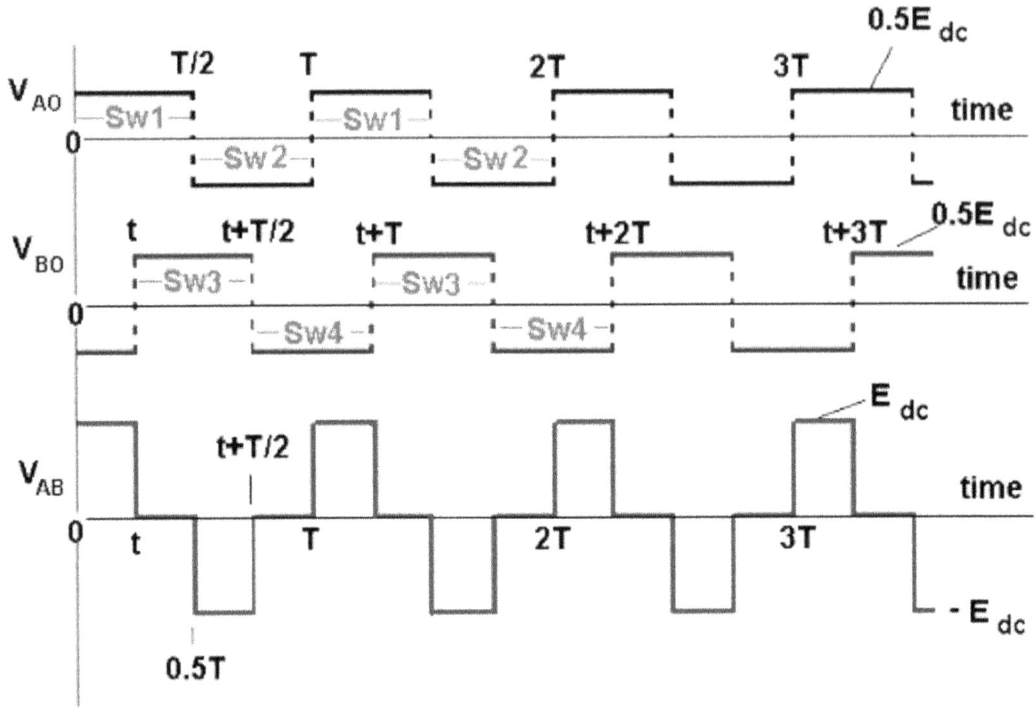

Figure 9.17 Pole and line voltages of single-phase inverter.

It may be more convenient to talk in terms of the phase displacement angle 'Φ' defined as:

$$\Phi = (2\pi/T)\, t \text{ radians} \qquad .....(9.14)$$

where 't' is the time by which the two pole voltages are staggered and 'T' is the time period of the square wave pole voltages.

The pole voltage VAO of the full bridge inverter may again be written considering the phase shift angle 'Φ' into account, the pole-B voltage may be written as

$$V_{BO} = \sum_{n=1,3,5,7,\ldots,\infty} \frac{2E_{dc}}{n\pi} \sin n(wt - \Phi) \qquad .....(9.15)$$

Difference of VAO and VBO gives the line voltage VAB. In full bridge inverter the single phase load is connected between points 'A' and 'B' and the voltage of interest is the load voltage VAB.

Taking difference of the voltage expressions given by Equations (9.14) and (9.15):

$$V_{AB} = \sum_{n=1,3,5,7,\ldots,\infty} \frac{2E_{dc}}{n\pi} \left[\sin nwt - \sin n(wt - \Phi)\right] \qquad .....(9.16)$$

The fundamental component of VAB may be written as

$$V_{AB,1} = \frac{2E_{dc}}{\pi}\left[\sin wt - \sin(wt - \Phi)\right] = \frac{4E_{dc}}{\pi}\cos(wt - \frac{\Phi}{2})\sin\frac{\Phi}{2} \quad .....(9.17)$$

The nth harmonic component in VAB may similarly be written as

$$V_{AB,n} = \frac{2E_{dc}}{n\pi}\left[\sin nwt - \sin n(wt - \Phi)\right] = \frac{4E_{dc}}{n\pi}\cos n(wt - \frac{\Phi}{2})\sin\frac{n\Phi}{2} \quad .....(9.18)$$

From the equation (9.17) the rms magnitude of the fundamental component of load voltage may be written as

$$(V_{AB,1})_{rms} = 0.9E_{dc}\sin\frac{\Phi}{2} \quad .....(9.19)$$

The rms magnitude of load voltage can be changed from zero to a peak magnitude of 0.9 Edc. The peak load voltage magnitude corresponds to Φ = 180 degrees and the load voltage will be zero for Φ = 0. For Φ = 180 degrees, the load voltage waveform is once again square wave of time period T and instantaneous magnitude E. As the phase shift angle changes from zero to 180 degrees the width of voltage pulse in the load voltage waveform increases. Thus, the fundamental voltage magnitude is controlled by pulse-width modulation.

Also, from equation (9.17) it may be seen that the line voltage distortion due to higher order harmonics for pulse width modulated waveform (except for Φ = 180 degrees) is less than the corresponding distortion in the square wave pole voltage. In fact, for some values of phase shift angle (Φ) many of the harmonic voltage magnitudes will drastically reduce or may even get eliminated from the load voltage. For example, for Φ = 60 degree the load voltage will be free from 3rd and multiples of third harmonics.

Three Phase Inverter: Figure 9.18 shows the power circuit of the three-phase voltage source inverter with square wave pole voltages, feeding a 3-phase balanced load. This circuit may be identified as three single-phase half-bridge inverter circuits put across the same dc bus. The individual pole voltages of the 3-phase bridge circuit are identical to the square pole voltages output by full bridge circuits.

The three pole voltages of the 3-phase square wave inverter are shifted in time by one third of the output time period. These pole voltages along with some other relevant waveforms have been plotted in Figure-9.19. The horizontal axis of the waveforms in this has been represented in terms of 'ωt', where 'ω' is the angular frequency (in radians per second) of the fundamental component of square pole voltage and 't' stands for time in second. In figure-9.19 the phase sequence of the pole voltages is taken as VAO, VBO and VCO. The numbering of the switches in Figure-9.18 has some special significance vis-à-vis the output phase sequence.

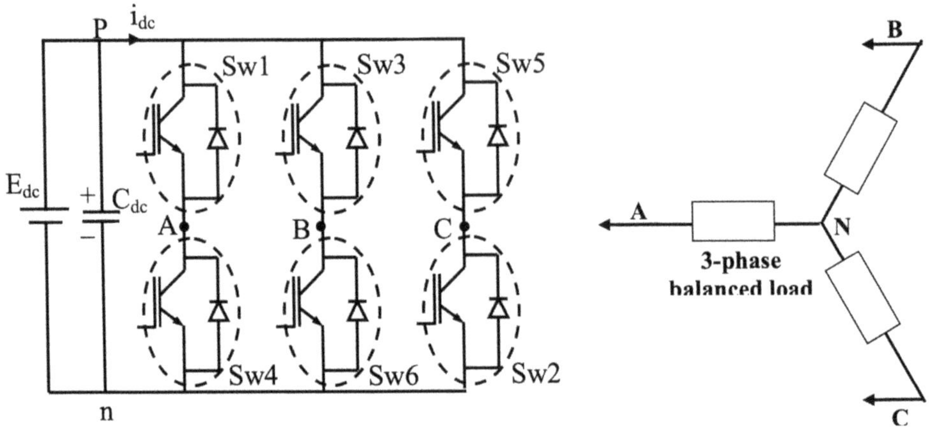

Figure 9.18 Three-phase voltage source inverter feeding a 3-phase balanced load.

To appreciate the particular manner in which the switches have been numbered, the conduction- pattern of the switches marked in Figure 9.18 may be noted. It may be seen that with the chosen numbering the switches turn on in the sequence: - Sw1, Sw2, Sw3, Sw4, Sw5, Sw6, Sw1, Sw2, ...and so on. Identifying the switching cycle time as 360 degrees (2π radians), it can be seen that each switch conducts for 180° and the turning on of the adjacent switch is staggered by 60 degrees. The upper and lower switches of each pole (leg) of the inverter conduct in a complementary manner. To reverse the output phase sequence, the switching sequence may simply be reversed. Considering the symmetry in the switch conduction pattern, it may be found that at any time three switches conduct. It could be two from the upper group of switches, which are connected to positive dc bus, and one from lower group or vice-versa (i.e., one from upper group and two from lower group). According to the conduction pattern i.e., there are six combinations of conducting switches during an output cycle: - (Sw5, Sw6, Sw1), (Sw6, Sw1, Sw2), (Sw1, Sw2, Sw3), (Sw2, Sw3, Sw4), (Sw3, Sw4, Sw5), (Sw4, Sw5, Sw6). Each of these combinations of switches conducts for 60° in the sequence mentioned above to produce output phase sequence of A, B, C.

The three load terminals are connected to the three output points (A, B, C) of the inverter. The load side phase voltages V_{AO}, V_{BO} and V_{CO} can be determined from the conduction pattern of the inverter switches. With reference to Figure-9.18 it may be seen that for $0 \le \omega t \le \pi/3$, switches Sw5, Sw6 and Sw1 conduct. Under the assumption of ideal switches Figure-9.20(a) will represent the equivalent inverter and load circuit during the time interval $0 \le \omega t \le \pi/3$. In the equivalent circuit representation, the non-conducting switches have been omitted, and a cross (**X**) sign is used to represent a conducting switch.

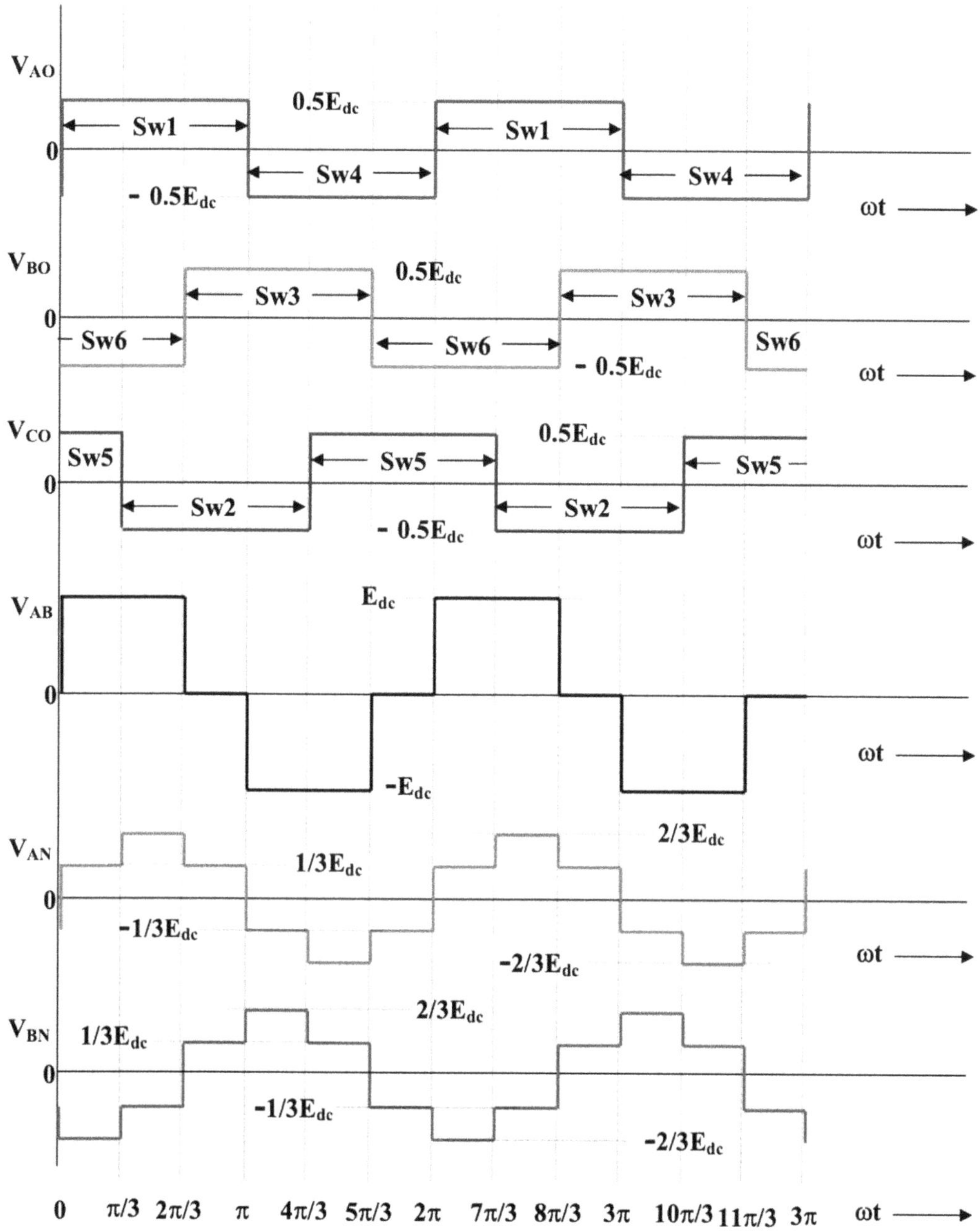

Figure 9.19 Some relevant voltage waveforms output by a 3-phase voltage source inverter.

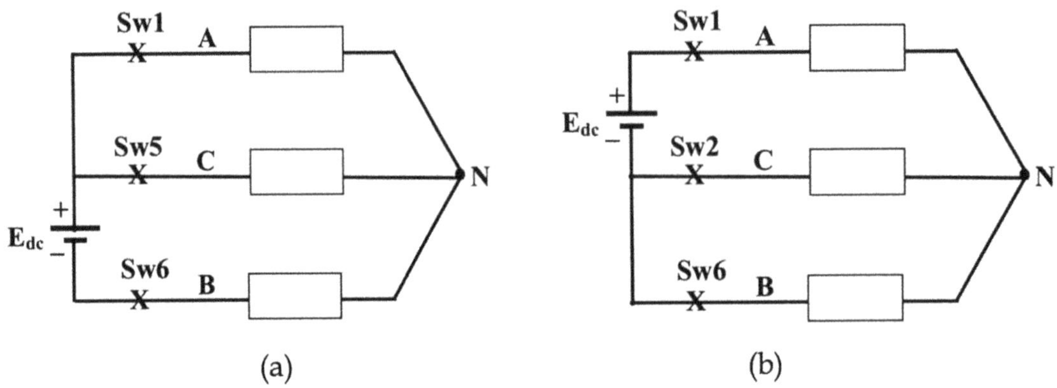

Figure 9.20 Schematic load circuit during conduction of (a) Sw5, Sw6 and Sw1 (b) Sw6, Sw1 and Sw2.

For a balanced 3-phase load the instantaneous phase voltage waveforms have been derived below for the following two cases (i) when the 3-phase load is purely resistive and (ii) when the load, in each phase, consists of a resistor in series with an inductor and a back e.m.f. In both the cases the equivalent circuit of Fig. 9.20(a) has been referred to derive the expression for load-phase voltage. For case (i), when the load is a balance resistive load, it is very easy to see that the instantaneous phase voltages, for $0 \leq \omega t \leq \pi/3$, will be given by $V_{AN} = 1/3\ Edc$, $V_{BN} = -2/3\ Edc$, $V_{CN} = 1/3\ Edc$. Figure-9.20(b) shows the equivalent circuit during $\pi/3 \leq \omega t \leq 2\pi/3$, when the switches Sw6, Sw1 and Sw2 conduct. The instantaneous load phase voltages may be found to be $V_{AN} = 2/3\ Edc$, $V_{BN} = V_{CN} = -1/3\ Edc$.

The load phase voltage waveforms for other switching combinations may be found in a similar manner. Two of the phase voltages, V_{AO} and V_{BO}, along with line voltage V_{AB} have been plotted over two output cycles in Figure-9.19. It may be seen that voltage V_{AN} is similar to V_{BN} but lags it by one third of the output cycle period. Further, it can be verified that the load phase voltage V_{CN} also has a waveform identical to the two other phase voltages, but time displaced by one third of the output time period. V_{AN} waveform leads V_{CN} by 120 degrees in the time (ωt) frame. It should be obvious that the fundamental component of the phase voltage waveforms will constitute a balanced 3-phase voltage having a phase sequence A, B, C. It may also be recalled that by suitably changing the switching sequence the output phase sequence can be changed. The phase voltage waveforms of Figure-9.19 show six steps per output cycle and are also referred as the six-step waveform.

The individual pole voltage waveforms output by the 3-phase square wave inverter are identical to the output waveform of a single-phase inverter. The expression for line voltage V_{AB} can be written as

$$V_{AB} = \sum_{n=1,3,5,7,\ldots,\infty} \frac{2E_{dc}}{n\pi}\left[\sin nwt - \sin n(wt - \frac{2\pi}{3})\right] \qquad \ldots..(9.20)$$

$$V_{BC} = \sum_{n=1,3,5,7,\ldots,\infty} \frac{2E_{dc}}{n\pi}\left[\sin n(wt - \frac{2\pi}{3}) - \sin n(wt - \frac{4\pi}{3})\right] \qquad \ldots..(9.21)$$

Accordingly, the fundamental magnitude of line voltages can be written as

$$V_{AB,1} = \frac{2E_{dc}}{\pi}\left[\sin wt - \sin(wt - \frac{2\pi}{3})\right] = \frac{2\sqrt{3}E_{dc}}{\pi}\sin(wt + \frac{\pi}{6})$$

$$V_{BC,1} = \frac{2\sqrt{3}E_{dc}}{\pi}\sin(wt - \frac{\pi}{2}), \quad V_{CA,1} = \frac{2\sqrt{3}E_{dc}}{\pi}\sin(wt - \frac{7\pi}{6})$$

The three fundamental line voltages are balanced (have identical magnitudes and are phase apart by 120 degrees). For most practical loads only the fundamental component of the inverter output voltage is of interest. However, the inverter output also contains significant amount of higher order harmonic voltages that cause undesirable distortion of the output waveform. It may, be noted that there are no even harmonics, and the line voltages are free from 3rd and multiples of 3rd order harmonics. Also, as the harmonic order (n) increases their magnitudes decrease inversely with the harmonic order. When expressed as a fraction of fundamental voltage magnitude, the line voltage distortions are mainly due to 20% of 5th harmonic, nearly 14% of 7th, nearly 9% of 11th and nearly 8% of 13th harmonic. Since most loads are inductive in nature with a low pass filter type characteristics the effect of very high order harmonics may be neglected. It may be noted that though the pole voltages have 3rd and multiples of 3rd order harmonic distortions, the line voltages are free from these distortions. Hence the load neutral point, rather than being connected to the mid-potential point of the input dc supply is deliberately left floating.

The floating neutral point does not allow a closed path for the 3rd and multiples of 3rd harmonic currents to flow (3rd or multiples of 3rd harmonic current, if present in the load phases, have identical instantaneous magnitudes in all the three phases and their algebraic sum needs to flow in or out of the load neutral point). By keeping the load neutral point floating, not only the need for bringing out the mid-potential point of DC supply is done away with, the triplen harmonic distortions of the load current is totally eliminated. Since there are no triplen harmonic currents in the load, the load phase voltages are also free from triplen harmonic distortions.

In fact, the six-stepped load-phase voltages shown in Figure-19.9. are found to be free from triplen harmonics. It turns out that by removing all triplen harmonics from the square-shaped pole voltage waveform one can arrive at the corresponding load-phase (six-stepped) voltage waveform. Accordingly, the load-phase voltages may be expressed in terms of its harmonic contents as shown below.

$$V_{AN} = \sum_{n=1,5,7,11,13...,\infty} \frac{2E_{dc}}{n\pi} \sin(nwt) \qquad .....(9.22)$$

$$V_{BN} = \sum_{n=1,5,7,11,13...,\infty} \frac{2E_{dc}}{n\pi} \sin n(wt - \frac{2\pi}{3}) \qquad .....(9.23)$$

$$V_{CN} = \sum_{n=1,5,7,11,13...,\infty} \frac{2E_{dc}}{n\pi} \sin n(wt + \frac{2\pi}{3}) \qquad .....(9.24)$$

For a balanced three-phase load, the instantaneous magnitude of any phase current can be determined by superposition of different harmonic currents of that phase. For a simple three- phase R-L load, the phase-A current (i_A) expression in terms of resistance (R) and inductance (L) of the load may be written as:

$$i_A = \sum_{n=1,5,7,11,13...,\infty} \frac{2E_{dc}}{n\pi\sqrt{R^2 + n^2\omega^2 L^2}} \sin[nwt - \tan^{-1}(\frac{n\omega L}{R})]. \qquad .....(9.25)$$

9.6.3 DC-DC Converters

DC-DC converters are widely used for traction motor control in electric automobiles, trolley cars, marine hoists, forklift trucks, and mine haulers. DC converters can be used in regenerative braking of dc motors to return energy back into the supply, and this feature results in energy savings for transportation systems with frequent stops. DC converters are used in dc voltage regulators; and also, are used in conjunction with an inductor, to generate a dc current source, especially for the current source inverter. The dc–dc converters are integral parts of energy conversion in the evolving area of renewable energy technology in smart grid.

Buck Converter: In the following topologies, the switching elements are represented as a simple switches. In reality, they can be a transistors, P-FETs or N-FETs, with or without drivers according to the detailed design requirements. As the name suggests, the step-down or buck converter converts a higher input voltage into a stabilized lower output voltage. A simplified circuit diagram and the main current and voltage waveforms are shown in Figure-9.21.

The simplest way to understand this circuit is to think of L_1 and C_1 forming a low pass filter. When switch S_1 is closed, the voltage across the load slowly ramps up as the capacitor C_1charges up through L_1. If S_1 is then opened, the energy stored in the magnetic field slowly ramp down. The average output voltage is then the mark/space ratio of the PWM control signal multiplied by the input voltage. In the following topologies, the switching elements are represented as a simple switches.

In reality, they can be transistors, P-FETs or N-FETs, with or without drivers according to the inductor is clamped to 0V at the switch end of the inductor by diode D_1, so the energy has no choice but to discharge into the capacitor and load, causing the voltage across the load.

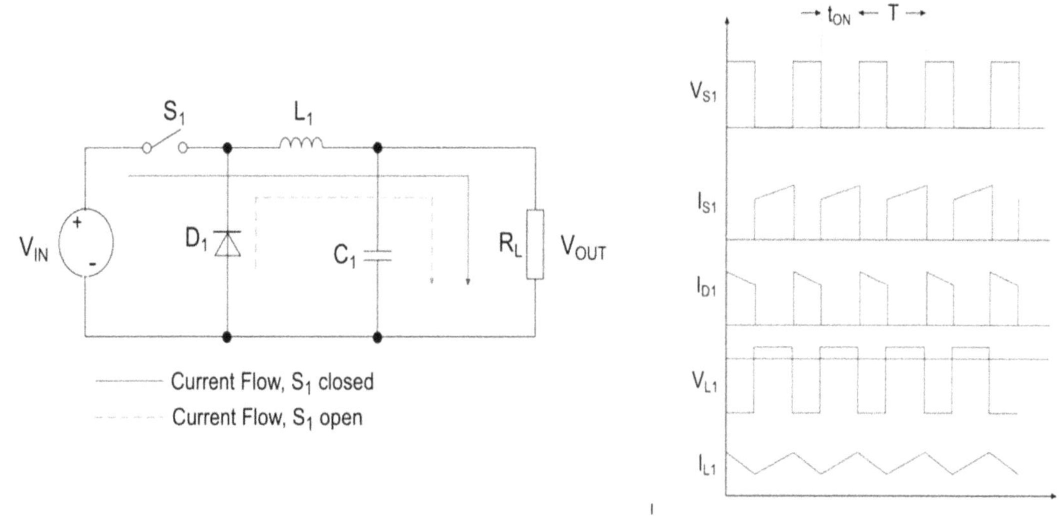

Figure 9.21 Simplified Schematic of Buck Regulator and related waveforms.

The transfer function can be derived by equating the voltage-time product of the inductance in the ON and OFF conditions. These two products must be the same because of the principle of energy conservation.

For the ON condition: $Energy_{IN} = (V_{IN} - V_{OUT})\ t_{ON}$ .....(9.26)

For the OFF condition: $Energy_{OUT} = (V_{OUT})\ t_{OFF}$ .... (9.27)

We know that $t_{ON} + t_{OFF} = T$ and the duty cycle δ= t_{ON}/T; where "T" switching time which is reciprocal of the switching frequency "f". i.e., T=1/ f.

Equating equation (9.26) and (9.27)

$$(V_{IN} - V_{OUT})\ t_{ON} = (V_{OUT})\ t_{OFF} \Rightarrow \quad {}^{V_{OUT}}/_{V_{IN}} = \delta \qquad .....(9.28)$$

The equation (9.28) is called the transfer function of the Buck converter.

The advantage of a buck converter is that the losses are very low (i.e. efficiency >97%) are readily achievable, especially in a synchronous design. Also, the switching frequency can be several hundreds of kHz to give a very compact construction with small inductors and a fast transient response. Finally, if the switching FET is disabled, the output is zero, so the no-load power consumption becomes negligible. For all of these reasons, the buck regulator makes a very attractive alternative to the linear regulator in many applications.

Boost Converter: As the name suggests, the step-up or boost converter converts a lower input voltage into a stabilized higher output voltage. A simplified circuit diagram and the main current and voltage forms are shown in Figure-9.22.

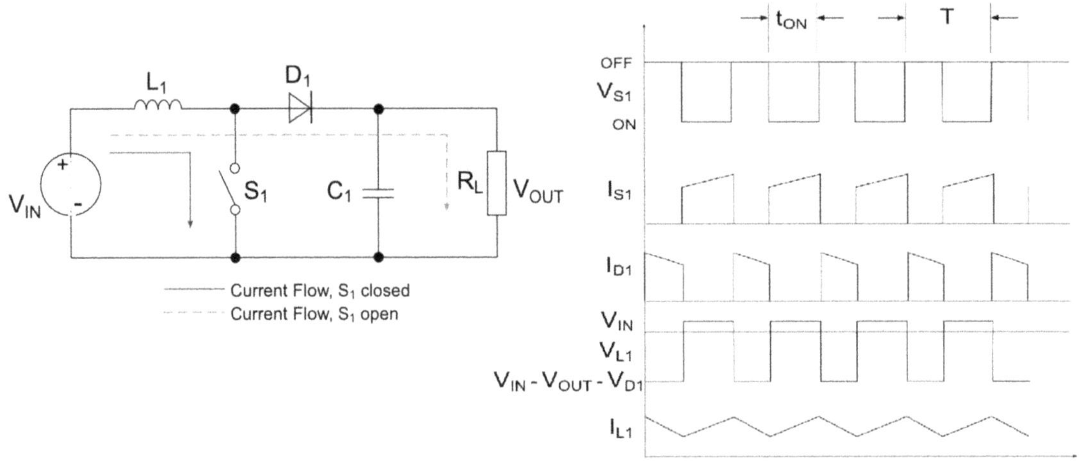

Figure 9.22 Simplified Schematic of Boost Converter and associated waveforms.

With S_1 closed, current flows through the inductor L_1 that increases linearly at a rate of (V_{IN}/L_1). During this period the load current is supplied from the stored energy in C_1. When the switch is opened again, the stored energy in the inductor causes high output voltage superimposed onto the input voltage. The resulting current flows via the freewheeling diode D_1 to supply the load and also re-charge C_1. The current through the inductor falls linearly and proportionally to $(V_{OUT} - V_{IN})/\ L_1$. The derivation of the transfer function is similar to that in the previous section, only the basic equations are rearranged.

For the ON condition: $Energy_{IN} = (V_{IN})\ t_{ON}$ .....(9.29)

For the OFF condition: $Energy_{OUT} = (V_{OUT} - V_{IN})\ t_{OFF}$ .....(9.30)

Equating the equations (9.29) and (9.30), we get

$$V_{OUT}/V_{IN} = 1/\ (1\text{-}\delta) \quad .....(9.31)$$

The advantage of the boost converter is that the output voltage can be varied with the mark-space ratio of the PWM signal to be equal to or above V_{IN}. This makes it especially suitable for increasing a low voltage battery output to a more useful higher voltage. However, in practice, a boost ratio of more than ×2 or ×3 makes the feedback stability difficult. Also, because the input current pulses increase proportionally to the boost gain, a converter that triples the input voltage draws triple the input current. This pulsed input current can cause EMI and voltage drop issues in the input leads. One further disadvantage with the boost converter is that the output cannot be switched off without adding a second switch in series with the input as disabling the PWM controller all one does not disconnect the load from the input. Finally, care must be taken not to allow the input voltage to rise above the output voltage. The PWM controller would then keep S_1 permanently open, and the input and output will be connected directly via L_1. and D_1 without regulation. Destructive currents can flow that will quickly destroy both the converter and the load. If this condition cannot be avoided, a topology that permits both buck and boost operation is needed.

Buck-Boost Converter: Buck-boost converter, converts an input voltage into a regulated negative output voltage that can be higher or lower than the absolute value of the input voltage. The simplified diagram in Figure-9.23 shows the basic circuit diagram and associated waveforms.

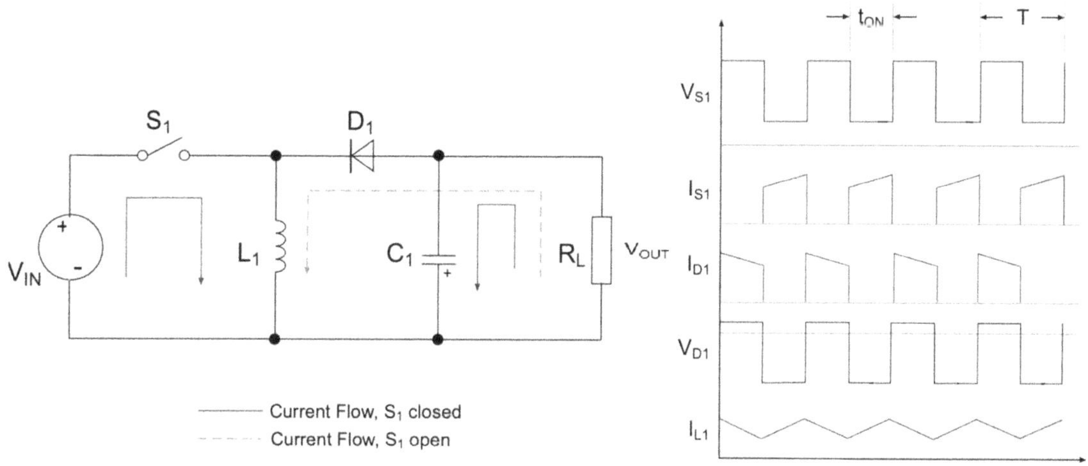

Figure 9.23 Basic circuit diagram of a Buck-Boost converter and associated waveforms.

In this circuit, when S_1 is closed, a current I_{L1}, which increases in proportion to V_{IN}/L_1 flows through L_1. Diode D_1 blocks any current flow into the load. During this time, the load current is supplied from the output capacitor C_1. When switch S_1 is opened, the energy stored in L_1 causes the switch end of the inductor to go negative (the other end of the inductor is grounded). The inverted current now flows into the load consisting of C_1 and R_L via D_1. This current decreases in proportion to V_{OUT}/L_1. Because of the direction of current flow, the output voltage is negative with respect to ground potential. Therefore, this topology is suitable for generating negative voltages only.

For the ON condition: $Energy_{IN} = (V_{IN})\ t_{ON}$ .....(9.32)

For the OFF condition: $Energy_{OUT} = -V_{OUT}\ t_{OFF}$ .....(9.33)

Equating the equations (9.31) and (9.32) we get

V_{OUT}/V_{IN}= -δ/ (1-δ) .....(9.34)

The advantage of a buck/boost converter is that the input voltage can be higher or lower than the regulated output voltage. For example, this is can be particularly useful in applications that need a stabilized 12 V output from a 12 V lead acid battery that can have a terminal voltage between 9 V when discharged to 14V when fully charged. Buck/Boosts are also very useful for stabilizing photovoltaic cell outputs. A solar cell can deliver very high voltage and current in bright sunlight, but low voltage and current when a cloud passes in front of the sun. As the voltage/current relationship changes, a buck/boost can be used for maximum power point tracking (MPPT) because the input/out- put voltage ratio can be continuously adjusted.

The biggest disadvantage is the inverted output voltage. Again, if used with a battery, then the output voltage inversion becomes irrelevant, because the battery supply can be left floating and the -V_{OUT} can then be connected to ground to give a positive-going output voltage. Another disadvantage is that the switch S_1 does not have a ground connection. This means that a level translator is needed in the PWM output circuit which can add cost and complexity to the design.

References

1. Ilhami Colak, Ersan Kabalci, Gianluca Fulli, Stavros Lazarou, A survey on the contributions of power electronics to smart grid systems, Renewable and Sustainable Energy Reviews, Volume 47, 2015, Pages 562-579, ISSN 1364-0321
2. W. Xu, W. Wang, Power electronic signaling technology-a new class of power electronics applications, IEEE Trans Smart Grid, 1 (3) (2010), pp. 332-339.
3. Alejandro J. del Real, Alicia Arce, Carlos Bordons Combined environmental and economic dispatch of smart grids using distributed model predictive control, Electr Power Energy Syst, 54 (2014), pp. 65-76
4. W. Chen, A.Q. Huang, C. Li, G. Wang, W. Gu, Analysis and comparison of medium voltage high power dc/dc converters for offshore wind energy systems, IEEE Trans Power Electron, 28 (4) (2013), pp. 2014-2023
5. David B. Richardson, Electric vehicles and the electric grid: a review of modelling approaches, Impacts, and renewable energy integration, Renew Sustain Energy Rev, 19 (2013), pp. 247-254
6. J.M. Guerrero, J.C. Vasquez, J. Matas, L.G. de Vicuña, M. Castilla, Hierarchical control of droop-controlled AC and DC microgrids — a general approach toward standardization, IEEE Trans Ind Electron, 58 (1) (2011), pp. 158-172

7. T.L. Vandoorn, J.D.M. De Kooning, B. Meersman, J.M. Guerrero, L. Vandevelde, Voltage-based control of a smart transformer in a microgrid, IEEE Trans Ind Electron, 60 (4) (2013), pp. 1291-1305
8. Ltera Inc. A network for the smart grid. 〈http://www.altera.com/technology/system-design/articles/2013/network-smart-grid.html〉
9. S. Paudyal, C.A. Canizares, K. Bhattacharya, Optimal operation of distribution feeders in smart grids, IEEE Trans Ind Electron, 58 (10) (2011), pp. 4495-4503
10. Lin Chen-Chun, Yang Chia-Han, Shyua Joseph Z. A comparison of innovation policy in the smart grid industry across the pacific: China and the USA. Energy Pol. Available online 7 March 2013, ISSN 0301-4215
11. Power Plus Communications AG. Smart grids communication – paving the way for change. http://www.ppc-ag.de/4-1-Smart+Grids.html〉
12. J.P. Lopes, C. Moreira, A.G. Madureira, Defining control strategies for MicroGrids islanded operation, IEEE Trans Power Syst, 21 (2) (2006), pp. 916-924
13. V.C. Gungor, Lu Bin, G.P. Hancke, Opportunities and challenges of wireless sensor networks in smart grid, IEEE Trans Ind Electron, 57 (10) (2010), pp. 3557-3564
14. B. Awad, J. Wu, N. Jenkins, Control of distributed generation, Elektrotech Inf, 125 (12) (2008), pp. 409-414
15. H. Jiayi, J. Chuanwen, X. Rong, A review on distributed energy resources and MicroGrid, Renew Sustain Energy Rev, 12 (9) (2008), pp. 2472-2483
16. Larsson M, Rehtanz C. Predictive frequency stability control based on wide-area phasor measurements. In: Proceedings of the IEEE power engineering society summer meeting, vol. 1; 2002. p. 233–8
17. R.M. Tumilty, I.M. Elders, G.M. Burt, J.R. McDonald, Coordinated protection, control & automation schemes for microgrids, Int J Distrib Energy Resour, 3 (3) (2007), pp. 225-241
18. T.L. Vandoorn, B. Renders, L. Degroote, B. Meersman, L. Vandevelde, Active load control in islanded microgrids based on the grid voltage, IEEE Trans Smart Grid, 2 (1) (2011), pp. 139-151.
19. J. De La Ree, V. Centeno, J.S. Thorp, A.G. Phadke, Synchronized phasor measurement applications in power systems, IEEE Trans Smart Grid, 1 (1) (2010), pp. 20-27
20. J. Ma, P. Zhang, H. Fu, B. Bo, Z. Dong, Application of phasor measurement unit on locating disturbance source for low-frequency oscillation, IEEE Trans Smart Grid, 1 (3) (2010), pp. 340-346
21. A.G. Phadke, J.S. Thorp, Synchronized phasor measurements and their applications, Springer, New York (2008)
22. Zhigang Liang, Rong Guo, Jun Li, A. Huang, A high-efficiency PV module-integrated DC/DC converter for PV energy harvest in FREEDM Systems, IEEE Trans Power Electron, 26 (3) (2011), pp. 897-909

23. Zhe Zhang, Riccardo Pittini, Michael A.E. Andersen, Ole C. Thomsena, A review and design of power electronics converters for fuel cell hybrid system applications, Energy Proc, 20 (2012), pp. 301-310
24. J. Arai, K. Iba, T. Funabashi, Y. Nakanishi, K. Koyanagi, R. Yokoyama, Power electronics and its applications to renewable energy in Japan, IEEE Circuits Syst Mag, 8 (3) (2008), pp. 52-66

A. Vaccaro, G. Velotto, A.F. Zobaa, A decentralized and cooperative architecture for optimal voltage regulation in smart grids, IEEE Trans Ind Electron, 58 (10) (2011), pp. 4593-4602

25. T. Funaki, T. Hikihara, Characterization and modeling of the voltage dependency of capacitance and impedance frequency characteristics of packed EDLCs, IEEE Trans Power Electron, 23 (3) (2008), pp. 1518-1525
26. Madureira A, Moreira C, Lopes J Peças. Secondary load-frequency control for micro-Grids in islanded operation. In: Proceedings of ICREPQ. Frankfurt, Germany; 2005. p. 1–4.
27. Colson CM, Nehrir MH. A review of challenges to real-time power management of microgrids. In: IEEE power & energy society general meeting; 2009. p. 1–8.
28. Estefanía Planas, Asier Gil-de-Muro, Jon Andreu, Iñigo Kortabarria, Iñigo Martínez de Alegría, General aspects, hierarchical controls and droop methods in microgrids: a review, Renew Sustain Energy Rev, 17 (2013), pp. 147-159
29. K. Kurohane, T. Senjyu, A. Yona, N. Urasaki, T. Goya, T. Funabashi, A hybrid smart AC/DC power system, IEEE Trans Smart Grid, 1 (2) (2010), pp. 199-204
30. M. Liserre, T. Sauter, J.Y. Hung, Future energy systems: integrating renewable energy sources into the smart power grid through industrial electronics, IEEE Ind Electron Mag, 4 (1) (2010), pp. 18-37

A. Vaccaro, G. Velotto, A.F. Zobaa, A decentralized and cooperative architecture for optimal voltage regulation in smart grids, IEEE Trans Ind Electron, 58 (10) (2011), pp. 4593-4602

31. Shao Zhang, King-Jet Tseng, D.M. Vilathgamuwa, TrongDuy Nguyen, Xiao-Yu Wang, Design of a robust grid interface system for PMSG-based wind turbine generators, IEEE Trans Ind Electron, 58 (1) (2011), pp. 316-328
32. S. Grillo, V. Musolino, L. Piegari, E. Tironi, C. Tornelli, DC islands in AC smart grids, IEEE Trans Power Electron, 29 (1) (2014), pp. 89-98
33. Zhe Zhang, Riccardo Pittini, Michael A.E. Andersen, Ole C. Thomsena, A review and design of power electronics converters for fuel cell hybrid system applications, Energy Proc, 20 (2012), pp. 301-310
34. Çolak, E. Kabalcı, R. Bayındır, Review of multilevel voltage source inverter topologies and control schemes, Energy Convers Manag, 52 (2011), pp. 1114-1128
35. Cao Hui, Zhang Hui, Jiang Wentao, Wei Santong. Research on PQ control strategy for PV inverter in the unbalanced grid. In: Proceedings of the Asia–Pacific power and energy engineering conference (APPEEC); 2012. p. 1–3.

36. Huajun Yu, Junmin Pan, Xiang An, A multi-function grid-connected PV system with reactive power compensation for the grid, Solar Energy, 79 (1) (2005), pp. 101-106
37. H. Kanchev, Di Lu, F. Colas, V. Lazarov, B. Francois, Energy management and operational planning of a microgrid with a PV-based active generator for smart grid applications, IEEE Trans Ind Electron, 58 (10) (2011), pp. 4583-4592
38. Ramon Zamora, Anurag K. Srivastava. Controls for microgrids with storage: Review, challenges, and research needs, Renew Sustain Energy Rev, 14 (7) (2010), pp. 2009-2018
39. M. Sechilariu, Baochao Wang, F. Locment, Building integrated photovoltaic system with energy storage and smart grid communication, IEEE Trans Ind Electron, 60 (4) (2013), pp. 1607-1618
40. Patarau T; Petreus D; Etz R; Cirstea M; Daraban S. Digital control of bidirectional DC–DC converters in Smart Grids. In: Proceedings of the 13th international conference on optimization of electrical and electronic equipment (OPTIM); 2012. p.1553–8.
41. O.C. Onar, A. Khaligh, A novel integrated magnetic structure based DC/DC converter for hybrid battery/ultracapacitor energy storage systems, IEEE Trans Smart Grid, 3 (1) (2012), pp. 296-307
42. M.H. Taghvaee, M.A.M. Radzi, S.M. Moosavain, Hashim Hizam, M. Hamiruce Marhaban, A current and future study on non-isolated DC–DC converters for photovoltaic applications, Renew Sustain Energy Rev, 17 (2013), pp. 216-227
43. Hua CC, Liao KA, Lin JR. Parallel operation of inverters for distributed photovoltaic power supply system. In: Proceedings of the IEEE PESC conference; 2002. p. 1979–83.
44. Çolak, E. Kabalcı, R. Bayındır, Review of multilevel voltage source inverter topologies and control schemes, Energy Convers Manag, 52 (2011), pp. 1114-1128
45. Cao Hui, Zhang Hui, Jiang Wentao, Wei Santong. Research on PQ control strategy for PV inverter in the unbalanced grid. In: Proceedings of the Asia–Pacific power and energy engineering conference (APPEEC); 2012. p. 1–3.
46. Huajun Yu, Junmin Pan, Xiang An, A multi-function grid-connected PV system with reactive power compensation for the grid, Solar Energy, 79 (1) (2005), pp. 101-106
47. H. Kanchev, Di Lu, F. Colas, V. Lazarov, B. Francois, Energy management and operational planning of a microgrid with a PV-based active generator for smart grid applications, IEEE Trans Ind Electron, 58 (10) (2011), pp. 4583-4592
48. Ramon Zamora, Anurag K. Srivastava, Controls for microgrids with storage: Review, challenges, and research needs, Renew Sustain Energy Rev, 14 (7) (2010), pp. 2009-2018
49. M. Sechilariu, Baochao Wang, F. Locment, Building integrated photovoltaic system with energy storage and smart grid communication, IEEE Trans Ind Electron, 60 (4) (2013), pp. 1607-1618
50. Kanchan Jha, Abdul Gafoor Shaik, A comprehensive review of power quality mitigation in the scenario of solar PV integration into utility grid, e-Prime - Advances in Electrical Engineering, Electronics and Energy, Volume 3,2023,100103, ISSN 2772-6711

51. Rashid, Muhammad H. *Power Electronics: Devices, Circuits, and Applications*. 4th ed., Pearson, 2013.

52. Steve Roberts M. DC/DC Book of Knowledge: Practical tips for the User. 3rd edition, RECOM Engineering GmbH & Co KG, Austria, 2016

Index

www.ingramcontent.com/pod-product-compliance
Ingram Content Group UK Ltd.
Pitfield, Milton Keynes, MK11 3LW, UK
UKHW061536190726
13845UKWH00006B/128

9 789349 562059